Magnetic Nanoparticles for Biomedical Applications

Edited by

Martin F. Desimone[1] and Rajshree B. Jotania[2]

[1]Cátedra de Química Analítica Instrumental, IQUIMEFA-CONICET, Facultad de Farmacia y Bioquímica, Universidad de Buenos Aires, (1113) Junin 956 Piso 3. Buenos Aires. Argentina

[2] Department of Physics, Electronics and Space science, University school of sciences, Gujarat University, Ahmedabad 380 009, India

Published by **Materials Research Forum LLC**
Millersville, PA 17551, USA

Published as part of the book series
Materials Research Foundations
Volume 143 (2023)
ISSN 2471-8890 (Print)
ISSN 2471-8904 (Online)

Print ISBN 978-1-64490-232-5
eBook ISBN 978-1-64490-233-2

Distributed worldwide by

Materials Research Forum LLC
105 Springdale Lane
Millersville, PA 17551
USA
https://www.mrforum.com

Manufactured in the United States of America
10 9 8 7 6 5 4 3 2 1

Table of Contents

Preface

Magnetic nanoparticles (MNPs) have found potential applications in the biomedical field because of their non-toxicity, high chemical stability, and biocompatibility. Over the last few years, MNPs have gained interest due to their unique structural and magnetic properties. Interestingly, the interactions of MNPs with biological media depend upon their crystal structure, shape, and size, too. In addition, the structural and magnetic properties of MNPs depend upon the synthesis method. MNPs are frequently used in DNA or protein separation, hyperthermia, tissue engineering, magnetic resonance imaging contrast enhancement agent, cancer therapy, drug delivery, bone, and dental repair, biosensors, etc. For all such uses, the selection of synthesis route to prepare MNPs is most challenging as it will determine the shape of the particles, the size distribution, and the surface morphology as well as magnetic properties. The topics covered in this book will focus on magnetic nanoparticles and coated nanoparticles (i.e.: ferrites nanoparticles, bimetallic-magnetic nanoparticles, magnetic fluid), as well as their associated synthesis, characterization, and in vivo or in vitro biomedical applications.

In chapter 1 Shivani R. Pandya and Harjeet Singh provide an overview of magnetic nanoparticles with their associated chemical and physical synthesis procedures. The chapter provides an in-depth analysis of surface modification of magnetic nanomaterials with polymeric, non-polymeric, inorganic and target-specific ligands with their effect on stability and magnetization for biomedical applications.

The synthesis, surface functionalization, modification, and coating also represent the underlying theme of chapter 2 written by M. V. Nikolic. She focused on magnetic spinel ferrite nanoparticles and nanocomposites with optimal properties for biomedical applications.

These initial two chapters are well connected with the topic of magnetic hyperthermia in chapter 3, where Robert Pullar illustrates the application of magnetic nanoparticles in this field. In particular, the chapter describes several properties of magnetic NPs that are of particular interest in biomedicine ranging from cancer therapy, hyperthermia, drug delivery to combined theragnostic systems.

In chapter 4, Marcela Van Raap and collaborators summarize and integrate the current state of knowledge on physical and cellular basis of oncologic magnetic thermotherapy. Interestingly, the authors analyze the cellular responses induced to counteract the various sources of stress associated with this technology.

Depending on particle size and morphology, magnetic nanomaterials have a wide variety of applications in various technological fields, as summarized in chapter 5 by

Kuzhichalil Peethambharan Surendran and coauthors. There, the authors provide an insightful panorama of recent trends in the synthesis of ferromagnetic Ni nanostructures through chemical reduction routes.

The increasing importance of electromagnetic properties is reflected in chapter 6. In this chapter Francisco E. Carvalho together with an international consortium of experts in the field analyze the effect of niobium pentoxide addition on the electromagnetic properties of cobalt ferrites.

In chapter 7, Ratiram G. Chaudhary and colleagues review methodologies for generating magnetic nanoparticles for diagnosis (i.e.: imaging, biosensing) and treatment (i.e.: hyperthermia, targeted drug delivery). The chapter also takes advantages of magnetic nanomaterials obtained by green synthesis procedures.

Magnetic nanoparticles are being developed for the localized drug delivery in patients with different pathologies. To facilitate the comprehension of this inherently sophisticated field, in chapter 8 Ayushi G. Patel, Rajshree B. Jotania and Martin F. Desimone review current work on the development of magnetic nanoparticle-based drug delivery systems.

The interactions of MNPs with the immune system as well as its applications in immunomodulatory therapies are described by Mauricio De Marzi´s team in chapter 9. It is pointed out that due to their properties, MNPs could be used with the aim of developing immunomodulatory therapies.

Finally, another field in which the application of MNPs is gaining attention is the development of DNA and RNA- based vaccines for immunization and immunotherapy. In chapter 10, Gisela Alvarez and colleagues summarize current approaches to use magnetofection for in vitro experiments as well as the study of in vivo vaccine assays or gene therapy using MNP as nucleic acid carriers.

We thank all the contributors for their high-quality cooperation and efforts to provide up-to-date manuscripts. We are thankful to reviewers of the chapters whose work significantly contributed to improving the quality of this book. We also thank Mr. Thomas Wohlbier and the entire team at Materials Research Forum LLC for their continuous cooperation during the publication of this book.

We hope this comprehensive book of a very dynamic field of research will provide the reader a thorough analysis of the great potential and diverse biomedical possibilities beyond the development of magnetic nanoparticles.

Martin F. Desimone and Rajshree B. Jotania

Materials Research Forum LLC
https://doi.org/10.21741/9781644902332-1

Chapter 1

Surface-Tailored Iron Oxide Magnetic Nanomaterials for Biomedical Applications

Shivani R. Pandya[1][*] and Harjeet Singh[2][**]

[1][*]Dept. of Forensic Science, PIAS, Parul University, Vadodara (Gujarat)

[2]Centre of Research for Development, Parul university, Vadodara (Gujarat)

*shivpan02@gmail.com, **mailgsbtm@gmail.com

Abstract

Iron Oxide Magnetic Nanomaterials (IOMNMs) are widely used biocompatible and FDA approved nanomaterials to develop numerous biomedical applications. However, bare IOMNMs have shown limited applications due to coulombic forces that increase the agglomeration, resulting in increased size. Thus, modifying the surface charge and design of IOMNMs are of much interest while talking about their applications in highly developed medical technologies and biotechnologies inclusive of MRI (Magnetic Resonance Imaging) contrast agents, magnetic separation and immobilization of different proteins, antibodies, enzymes and several other biological substances. Usually, IOMNMs are modified with biocompatible functional groups like amine, carboxylic acid, hydroxyl group to enhance their bioavailability. The present article emphasizes possible synthetic approaches for tailored iron oxide nanoparticles and their surface chemistry, allowing both therapeutic and diagnostic applications (theranostic).

Keywords

Magnetic Nanomaterial, Biomedical Application, Theranostic, Synthesis, Drug Delivery

Contents

Materials Research Forum LLC
https://doi.org/10.21741/9781644902332-1

1. Nanotechnology: An overview

"There is plenty of room at the bottom"

In 1959 Richard Feynman, father of nanotechnology and an American Physicist thought up these words explaining the top-down approach in nanotechnology. From then until now, for almost half a century, the 10^{-9} factor has remained a very significant field and a challenge that lies ahead in terms of smart materials and application in various scientific fields. The applications come from all sectors of public concern, such as industries, building materials, catalysis and biomedical technology [1]. This chapter is centred on the synthesis of MNPs (magnetic nanoparticles), their bio-functional modifications as well as their multi-applications.

Nanotechnology provides the equivalent of a single nano object as well as the building materials and divisions built by them and the process takes place in the nanometer system. Nanomaterial is a substance with its essential properties that are expressed by the nano-object it contains [2]. These materials can be divided into (i) nanostructured materials and (ii) nanodispersion materials, where nanostructured materials consist of a recurring isotropic unit of solid nanometer sized particles and nanodispersion materials having nanosized particles containing homogeneous solution. Nanoparticles are those of a size of <100 nm, also known as 0-D nanoobjects. However, well organised arrangements of atoms in nanoparticles are known as nanocrystallites [3].

Nanocrystallite materials are comprised of either metal or metal oxide. Like metal, nanoparticles are extremely unstable and are toxic. Therefore, metal oxides are of choice in the biomedical field. Metal oxide nanocrystallites are often composed of transition metal ions due to their widespread use in industries and medical practices, in which, iron oxide nanoparticles of different sizes and shapes have been widely used as memory chip devices, catalysts, carriers of drug and MRI agents [4].

1.1 Magnetic nanoparticles

Contrary to nanoparticles, magnetic clusters are entirely indistinguishable MNPs and their magnetism can be expressed in terms of exchange modified paramagnetism. Iron oxide is found as rust in nature and widely used, as it is inexpensive and plays crucial roles in various geological as well as biological processes. Humans have iron as indispensable core of haemoglobin and they use it as an iron ore in thermite, a catalyst, along with durable pigments in coating, paint and coloured concretes (Fig. 1).

On the basis of different oxidation states iron oxides can be divided as:

- Iron(II) ferrous oxide (wüstite-FeO)
- Iron (ferrous ferric) oxide (magnetite-Fe_3O_4)
- Iron (ferric) oxides Fe_2O_3
  - Hematite (α-Fe_2O_3)
  - β-Fe_2O_3

- ○ Maghemite (γ-Fe_2O_3)
- ○ ϵ-Fe_2O_3, amorphous and high-pressure form

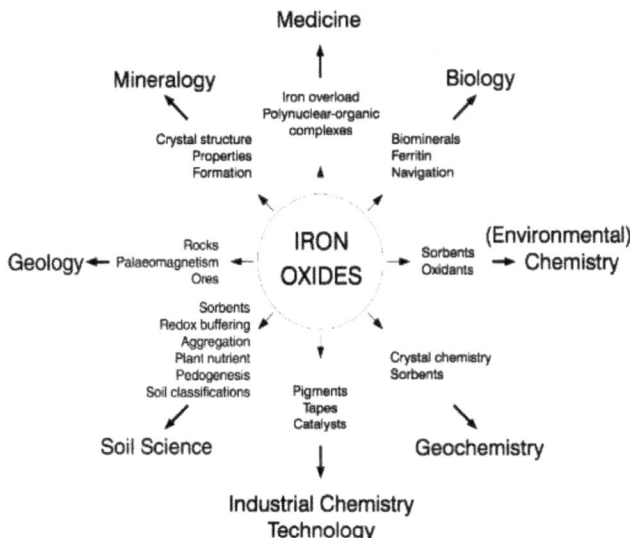

Fig. 1. Multi-dimensional functions of iron oxide NPs [5].

Iron oxides based MNPs are being discussed in detail in literature respectively [6, 7], where Magnetite and maghemite are closely related forms of iron oxide. Particles having a size between <10 to 20 nm displaying an exclusive arrangement of magnetism, *i.e.*, superparamagnetism. Therefore, magnetic iron oxide (Fe_2O_3 and γ-Fe_2O_3) attracted a great deal of attention as they are less toxic, superparamagnetic in nature and are easy to separate. Also, iron oxide proved to be helpful in diagnosis and treatment in relation to drug delivery and magnetic resonance imaging [8].

1.1.1 Elementary properties of magnetite and maghemite

In iron oxide's magnetite (Fe_3O_4) form, Fe (II) and Fe (III) lattices are arranged irregularly and are estranged by oxygen atoms leading to electronic connections. So, it leads to a structure which is reverse spinal, having ferromagnetic properties. Likewise, maghemite (γ-Fe_2O_3) show ferrimagnetic properties having a reverse spinal structure, due to lattice vacancies, rising unremunerated electron-spins [9]. Both the aforementioned forms illustrate parallel crystallographic properties having similar structure and minor differences in lattice structure, where a=8.3515 Å for γ-Fe_2O_3 and a=8.396 Å for Fe_3O_4. So, it is tricky to distinguish between magnetite and maghemite particles effortlessly (Table 1). Though, certain analytical techniques *viz.* FT-IR spectroscopy, Raman spectroscopy, XPS (X-ray photoelectron spectroscopy), Mössbauer, magnetometer and especially X-ray Diffraction could help to identify differences in magnetite and maghemite.

Table 1. Significant properties of γ-Fe_2O_3 (maghemite) and Fe_3O_4 (magnetite) [10-12].

Feature	Maghemite	Magnetite
Colour	Brown to brownish–red	Black
Type of structure	Defective inverse spinal	Inverse spinal
Space group	Cubic (P4$_3$32)	Cubic (Fd$_3$m)
Unit cell a [Å]	8.3474[8]/8.3515[9]/8.351[10]	8.396
Space group	Tetragonal (P4$_1$2$_1$2)	
Unit cell a [Å]	8.347[8]/8.349[10]	
Unit cell a [Å]	25.01[8]/24.996[10]	
Band gap [eV]	2.03	0.1
Néel temperature TN [K]	820–986 (estimated)	850
Saturation magnetisation [Am^2 kg^{-1}]	72	98

1.2 Methods of MNPs synthesis

Synthesizing magnetic nanoparticles having a desired shape and size for targeted applications remains challenging both scientifically and technologically. In a comprehensive sense, MNPs can be synthesized using physical, chemical and biological means (Fig. 2 and 3).

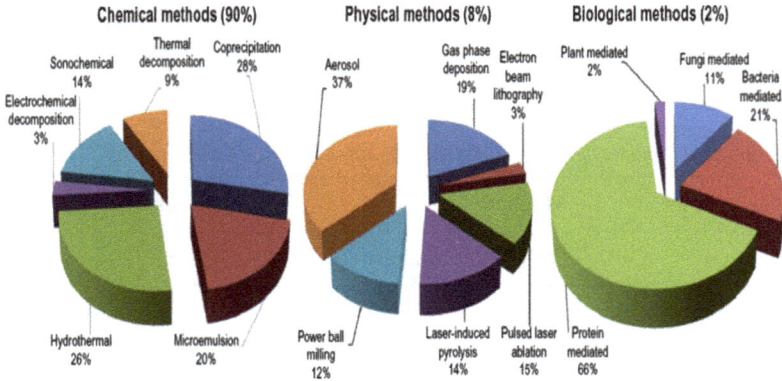

Fig. 2. Comparison between syntheses of MNPs by three different courses [13].

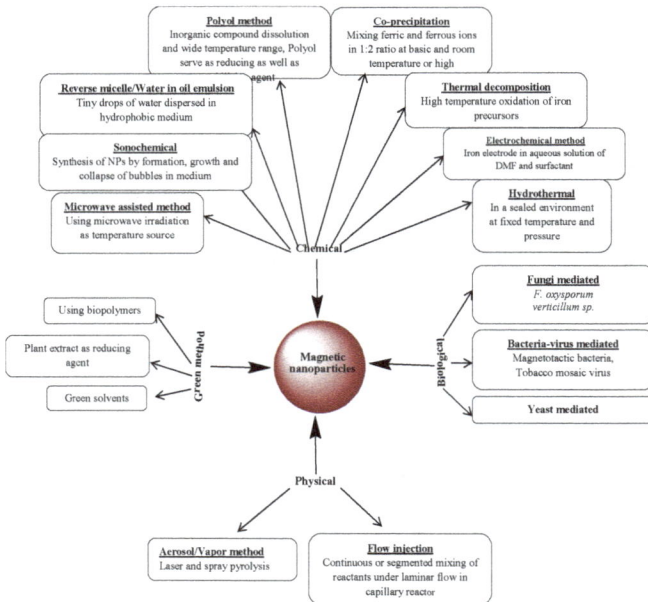

Fig. 3. Scheme illustrating syntheses of MNPs by Chemical, Biological, and Physical methods.

1.2.1 Chemical route of synthesis

A number of chemical approaches are in use for the synthesis of MNPs on the basis of desirable applications, *viz.* microemulsion method [14], sol-gel technique [15], sonochemical method [16], hydrothermal method [17], hydrolysis and thermolysis procedures [18]. Synthesizing superparamagnetic nanoparticles is a complex process because of their colloidal properties. Optimization of experimental conditions for making monodispersed magnetic grains of the desired size and shape is foremost important. As well as choice of reproducible procedure without having any complex purification process like ultracentrifugation [19], size exclusion chromatography [20], and magnetic filtration [21] is equally important. These methods are in extensive use for the preparation of particles having a desirable composition and constricted size distribution. However, the most frequently used method for MNPs production is the co-precipitation (chemical) method using different salts of iron [22-24].

1.2.1.1 Co-precipitation method

This is one of the oldest, facile and most competent methods for obtaining MNPs. Generally, preparation of iron oxides ($Fe_3O_4/ \gamma\text{-}Fe_2O_3$) can be achieved if the stoichiometric ratio of Fe^{2+} and Fe^{3+} salts is mixed and precipitated in the aqueous medium. Chemical reaction for synthesis of MNPs is mentioned below in equation 1,

$$Fe^{2+} + 2Fe^{3+} + 8OH^- \longrightarrow Fe_3O_4 + 4H_2O \qquad (1)$$

If thermodynamics for the above reaction is considered, the absolute precipitation of Fe_3O_4 could be obtained between pH 8 to 14 having a 2:1 Fe^{+3} and Fe^{+2} stoichiometric mixture ratio in an inert environment [25].

MNPs are unstable and are sensitive to oxidation. For this reason, Magnetite gets converted to maghemite when incident to oxygen ($\gamma\text{-}Fe_2O_3$).

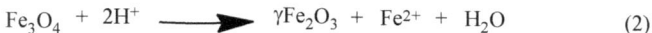

$$Fe_3O_4 + 2H^+ \longrightarrow \gamma Fe_2O_3 + Fe^{2+} + H_2O \qquad (2)$$

Though, oxidation is not a reason alone for the conversion of Fe_3O_4 (Magnetite) into $\gamma\text{-}Fe_2O_3$ (maghemite). As shown in equation 2, at a particular pH, transfer of electrons and ions is also responsible for the conversion of Magnetite to maghemite. It has been observed that in acidic conditions when oxygen is absent, Fe^{2+} gets easily desorbed, forming a hexa-aqua complex within solution, whereas, having basic conditions it is found involved in oxidation and reduction processes on the surface of MNPs and oxidation of Fe^{2+} is linked with repositioning of cations from the framework of lattice which generate the cationic vacancies for balancing the charge and expound maghemite phase. As shown in Formula 1, for magnetite, ions of Fe scattered in octahedral (Oh) and tetrahedral (Td) sites of the spinal geometry leading to blockage of the cationic vacancies of octahedral sites that are observed in maghemite form. The scheme of vacancy ordering is sample preparation dependent, which causes symmetry sinking and possible superstructures. These may be completely or partially random or ordered. Also, it could be completely ordered. It has been observed that particles having sizes less than 5 nm can display ordered vacancies and might be observed using FTIR/XRD.

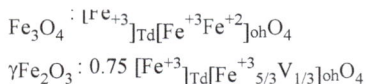

$$Fe_3O_4 : [Fe^{+3}]_{Td}[Fe^{+3}Fe^{+2}]_{oh}O_4$$

$$\gamma Fe_2O_3 : 0.75\ [Fe^{+3}]_{Td}[Fe^{+3}_{5/3}V_{1/3}]_{oh}O_4$$

Formula 1: Magnetite and maghemite Geometry

Co-precipitation routes higher yield of nanoparticles, but it is extremely difficult to control the size of the particles because it is completely dependent on the kinetic parameters which control the growth of crystals. Nucleation and growth are the two major phases which are involved in co-precipitation [26-29] producing monodispersed MNPs (Fig. 4). It is important to keep both the mentioned stages separated by evading nucleation during the course of growth [30]. The cause relies on the fact that as the concentration of medium arrives at decisive supersaturation, a very petite nucleation burst occurs and growth rate of nuclei gets sluggish, which causes the solutes to get diffused on the nuclei's surface.

The LaMer diagram [31] behaves as a model for the genesis of monodispersed nano or micro-particles *via* crystal development mechanism and nucleation. If we talk about supersaturated solution, formation of nuclei leads to the formation of a narrow size distribution on account of successive growth of nuclei [32]. Therefore, it can be concluded that monodispersed particle size-control could be attained by short nucleation periods. After nucleation gets completed and a sum of particles is formed, it remains unchanged during the entire particle growth process. Thus, a variety of factors could be tailored to synthesize MNPs having specific size, magnetic properties and surface functionalities [33-37].

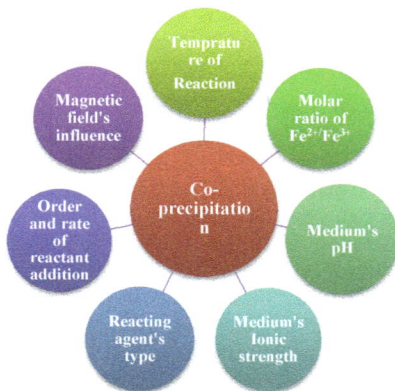

Fig. 4. Representation of major factors concerned with co-precipitation method

In short, we can conclude that particles having a size range of 2 to 17 nm could be acquired by processing the shape and size of MNPs by regulating the ionic strength, nature of the salts, temperature, pH, for instance, sulfatesor nitrates, chlorides, perchlorates and Fe(II) or Fe(III) ratios [21]. These days the various procedures such as coating, capping, functionalisation with organic/inorganic/polymeric agents are in use to control the size of nanoparticles as well as stability.

Massart (1981) carried out the synthesis of superparamagnetic MNPs in a controlled manner following alkaline precipitation of $FeCl_3$ and $FeCl_2$ [38]. In the inventive synthesis, the shape of synthesized magnetic nanoparticles was spherical and their diameter (8nm) was measured by means of XRD [39].

The function of different bases such as CH_3NH_2, ammonia, and NaOH [29, 39], at a particular pH were observed along with the addition of cations, viz. Na^+, Li^+, $N(CH_3)_4^+$, K^+, NH_4^+ and the ratio of Fe^{+2} to Fe^{+3} on the yield and size of the magnetic nanoparticles. Therefore, if aforementioned parameters are modified, there are chances to have particles with a narrow size range (between 16.6-4.2 nm) [39]. But, a number of studies have revealed that by restricting acidity as well as ionic strength, it could form engineered magnetic nanoparticles having sizes between 2 to 15 nm and, due to dissimilarity in electrostatic charges of the MNPs, variation in shapes could be observed [40-41].

MNPs are able to disperse in both aqueous as well as non-polar medium like oil/organic solvents on the basis of specific application such as preparation of capsules, vesicles and magnetic-emulsions [13, 42-44]. Babes et al. (1999) has investigated the effect of a range of parameters in relation to iron medium and concentration, especially Fe^{+2} to Fe^{+3} molar concentrations. The results of the study revealed that if Fe^{+2}/Fe^{+3} molar ratios increase and preparation yield decreases, mean size also increases [39, 45-46], and they established that 0.4 and 0.6 ratios are effective in synthesis of particles for using them as contrast agents [1].

Moreover, it cannot be deserted that the mean size of magnetic nanoparticles is highly reliant on the ionic strength and pH of the reaction medium [47-48]. Therefore, it can be concluded that a rise in pH and ionic strength could form particles with small size and having a narrow size distribution.

Qui et al. (2000) has also worked on the effect of ionic strength on the formation of magnetite particles in a reaction medium [49]. They found that addition of 1M NaCl in a reaction mixture could form nanoparticles of iron oxide having size of 1.5 nm, which is much smaller than could be achieved with traditional methods. Moreover, at higher ionic strength, synthesis of nanoparticles showed a lower Ms value (63 emu/g) while in the absence of NaCl the value increased up to 71 emu/g. As ionic strength was increased and size was decreased, magnetization also decreased. Apart from all the factors listed above, there are a few additional factors that could influence the size of nanoparticles. The high degree of mixing reduces particle size. Similarly, a decrease in polydispersity and size is found when a precipitant is added to reactants [39]. Injection rate flux does not have noteworthy consequences on synthesis of nanoparticles [21].

Temperature is one of the other important elements that could direct the dimensions and shape of MNPs. Various researchers have reported its significance in selective crystal formation. It is an established fact that a rise in temperature may lessen magnetite particle formation [21, 29]. This finding is in support of the nucleation theory and particle growth. According to the literature, magnetite nanoparticle synthesis should be carried out at higher temperatures (above 80°C) [50] because at low temperatures (below 60°C) amorphous hydrated oxyhydroxides could be formed and they get easily converted to Fe_2O_3 [51]. Various investigators also found that nitrogen gas bubbling have an impact on formation of magnetite particles where oxygen removal by nitrogen gas bubbling could discontinue over critical oxidation of particles of magnetite with decrease in size of the particles [52].

1.2.1.2 High temperature thermal decomposition and Hydrothermal Technique

This method is an extremely high-temperature dependent chemical method. In previous literature, MNPs synthesis using hydrothermal method and ultrafine powders has already been described [53-55]. As the name suggests, these are the reactions which are performed generally in aqueous medium in reactors or autoclaves which are teflon lined at an extremely high pressure (about 2000 psi) and temperature (more than 200°C) for a given period of time. There are generally two different hydrothermal and thermal decomposition pathways which are in use for the synthesis of ferrites in hydrothermal conditions, *viz.* neutralization or hydrolysis and oxidation of the intermediary hydroxides. More or less, both the given methods are the same; disregarding those ferrous salts that are in use for the earlier one [56].

Pressure, temperature and time are the key factors which govern the size of particles and the type of product that yields [57]. Studies have shown that if reaction time and water content is increased, it could lead to the formation of larger magnetic nanoparticles.

It leads to the fact that the size of particles at the time of crystallization is prompted by the nucleation rate and growth of grains of a particular type and, eventually, it relies on the temperature of the reaction and continuous parameters [57]. Interestingly, at elevated temperatures, the rate of nucleation increases rather than the grain growth, forming smaller sized particles. Dissimilar to this, if reaction time is increased it would favour grain growth and particles would have bigger size [58].

Zheng *et al.* (2006) carried out a hydrothermal method for the synthesis of $Fe_3O_4@C$ and Fe_3O_4 using $FeCl_3$ (as precursor), glucose (as reducing agent) and sodium acetate or NaOH (as base) at 160°C for 6h [59]. Furthermore, sodium citrate acts as a carbon source in the reaction and, using electrostatic interactions, can dispose of the surroundings and form MNPs. Using this method, it has been found that magnetic nanoparticles having sizes smaller than 10 nm and 120 nm of $Fe_3O_4@C$ were obtained. As the size of magnetic nanoparticles is very small, they show superparamagnetism due to the absence of remanence, while $Fe_3O_4@C$ shows superferromagnetic properties. These magnetic nanoparticles are potent dye degraders and show good adsorption properties [60].

Controlled sized MNPs having monodispersity could be achieved by corrosion of precursors (organic) of Fe like $Fe(Cup)_3$, $Fe(CO)_5$ and $Fe(acac)_3$ at a high temperature in organic solvents. At a high temperature of $100°C$ Fe carbonyl decomposition in solvent medium having oleic acid and octylether generally forms iron carbonyl [61]. In an additional two-step procedure, thermal decomposition was achieved at a temperature of $100°C$ using $Fe(CO)_5$ (substrate) and oleic acid (solvent) producing highly crystalline monodispersed maghemite crystals. Following this procedure, a size range of 4 to 16 nm could be achieved [62].

Tartaj *et al.* (2006) and Sato *et al.* (1994) reviewed detailed chemistry and mechanism of synthesis of magnetic nanoparticles [30, 63]. They summarized that dimension and morphology of the nanoparticles could be controlled by temperature, pressure, ratio of precursor's concentration, solvent medium and complexation strength. Colloidal media can be stabilized using stabilizers and surfactants by the phenomenon of adsorption on the surface of magnetic nanoparticles. Moreover, it is significant for manufacturing on an industrial scale.

Magnetite nanoparticles which are hydrophobic in nature could be changed to hydrophilic using bipolar surfactant, where hydrophilic aminopropyl groups could adhere to the surface *via* silica shell [64]. Sun *et al.* (2004) reported the use of $Fe(ocaac)_3$ and 1,2-hexadecanediol in oleic acid and oleyl amine respectively, for the synthesis of extremely monodispersed MNPs. Following this method, particles having a size range of 4-20 nm could be obtained and their hydrophobic nature could be transformed to hydrophilic as bipolar surfactants are added [65].

Nanoparticles with hydrophobicity having a size range of 4-11 nm might be obtained employing a method of thermal decomposition of ironoleate as well as iron-pentacarbonyl over a wide range of temperatures and they can be easily dispersed in different organic solvents [66]. Several researchers are operational for the synthesis of nanoparticles using iron chloride as precursor, as it is non-toxic in nature [67, 68]. Magnetic nanoparticles are usually dispersed in solvents which are organic in nature. Whereas they are non-dispersible in water as they are hydrophobic in nature. Post-synthetic procedures are needed to make nanocrystals hydrophilic. Some workers synthesized hydrophilic magnetic nanoparticles by thermal decomposition of iron chloride and iron acetylacetone in both acidic and basic media [69-70]. They used 2-pyrollidone as solvent, which is an excellent stabilizer and it forms co-ordination bonds with metal ions [71]. It is fascinating that with reflux time alteration, the morphology of MNPs alters from spherical form to cubic. PEG coated MNPs which are soluble in water could be synthesized under a similar experimental setup by means of surface capping agents such as mono-carboxyl-terminated PEG [72] or else α, ω-dicarboxyl-terminated PEG [73].

Wan *et al.* (2007) synthesized water-soluble MNPs *via* thermal decomposition of $Fe(acac)_3$ using triethylene glycol as solvent [74]. To a huge degree, they were capable of preventing aggregation of particles as well as growth. Maity *et al.* (2009) extended this work by anticipating solubility of magnetic nanoparticles in aqueous medium by changing factors such as charge on the surface of MNPs and their coating. Astonishingly, the outcome represented the development of MNPs which are crystalline, monodispersed and superparamagnetic with 100K blocking temperature [75].

1.2.1.3 Sol-Gel/Polyol technique

It is a wet synthetic method used for the synthesis of nanoparticles in a reaction medium by hydroxylation and condensation of substrate. It forms a "sol" of nanoparticles. If the "sol" is condensed and polymerized, it forms 3-D nanostructured wet gels [76, 77]. Previous studies revealed that synthesized gel properties are solely dependent on the structure formed during the sol-stage and formation of crystal structure [76, 78]. Precursor salt concentration, solvent, pH and temperature are key factors that administer the reaction kinetics, nucleation and structural properties of the formed gel [79, 80].

At a low room temperature, sol-gel methods could be employed for the synthesis of MNPs [81-84] using precursors like alkoxide or non-alkoxide. Hasanpour *et al.* (2013) and Cui *et al.* (2013) have synthesized magnetic nanoparticles utilizing precursors such as ferric nitrate nonahydrate and solvents such as ethanol or ethylene glycol under optimized conditions of temperature [85, 86]. The sol-gel method is extensively exploited for the synthesis of nanocomposites in combination with different inorganic materials [87].

1.2.1.4 Microwave assisted technique

In co-precipitation and hydrothermal environments, this method is explored for the synthesis of MNPs [88-90]. This process is extremely significant because of its very short reaction time. In this method, ethylene or diethylene glycol and hydrazine-N_2H_4 are used as reducing agents and capping agents. This method is employed to achieve biologically compatible nanoparticles in a very petite reaction time. Xiao *et al.* (2012) used this technique for synthesizing magnetic nanoparticles having a fine size distribution and used those nanoparticles as contrast agent in MR blood pool [91]. Same nanoparticles of the size range between 40-300 nm were formed following the same parameters using the "Polyol method". Komarneni *et al.* (2012) and Nan *et al.* (2009) followed this technique to construct functionalized magnetic nanoparticles by altering factors like temperature, precursors, time and synthetic conditions [92-93].

1.2.1.5 Oil-water emulsion method

For the last three decades, it has been well liked traditional synthetic wet method to synthesize MNPs [94]. In this method, the microemulsion having the reverse aqueous micellar phase is diffused into a continuous oil phase. It is also employed to synthesize core shell magnetic nanoparticles. Lu *et al.* (2013) synthesized MNPs following this technique and prepared nanoparticles having very small size with the help of surfactants such as DTAB, Brij30, SDS, DEAB, DBAB, CTAB, and Gemini surfactant [95]. The entire preparation was carried out in an atmosphere having nitrogen to avoid oxidation.

The MNPs prepared were aged, washed and subsequently dried for approximately 8 hours at 80°C. Following given parameters, MNPs having a size of 13-16 nm were recovered.

1.2.1.6 Electrochemical technique

This method was instituted by Reetz *et al.* (1994) and Pascal *et al.* (1999) for the synthesis of MNPs having a size range of 3 to 8 nm by means of an iron electrode in aqueous cationic surfactant as well as DMF. It was observed that current density is directly linked to change on MNPs [96-97].

1.2.2 Physical technique

Physical method employs different techniques for the synthesis of MNPs *viz.* aerosol method, gas phase deposition, electron beam lithography, power ball milling, and pulsed laser ablation. Among all the aforesaid methods, aerosol-based preparation is commonly used, which is discussed below in detail.

1.2.2.1 Aerosol/Pyrolysis/vapour method

In this method, aerosols are prepared by means of diluted precursors using an ultrasound generator. Aerosols with the assistance of carrier gas are carried to a quartz tube furnace where post-synthetic processes such as calcination, evaporation, and densifications of powder materialize. The prepared sample is recovered using high voltage mo-wire at the end of the tube [98]. In contrast to previous discussed methods, this method is much more economical as it involves less investment and other operational expenses [99]. This process is in use for the preparation of superparamagnetic nanoparticles having diverse shapes and sizes with or without silica [100].

1.2.3 Biological technique

This method is the least used method to synthesize nanoparticles as it yields fewer amounts of nanoparticles and requires more time. Stability of synthesized nanoparticles is also less. This method is adopted for the green synthesis of nanoparticles concerning toxicity associated with them. Method is discussed as below.

1.2.3.1 Biosynthesis by means of fungi, bacteria and yeast

MNPs can be synthesized using biosynthetic methods using actinomycetes, bacteria, yeast, fungi, and viruses. Though synthesized nanoparticles are not monodispersed but the rate of synthesis is very sluggish. Magnetotactic bacteria are widely used to synthesize magnetic nanoparticles (magnetite form) [101]. Certain bacteria such as (i) *A. magnetotacticum* [102], (ii) *Magnetotactic bacterium* (MV1) [103], (iii) Sulfate-reducing bacteria [104], (iv) *Magnetospirillum magnetotacticum* [105], and (v) *M. magnetotacticum* (MS-1) [106] are extensively used for the synthesis of MNPs.

1.3 Surface modification of magnetic nanoparticles for biomedical applications and their effect on stability and magnetization

As conversed in the preceding sections it has been observed that synthesized metallic nanoparticles are usually hydrophobic in nature. This problem raises stability as well as dispersion issues. Although nanoparticles may show an exceptional range of shape, size and magnetization, their targeted applicability due to hydrophobic nature does not match expectations. Therefore, modification of nanoparticles with respect to their surface becomes an important parameter to enhance their targeted applicability in various fields, such as targeted drug delivery, MRI, bio-separation, in addition to numerous other biomedical applications. Details of surface modification of nanoparticles using different methods are discussed below in the coming subsections. Figure 5 depicts a broad schematic illustration by which modification of the surface of MNPs could be done.

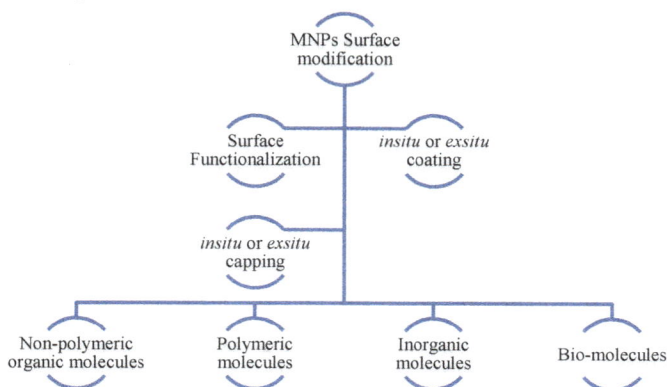

Fig. 5. Scheme showing mechanisms of MNPs surface modification and materials used.

1.3.1 Non-polymeric organic molecule-based surface modification

At the outset, Yee *et al.* (1999) found that amorphous MNPs being surface adsorbed by alkanephosphonic and alkanesulphonic acid stabilized the colloidal dispersion. They found that there are two possible bonding strategies (i) where only one O atom could bind onto the surface of Fe^{+3} ions and (ii) where two O atoms could bind onto the surface of Fe^{+3} ions [107]. Sahoo *et al.* (2001) surface functionalized magnetite using lauric acid, oleic acid and dodecylphosphonic acid for stabilizing nanoparticles in organic solvent system. The results demonstrated that alkyl phosphonates and phosphonates are excellent agents for derivatization, for obtaining a system which is thermodynamically stable. The bonding mechanism on the surface of nanoparticles is explained by quasi-bilayer structure formation by organic non-polymeric molecules [108]. Since polymeric molecules direct the formation of a very thick surface layer, Porter *et al.* (2001) exploited monomeric organic molecules as coating agents for iron oxides which is adequate to prevent protein absorption [109]. Notably, pure Phosphorylcholine in comparison to derived Phosphorylcholine polymer found acting as protection of prosthesis by protein contaminant inhibition [110].

1.3.2 Polymeric molecules-based surface modification

MNPs surface coated with polymeric materials exhibit high potential applications in different fields of science. The synthesis is carried out by precipitation of nanoparticles in gel matrix or polymer [111]. For the first time Ugelstad *et al.* (1993) carried out iron oxide's direct precipitation within the polystyrene pores having 2.8 and 4.5 μm particle size using acrylamide as monomer. But the nature of prepared particles was hydrophobic [112]. Interestingly, Kawaguchi *et al.* (1996) were the first to report magnetic latexes having hydrophilic nature utilizing acrylamide as monomer [113]. Sauzedde *et al.* (1999) further carried similar work following methodology suggested by Furusawa *et al.* (1994). They found that these latexes are sensitive to temperature, therefore, exhibits temperature sensitive properties [114-115].

Lee *et al.* (1996) reported PVA modified MNPs *via* iron salt's precipitation with PVA. They found that an increase in PVA concentration leads to the formation of nanoparticles having crystalline nature [116]. In another method, *i.e., "ex-situ* method", MNPs were modified following precipitation to obtain better dispersion stability [117]. In comparison to surface engineered nanoparticles, uncoated particles align freely to an applied magnetic field and in turn inhibit agglomeration [118]. Generally, Coating reduces the magnetic moments because of the "non-collinear spin structure" at the nanoparticle's interface and coated surfactants.

Nanoparticles must be extremely hydrophilic for drug delivery applications. Their size must range to less than 100 nm to escape barriers of biological membranes. Therefore, to enhance the time period of circulation of these nanoparticles in the stream of human blood, it becomes important to make them hydrophilic through coating with hydrophilic polymers such as poly ethylene glycol, which could inhibit absorption of proteins and cause agglomerated clusters of macrophages inside the system [119].

1.3.3 Inorganic molecules-based surface modification

Nanoparticles of the metallic core shell consists of iron oxide (inside) and metallic shell (outside) of inorganic material. Magnetic nanoparticles treated with silica, gold and gadolinium have already been reported with potential application in different fields of biomedical sciences. Chen *et al.* (2003) developed two different types of gold shell magnetic nanoparticles (i) acicular and (ii) spherical, and, studied their capacity of heat treatment, co- reactivity and magnetization saturation [120].

Gold shell proved to be an exceptional surface for consequential functionalization by means of biological and chemical molecules [121]. Coating with gold was also carried out by Carpenter (2001) using microemulsion [122]. Surface gold coatings affect core oxidation and result in biocompatibility for biomedicine applications. Zhou *et al.* (2001) synthesized an iron core coated with gold nanoparticle using reverse micellar technique and attained a core size of approximately 6 nm and 1-2 nm shell size [123]. A lot of research has also been carried out on silica coated magnetic nanoparticles [124-126]. Moreover, the barriers related to the application of magnetic nanoparticles in Magnetic Resonance Imaging have also been resolved by Gadolinium based surface functionalization [127-128].

1.3.4 Target specific ligands and biomolecules-based surface functionalization

Biomolecules-based ligands such as vitamins, proteins, antibodies and amino acids could be bound on the surface of nanoparticles or polymers using linkers such as amide esters for targeted applications. Amino acid-based fictionalization of magnetic nano-particles is also important in introducing minor doses of amino acids to overcome the deficit of desired amino acid in the system [129]. In this context, used amino acids are essential and it is vital to additionally supply these amino acids from outside to supplement the system (Table 2).

Table 2. Coating agents with tailored functional group after surface functionalization.

Functional group	Functional moieties	References
-OH	Polyetyleneglycol	[133]
	Dextran	[134]
	Polyvinylalcohol	[135]
-NH$_2$	Di-amine	[136]
	Polyethyleneglycol with terminal amine group	[137]
	Poly (L-lysine)	[138]
	Polyetylenemine	[139]
	Chitosan	[140]
-COOH	Citratic acid	[141]
	Polymethacrylic acid	[142]
	Alginate	[143]
	Polyethyleneglycol with terminal carboxyl group	[144]
	Carboxymethyl cellulose	[145]
	Polyacrylic acid	[146]

Also, it has been testified that excess of amino acids, for instance, leucine and arginine increase shrinkage of cells, leading to cell death by malignancy [130]. Cysteine sulfhydryl (-SH) group is reported as an excellent antioxidant for repairing immune system defects in patients suffering from AIDS, as well as cancer [131-132]. So, it can be concluded that magnetic nanoparticles having a coating of amino acids have tremendous potential for cancer diagnosis and combating the disease.

1.4   Applications of MNPs

Since the early 40's, magnetic nanoparticles have been well known for their application as carrier-systems and have been adapted for waste-water-treatment for the very first time. Since then, magnetic nanoparticles having diverse shapes, sizes, and properties are in use for different biological and industrial applications (Fig. 6).

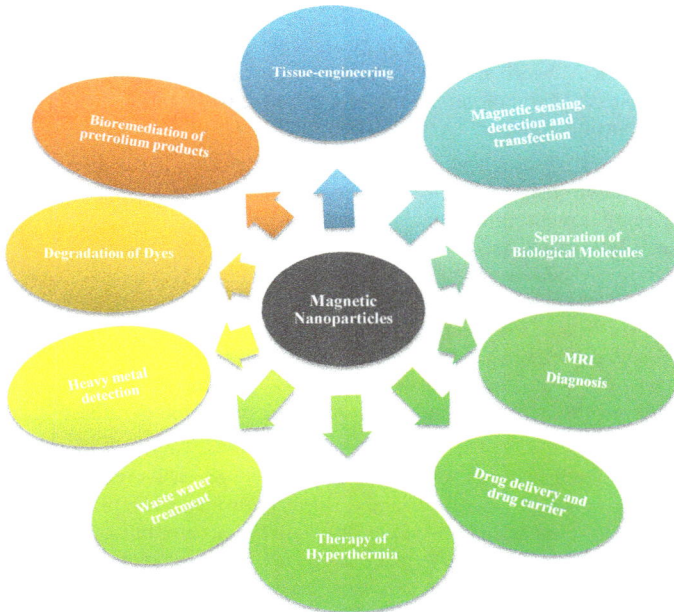

Fig. 6. Diverse environmental and biomedical applications of magnetic nanoparticles

1.4.1 Applications in biological systems

Nano-medicine is one of the approaching fields offering answers to a variety of problems related to biomedical science. For instance, antibiotic discovery was a major historical event, but now resistance in microbes to antibiotics is becoming a major problem [147-148]. Different research groups are practicing to develop materials using nano-science to combat such problems and prepare certain materials having antimicrobial and anticancer biological properties. Some of the important applications of MNPs are discussed below in detail.

1.4.1.1 Application in drug delivery

Magnetic nanoparticles have splendid properties, making them excellent carriers in drug delivery systems which are very targeted and specific to an organ. Widder *et al.* (1970) explained magnetic nanoparticle's applications as a carrier in drug delivery [149]. Magnetic nanoparticles sustainability is exclusively size and surface dependent [150]. The drug is released and absorbed by the targeted cells as it concentrates on a targeted location. The rate of release of the targeted drug is dependent on factors such as morphology and nature of supporting materials, mechanism of delivery, temperature of release and pH.

Both complex systems such as core shell and composites as well as magnetic loaded systems have been tested for drug delivery. The test drug is chemically bound on the surface of iron *via* several functional groups such as carboxyl, amino and thiol [151-154]. Yuan *et al.* (2008) used magnetic nanoparticles carriers for the drug doxorubicin. The magnetic nanoparticles served as core which are bound to doxorubicin through an acid-labile hydrozone bond and are a polymer chitosan derivative encapsulated which is temperature sensitive. This practice was carried out at pH 5.3 for cancerous cells and also at pH 7.4 for blood cells. Also, temperatures of 20°C, 37°C as well as 40°C (hyperthermal temperature) were evaluated for testing the efficacy of the nanoparticles. The expressed results displayed that pH 5.3 is more favorable for doxorubicin release in comparison to pH 7.4 [155].

Also, magnetic nanoparticles coated with chitosan are extensively tested and used for the release of drugs (for *e.g.,* Doxorubicin and Bortezomib) based on pH [156-157]. Cisplatin has also been exploited as an anti-cancer drug delivery carrier but as of its huge toxicity it is necessary to target drug very specifically to the target cells [158-159]. Magnetic nanoparticles have also been studied as drug carrier for ibuprofen (anti-inflammatory drug) for the targeted delivery of drug [160]. Table 3 depicting different types of MNP-based drug delivery systems along with of mechanism of drug release.

Table 3. Reported studies on MNPs functionalization and their potential use as drug carrier for drug delivery.

Functionalization agent	Drug	Drug release mechanism	Ref.
Arabic acid, Silica & polysaccharide cross linker	Antibody	Enzymatic cleavage	[161]
Chitosan and Poly acrylic acid Multilayer	Cefradine	pH mediated	[162]
Chitosan	Cefradine	pH mediated	[163]
PAH@Fe$_3$O$_4$	Fluorescein isothiocyanate (FITC)-Dextran	Magnetic field facilitated	[164]
PolyamidoaminePoly (methyl methacrylate) PMMA	Fluorescein iso-thiocyanate (FITC)	Thermo facilitated	[165]
Poly-L-lysine hydrochloride (PLL), poly-L- glutamic acid (PGA)	DNA	pH mediated	[166]
Alginate@chitosan	Insulin	Magnetic field facilitated	[167]
2-hydroxypropyl- cyclodextrin (HCD)@Gum Arabic@MNPs (GAMNP)	Ketoprofen	Undefined	[168]
PEG@SiO$_2$	Doxorubicin	Enzymatic cleavage / Osmosis mediated	[169]
Poly-N-isopropylacrylamide (PNIPAAM) and poly-D, L-lactide-co glycolide (PLGA)	Bovine Serum Albumin (BSA) and Curcumin	Temperature Facilitated	[170]
Poly(N-isopropylacrylamide) (PNIPAM)	Doxorubicin	Thermo Facilitated	[171]
Poly-aniline-co-sodium N-(1-onebutyric acid)-aniline (SPAnNa)	1,3-bis 2-chloroethyl -1-nitrosourea	Ultrasound & Magnetic field	[172]

Poly (sodium 4-styrenesulfonate) (PSS) and poly (allylamine hydrochloride) (PAH)	Dye	Magnetic heating	[173]
Polyethyleneglycol (PEG)	Doxorubicin	enzymatic cleavage	[174]
Multiwalled carbon nanotubes (MWNTs)	Doxorubicin	pH mediated	[175]
Cyclodextrin with pluronic polymer (F-127)	Curcumin	Dissociation of surface bound curcumin molecules that exist on the CD or F127 polymer matrix. The remaining sustained drug release was due to the slow release of the drug entrapped inside CD and/or F127 polymer layers.	[176]
Polyaniline-co-N-1-one-butyric acid aniline shell with biocompatible O-(2-aminoethyl) PEG (EPEG)	Doxorubicin, Paclitaxel	Temperature Facilitated	[177]
PEG-b-PCL deblock copolymers	Triamterene & doxorubicin	Temperature Facilitated	[178]

1.4.1.2 Hyperthermia therapy

Hyperthermia is a method of treating cancer cells in different regions of the body by elevating temperature. The temperature range lies between 41 to 43°C for approximately 20 to 60 minutes for treating malignant cells of the tissues [150, 179-180]. This temperature range is attained by different sources such as microwave, radiofrequency, infrared irradiation and ultrasonication [181-182].

Magnetic hyperthermia is an evolving and one of the innovative approaches for treating cancer tissues [183]. Applied magnetic field ranging between 50-300 kHz is generally practiced for generation of heat. According to the available literature, ferrite ions (*e.g.,* cobalt ferrite) show minor toxicity and higher heating are generally preferred for magnetic hyperthermia therapy [184, 187]. Qu *et al.* (2010) used Fe_3O_4-chitosan based nanoparticle size of around 10.5 nm size for hyperthermia. This study revealed that a Fe_3O_4-chitosan based system has tremendous potential for generation of heating for hyperthermia [188].

1.4.1.3 Applications in bio-separation

Among the various materials available, magnetic nanoparticles are one of the extensively used material employed for separation of different biomolecules, as they are easy to remove under an applied magnetic field. This particular property emerges due to the presence of hydroxyl groups present on the surface of MNPs, which usually form hydrogen as well as covalent bonds with different biomolecules [189-191]. MNPs could be used in their different available forms *viz.* composite, core or oxide forms. They are feasible in both active and passive separations.

Aguilar-Arteaga *et al.* (2010) summarized a review describing magnetite analytical application. In this review they emphasized functionalized solid magnetic matrix with certain agents and detailed their use in separation and enzyme immobilization [192]. Also, Liu *et al.* (2009) carried out synthesis of $Fe_3O_4@NiSiO_3$ following hydrothermal technique and carried out magnetic separation of HIS-tagged proteins as well as of biomolecules having very low molecular weight [193].

1.4.1.4 Applications in tissue engineering

In the tissue engineering field, MNPs have enormous potential applications because of their high level of susceptibility and target specific nature. They could be guided to specific cells, especially to osteoblast [194]. Panseri *et al.* (2012) examined the effect of iron on bone remodeling. They carried out the synthesis of HA/Fe_3O_4 using various ferrite salts and examined their *in vitro* as well as *in vivo* activities under a constant applied magnetic field. The results revealed excellent viability of human osteoblast cells [194]. Kito *et al.* (2013) exploited magnetite particles for constructing multilayered sheets of pluripotent stem cells [195].

A number of studies have found that magnetic nanoparticles could inhibit damage to DNA especially in epithelial cells of the lungs and hemopatoma [196-197], while other researchers have worked on cell apoptosis [198]. Ito *et al.* (2004; 2005 and 2006) worked on magnetically driven tissue mimics such as multi-layered cell sheets and heterotypic cell sheets [199-201].

1.4.1.5 Applications in MRI imaging

Among all biomedical applications, MNPs exhibit an important application in the field of MRI (magnetic resonance imaging). Barick *et al.* (2014) worked on carboxyl based magnetic nanoparticles for MRI. This study found that magnetic nanoparticles having a size of about 10 nm provide an excellent magnetization, cyto-compatibility and good stability with cell lines [202]. While, Kim *et al.* (2005) prepared magnetic nanoparticles having oleic acid as a surfactant showing biomedical application as MRI contrast agent [203].

1.4.1.6 Applications as catalyst

Magnetite and hematite magnetic nanoparticles could be used as catalysts for various industrial reactions such as high temperature water gas shift, Haber's process, Fisher-Tropsch synthesis, desulfurization reaction, large-scale manufacturing of butadiene and alcohol oxidation [204-205].

Concluding remarks

Considerable progress has been made in the development of stable and monodispersed magnetic nanoparticles for developing potent applications in the area of nano-biotechnology. As MNPs show wider biological application, the colloidal and water stable variants are in the main stream of consideration when opting for methods of synthesis. Among all the available methods, the co-precipitation method is found as the most efficient as well as cost effective method for the synthesis of MNPs and organometallic nanomaterials. Though the co-precipitation method is the most effective method for one pot synthesis of water compatible MNPs, it is found incompetent for manufacturing uniform-sized distributed particles. However, a number of coating agents have been examined for providing stability to magnetic nanoparticles by breaking inter-columbic forces and restricting cluster formation. Biomolecules are of great interest in connection to their higher biocompatibility. Also, various proteins, antibodies, DNA and vitamins have been reported as functional ligands on the surface of magnetic nanoparticles for certain targeted applications.

References

[1] S.M. Moghimi, A.C.H. Hunter, J.C. Murray, Long-circulating and target-specific nanoparticles: theory to practice, Pharm Rev. 53 (2001) 283-318.

[2] J. Jeevanandam, A. Barhoum, Y.S. Chan, A. Dufresne, M.K. Danquah, Review on nanoparticles and nanostructured materials: history, sources, toxicity and regulations, Beilstein J. Nanotechnol. 9 (2018) 1050-1074. https://doi.org/10.3762/bjnano.9.98

[3] U. Riaz, T. Mehmood, S. Iqbal, M. Asad, R. Iqbal, U. Nisar, M.M. Akhtar, Historical Background, Development and Preparation of Nanomaterials, in: M.B. Tahir, M. Rafique, M. Sagir (Eds.), Nanotechnology and photocatalysis for environmental applications Nanotechnology, Springer, Singapore, 2021, https://doi.org/10.1007/978-981-15-9437-3_1

[4] I. Khan, K. Saeed, I. Khan, Nanoparticles: Properties, applications and toxicities, Arabian Journal of Chemistry. 12 (2019) 908-931. https://doi.org/10.1016/j.arabjc.2017.05.011

[5] A.R.L. Pimenta, NanopartículasMagnéticas para Nanomedicina, Dissertação para obtenção do Grau de Mestre emEngenharia de Materiais (2010), Thesis.

[6] W. Smith, J. Hashemi, Foundations of Materials Science and engineering, fourth ed., McGraw-Hill Higher Education, 2005.

[7] U. Wertmann, R. Cornell, Iron Oxides in the Laboratory, second ed., John Wiley & Sons, 2000.

[8] S.M. Dadfar, K. Roemhild, N.I. Drude, S. von Stillfried, R. Knüchel, F. Kiessling, T. Lammers, Iron oxide nanoparticles: Diagnostic, therapeutic and theranostic applications, Advanced Drug Delivery Reviews. 138 (2019) 302-325. https://doi.org/10.1016/j.addr.2019.01.005

[9] Z. Peter, Magnetic nanoparticles: production and applications, Masaryk University, 2013.

[10] A. Bahari, Characteristics of Fe_3O_4, α -Fe_2O_3, and γ- Fe_2O_3 Nanoparticles as Suitable Candidates in the Field of Nanomedicine, J. Supercond. Nov. Magn. 30 (2017) 2165-2174. https://doi.org/10.1007/s10948-017-4014-8

[11] L.S. Ganapathe, M.A. Mohamed, R.M. Yunus, D.D. Berhanuddin, Magnetite (Fe_3O_4) Nanoparticles in Biomedical Application: From Synthesis to Surface Functionalisation, Magnetochemistry 6 (2020) 68. https://doi.org/10.3390/magnetochemistry6040068

[12] N. Ajinkya, X. Yu, P. Kaithal, H. Luo, P. Somani, S. Ramakrishna, Magnetic Iron Oxide Nanoparticle (IONP) Synthesis to Applications: Present and Future, Materials (Basel).13 (2020) 4644. https://doi.org/10.3390/ma13204644

[13] A. Ali, H. Zafar, M. Zia, I. ul Haq, A.R. Phull, J.S. Ali, A. Hussain, Synthesis, characterization, applications, and challenges of iron oxide nanoparticles, Nanotechnol Sci. Appl. 19 (2016) 49-67. https://doi.org/10.2147/NSA.S99986

[14] M. Salvador, G. Gutiérrez, S. Noriega, A. Moyano, M.C. Blanco-López, M. Matos, Microemulsion Synthesis of Superparamagnetic Nanoparticles for Bioapplications, Int. J. Mol. Sci. 22 (2021) 427. https://doi.org/10.3390/ijms22010427

[15] S. Majidi, F.Z. Sehrig, S.M. Farkhani, M.S. Goloujeh, A. Akbarzadeh, Current methods for synthesis of magnetic nanoparticles, Artif. Cells Nanomed. Biotechnol. 44 (2016) 722-34. https://doi.org/10.3109/21691401.2014.982802

[16] J.A. Fuentes-García, A. Carvalho Alavarse, A.C. Moreno Maldonado, A. Toro-Córdova, M.R. Ibarra, G.F. Goya, Simple Sonochemical Method to Optimize the Heating Efficiency of Magnetic Nanoparticles for Magnetic Fluid Hyperthermia, ACS omega. 5 (2020) 26357-26364. https://doi.org/10.1021/acsomega.0c02212

[17] A. Ali, T. Shah, R. Ullah, P. Zhou, M. Guo, M. Ovais, Z. Tan, Y. Rui, Review on Recent Progress in Magnetic Nanoparticles: Synthesis, Characterization, and Diverse

Applications, Front Chem. 9 (2021) 629054.
https://doi.org/10.3389/fchem.2021.629054

[18] S. Shukla, R. Khan, A. Daverey, Synthesis and characterization of magnetic nanoparticles, and their applications in wastewater treatment: A review, Environmental Technology & Innovation. 24 (2021) 101924. https://doi.org/10.1016/j.eti.2021.101924

[19] A. Baki, N. Löwa, A. Remmo, F. Wiekhorst, R. Bleul, Micromixer Synthesis Platform for a Tuneable Production of Magnetic Single-Core Iron Oxide Nanoparticles, Nanomaterials. 10 (2020) 1-25. https://doi.org/10.3390/nano10091845

[20] A. Mota-Cobián, C. Velasco, J. Mateo, S. España, Optimization of purification techniques for lumen-loaded magnetoliposomes, Nanotechnology. 31 (2020) 145102. https://doi.org/10.1088/1361-6528/ab5f80

[21] L. Babes, B. Denizot, G. Tanguy, J.J Le Jeune, P. Jallet, Synthesis of Iron Oxide Nanoparticles Used as MRI Contrast Agents: A Parametric Study, J. Colloid Interface Sci. 212 (1999) 474-482. https://doi.org/10.1006/jcis.1998.6053

[22] H. Mohammadi, E. Nekobahr, J. Akhtari, M. Saeedi, J. Akbari, F. Fathi, Synthesis and characterization of magnetite nanoparticles by co-precipitation method coated with biocompatible compounds and evaluation of in-vitro cytotoxicity, Toxicology reports, 8 (2021) 331-336. https://doi.org/10.1016/j.toxrep.2021.01.012

[23] N.A. Yazid and Y.C. Joon, Co-precipitation Synthesis of Magnetic Nanoparticles for Efficient Removal of Heavy Metal from Synthetic Wastewater, 6th International Conference on Environment, 6 (2019) 1-11. https://doi.org/10.1063/1.5117079

[24] Y.K. Sun, M. Ma, Y. Zhang, N. Gu, Fe_3O_4@Pt nanoparticles with enhanced peroxidase-like catalytic activity, Colloids Surf. A. 105 (2013) 36-39. https://doi.org/10.1016/j.matlet.2013.04.020

[25] J.P. Jolivet, C. Chaneac, E. Tronc, Iron oxide chemistry. From molecular clusters to extended solid networks, Chem. Commun. 5 (2004) 481-483. https://doi.org/10.1039/B304532N

[26] R.K. Gautam, M.C. Chattopadhyaya, Functionalized Magnetic Nanoparticles: Adsorbents and Applications, in: R.K. Gautam, M.C. Chattopadhyaya (Eds.), Nanomaterials for Wastewater Remediation, Butterworth-Heinemann, 2016, pp. 139-159. https://doi.org/10.1016/B978-0-12-804609-8.00007-8

[27] U. Manzoor, F.T. Zahra, S. Rafique, MT. Moin, M. Mujahid, Effect of Synthesis Temperature, Nucleation Time, and Postsynthesis Heat Treatment of ZnO Nanoparticles and Its Sensing Properties, Journal of Nanomaterials. 18 (2015) 1-6. https://doi.org/10.1155/2015/189058

[28] R.M. Cornell, U. Schertmann, Iron Oxides in the Laboratory: Preparation and Characterization, VCH Publishers, Weinheim, Germany, 1991.

[29] N.M. Gribanow, E.E. Bibik, O.V. Buzunov, V.N. Naumov, Physico-chemical regularities of obtaining highly dispersed magnetite by the method of chemical condensation, J. Magn. Magn. Mater. 85(1990) 7-10. https://doi.org/10.1016/0304-8853(90)90005-B

[30] P. Tartaj, M.P. Morales, S. Veintemillas-Verdaguer, T. Gonzalez-Carreno, C.J. Serna, Synthesis, properties and biomedical applications of magnetic nanoparticles, Handbook of Magnetic Materials, 2006. https://doi.org/10.1016/S1567-2719(05)16005-3

[31] V.K. LaMer, R.H. Dinegar, Theory, production and mechanism of formation of monodispersed hydrosols, J. Am. Chem. Soc. 72 (1950) 4847. https://doi.org/10.1021/ja01167a001

[32] F. Tourinho, R. Franck, R. Massart, R. Perzynski, Synthesis and mangeitc properties of managanese and cobalt ferrite ferrite ferrofluids, Prog. Colloid Polym. Sci. 79 (1989) 128-134. https://doi.org/10.1007/BFb0116198

[33] R. Weissleder, U.S. Patent, 5,492,814, 1996.

[34] H. Itoh, T. Sugimoto, Systematic control of size, shape, structure, and magnetic properties of uniform magnetite and maghemite particles, J. Colloid. Interface Sci. 265 (2003) 283-295. https://doi.org/10.1016/S0021-9797(03)00511-3

[35] D. Thapa, V.R. Palkar, M.B. Kurup, S.K. Malik, Properties of magnetite nanoparticles synthesized through a novel chemical route, Mater. Lett. 58 (2004) 2692-2694. https://doi.org/10.1016/j.matlet.2004.03.045

[36] H. Pardoe, W. Chua-anusorn, T.G. St. Pierre, J. Dobson, Structural and magnetic properties of nanoscale iron oxide particles synthesized in the presence of dextran or polyvinyl alcohol, J. Magn. Magn. Mater. 225 (2001) 41-46. https://doi.org/10.1016/S0304-8853(00)01226-9

[37] S.E. Khalafalla, G.W. Reimers, Preparation of dilution-stable aqueous magnetic fluids, IEEE Trans. Magn. 16 (1980), 178-183. https://doi.org/10.1109/TMAG.1980.1060578

[38] R. Massart, Preparation of aqueous magnetic liquids in alkaline and acidic media, IEEE Trans. Magn. 17 (1981) 1247-1248. https://doi.org/10.1109/TMAG.1981.1061188

[39] R. Massart, V. Cabuil, Synthèseen milieu alcalin de magnétitecolloïdale: contrôle du rendement et de la taille des particules: Synthesis of colloid dal magnetite in alkaline medium: yield and particle size control, J. Chim. Phys. 84 (1987) 967-973. https://doi.org/10.1051/jcp/1987840967

[40] C. Iacovita, I. Fizeşan, A. Pop, L. Scorus, R. Dudric, G. Stiufiuc, N. Vedeanu, R. Tetean, F. Loghin, R. Stiufiuc, C.M. Lucaciu, In Vitro Intracellular Hyperthermia of Iron Oxide Magnetic Nanoparticles, Synthesized at High Temperature by a Polyol

Process, Pharmaceutics. 12 (2020) 424.
https://doi.org/10.3390/pharmaceutics12050424

[41] F. Arteaga-Cardona, N.G. Martha-Aguilar, J.O. Estevez, U. Pal, M.Á. Méndez-Rojas, U. Salazar-Kuri, Variations in magnetic properties caused by size dispersion and particle aggregation on CoFe$_2$O$_4$, SN Appl. Sci. 1 (2019) 412. https://doi.org/10.1007/s42452-019-0447-y

[42] B.I. Kharisov, H. Dias, O.V. Kharissova, A. Vázquez, Y.F. Peña, I. Gómez, Solubilization, dispersion and stabilization of magnetic nanoparticles in water and non-aqueous solvents: recent trends, RSC Advances. 4 (2014) 45354-45381. https://doi.org/10.1039/C4RA06902A

[43] R. Massart, J. Roger, V. Cabuil, New Trends in Chemistry of Magnetic Colloids: Polar and Non-Polar Magnetic Fluids, Emulsions, Capsules and Vesicles, Braz. J. Phys. 25 (1995) 135-141.

[44] S. Neveu-Prin, V. Cabuil, R. Massart, P. Escaffre, J. Dussaud, Encapsulation of magnetic fluids, J. Magn. Magn. Mater. 122 (1993) 42-45. https://doi.org/10.1016/0304-8853(93)91035-6

[45] J.P. Jolivet, P. Belleville, E. Tronc, J. Livage, Influence of Fe(II) on the formation of the Spinel iron oxide in alkaline medium, J. Clays Clay Miner. 40 (1992) 531-539. https://doi.org/10.1346/CCMN.1992.0400506

[46] E. Tronc, P. Belleville, J.P. Jolivet, J. Livage, Transformation of ferric hydroxide into spinel by iron(II) adsorption, Langmuir 8 (1992) 313-319. https://doi.org/10.1021/la00037a057

[47] M. Bustamante-Torres, D. Romero-Fierro, J. Estrella-Nuñez, B. Arcentales-Vera, E. Chichande-Proaño, E. Bucio, Polymeric Composite of Magnetite Iron Oxide Nanoparticles and Their Application in Biomedicine: A Review, Polymers, 14 (2022) 752. https://doi.org/10.3390/polym14040752

[48] J.P. Jolivet, Metal Oxide Chemistry and Synthesis from Solution to Solid State, Wiley: Chichester, U.K., 2000.

[49] X. Qui, Synthesis and characterization of magnetic nanoparticles. Chin. J. Chem. 18 (2000) 834-837. https://doi.org/10.1002/cjoc.20000180607

[50] R.F. Ziolo, E.P. Giannelis, B.A. Weinstein, M.P. O'Horo, B.N. Ganguly, V. Mehrotra, M.W. Russell, D.R. Huffman, Matrix-Mediated Synthesis of Nanocrystalline γ- Fe$_2$O$_3$: A New Optically Transparent Magnetic, Mater. Sci. 257 (1992) 219-223. https://doi.org/10.1126/science.257.5067.219

[51] L.F. Shen, P.E. Laibinis, T.A. Hatton, Bilayer surfactant stabilized magnetic fluids: Synthesis and interactions at interfaces, Langmuir. 15 (1999) 447-453. https://doi.org/10.1021/la9807661

Materials Research Forum LLC
https://doi.org/10.21741/9781644902332-1

[52] A. Ali, H. Zafar, M. Zia, I. ul Haq, A.R. Phull, J.S. Ali, A. Hussain, Synthesis, characterization, applications, and challenges of iron oxide nanoparticles, Nanotechnology, Science and Applications. 9 (2016) 49-67. https://doi.org/10.2147/NSA.S99986

[53] B. Mao, Z. Kang, E. Wang, S. Lian, L. Gao, C. Tian, C. Wang, Synthesis of magnetite octahedrons from iron powders through a mild hydrothermal method, Mater. Res. Bull. 41 (2006) 2226-2231. https://doi.org/10.1016/j.materresbull.2006.04.037

[54] H. Zhu, D. Yang, L. Zhu, Hydrothermal growth and characterization of magnetite (Fe$_3$O$_4$) thin films, Surf. Coat. Technol. 201 (2007) 5870-5874. https://doi.org/10.1016/j.surfcoat.2006.10.037

[55] S. Giri, S. Samanta, S. Maji, S. Ganguli, A. Bhaumik, Magnetic properties of α-Fe$_2$O$_3$ nanoparticle synthesized by a new hydrothermal method, J. Magn. Magn. Mater. 285 (2005) 296-302. https://doi.org/10.1016/j.jmmm.2004.08.007

[56] M.A. Willard, L.K. Kurihara, E.E. Carpenter, S. Calvin, V.G Harris, H. Nalwa, Encyclopedia of Nanoscience and Nanotechnology, fifth ed., vol. 1, American Scientific Publishers, Valencia, CA, 2004, 815.

[57] D. Chen, R. Xu, Hydrothermal synthesis and characterization of nanocrystalline Fe$_3$O$_4$ powders, Mater. Res. Bull. 33 (1998) 1015-1021. https://doi.org/10.1016/S0025-5408(98)00073-7

[58] Y.H. Zheng, Y. Cheng, F. Bao, Y.S. Wang, Synthesis and magnetic properties of Fe$_3$O$_4$ nanoparticles, Mater. Res. Bull. 41 (2006) 525-529. https://doi.org/10.1016/j.materresbull.2005.09.015

[59] F.C. Meldrum, B.R. Heywood, S. Mann, Magnetoferritin: in vitro synthesis of a novel magnetic protein, Science 257 (1992) 522. https://doi.org/10.1126/science.1636086

[60] D.P.E. Dickson, S.A. Walton, S. Mann, K. Wong, Properties of magnetoferritin: a novel biomagnetic nanoparticle. Nanostruct. Mater. 9 (1997) 595-598. https://doi.org/10.1016/S0965-9773(97)00133-5

[61] A.M. Abu-Dief, S.M. Abdel-Fatah, Development and functionalization of magnetic nanoparticles as powerful and green catalysts for organic synthesis, Beni-Suef University Journal of Basic and Applied Sciences, 7 (2018) 55-67. https://doi.org/10.1016/j.bjbas.2017.05.008

[62] T. Hyeon, S.S. Lee, J. Park, Y. Chung, H.B. Na, Synthesis of Highly Crystalline and Monodisperse Maghemite Nanocrystallites without a Size-Selection Process, J. Am. Chem. Soc. 123 (2001) 12798-12801. https://doi.org/10.1021/ja016812s

[63] S. Sato, T. Murakata, H. Yanagi, F. Miyasaka, S. Iwaya, Hydrothermal synthesis of fine perovskite PbTiO$_3$ powders with a simple mode of size distribution, J. Mater. Sci. 29 (1994) 5657-5663. https://doi.org/10.1007/BF00349961

[64] K. Woo, J. Hong, J.P. Ahn, Synthesis and surface modification of hydrophobic magnetite to processible magnetite@silica-propylamine, J. Magn. Magn. Mater. 293 (2005) 177-181. https://doi.org/10.1016/j.jmmm.2005.01.058

[65] S. Sun, H. Zeng, D.B. Robinson, S. Raoux, P.M. Rice, S.X. Wang, G. Li. Monodisperse MFe_2O_4 (M= Fe, Co, Mn) Nanoparticles, J. Am. Chem. Soc. 126 (2004), 273-279. https://doi.org/10.1021/ja0380852

[66] J. Park, E. Lee, N.M. Hwang, M. Kang, S.C. Kim, J.G. Hwang, G. Park, H.J. Noh, J.H. Kim, J. Park, H. Hyeron, One-nanometer-scale size-controlled synthesis of monodisperse magnetic iron oxide nanoparticles, Angew. Chem. Int. Ed. 44 (2005) 2873-2877. https://doi.org/10.1002/anie.200461665

[67] N.R. Jana, Y. Chen, X. Peng, Size- and Shape-Controlled Magnetic (Cr, Mn, Fe, Co, Ni) Oxide Nanocrystals via a Simple and General Approach, Chem. Mater. 16 (2004) 3931-3935. https://doi.org/10.1021/cm049221k

[68] J. Park, K. An, Y. Hwang, J.G. Park, H.J. Noh, J.Y. Kim, J.H. Park N.M. Hwang, J.H. Hyeron, Ultra-large-scale syntheses of monodisperse nanocrystals, Nat. Mater. 3 (2004) 891-895. https://doi.org/10.1038/nmat1251

[69] Z. Li, H. Chen, H.B. Bao, M.Y. Gao, One-Pot Reaction to Synthesize Water-Soluble Magnetite, Nanocrystals Chem. Mater. 16 (2004) 1391-1393. https://doi.org/10.1021/cm035346y

[70] J. Wan, W. Cai, J. Feng, X. Meng, E. Liu, In-situ decoration of carbon nanotubes with nearly monodisperse magnetite nanoparticles in liquid polyols, J. Mater. Chem. 17 (2007) 1188. https://doi.org/10.1039/b615527h

[71] Z. Li, Q. Sun, M. Gao, Preparation of Water-Soluble Magnetite Nanocrystals from Hydrated Ferric Salts in 2-Pyrrolidone: Mechanism Leading to Fe_3O_4, Angew. Chem. Int. Ed. 44 (2004) 123-126. https://doi.org/10.1002/anie.200460715

[72] Z. Li, L. Wei, M. Gao, H. Lei, One-Pot Reaction to Synthesize Biocompatible Magnetite Nanoparticles, Adv. Mater. 8 (2005) 1001-1005. https://doi.org/10.1002/adma.200401545

[73] Y.W. Jun, Y.M. Huh, J.S. Choi, J.H. Lee, H.T. Song, S. Kim, S. Yoon, K.S. Kim, J.S. Shin, J.S. Suh, J. Cheon, Nanoscale Size Effect of Magnetic Nanocrystals and Their Utilization for Cancer Diagnosis via Magnetic Resonance Imaging, J. Am. Chem. Soc. 127 (2005) 5732-5733. https://doi.org/10.1021/ja0422155

[74] J. Wan, W. Cai, X. Meng, E. Liu, Monodisperse water-soluble magnetite nanoparticles prepared by polyol process for high-performance magnetic resonance imaging, Chem. Commun. 47 (2007) 5004-5006. https://doi.org/10.1039/b712795b

[75] D. Maity, S.N. Kale, R. Kaul-Ghanekar, J. Xue, J. Ding, Studies of magnetite nanoparticles synthesized by thermal decomposition of iron (III) acetylacetonate in tri

Materials Research Forum LLC
https://doi.org/10.21741/9781644902332-1

(ethylene glycol), J. Magn. Magn. Mater. 321 (2009) 3093-3098.
https://doi.org/10.1016/j.jmmm.2009.05.020

[76] X.Q. Liu, S.W. Tao, Y.S. Shen, Preparation and characterization of nanocrystalline
α- Fe_2O_3 by a sol-gel process, Sens. Acuators. A. 40 (1997) 161-165.
https://doi.org/10.1016/S0925-4005(97)80256-0

[77] K.M.M. Kojima, F. Mizukami, K. Madea, Selective formation of spinel iron oxide in
thin films by complexing agent-assisted sol-gel processing, J. Sol-Gel Sci. Technol. 8
(1997) 77-81. https://doi.org/10.1007/BF02436821

[78] C. Ortiz, G. Lim, M.M. Chen, G. Castillo, Physical properties of spinel iron oxide
thin films, J. Mat. Res. 3 (1988) 344-350. https://doi.org/10.1557/JMR.1988.0344

[79] C. Cannas, D. Gatteschi, A. Musinu, G. Piccaluga, C. Sangregorio, Structural and
magnetic properties of Fe_2O_3 nanoparticles dispersed over a silica matrix, J. Phys.
Chem. 102 (1998) 7721-7726. https://doi.org/10.1021/jp981355w

[80] C.J. Brinker, G.W. Sherrer, Sol-Gel Science. Academic Press, New York, 1990.

[81] U. Schwertmann, R.M. Cornell, Iron oxides in the laboratory: preparation and
characterization, VCH, New York, 1991.

[82] H.Z. Qi, B.A. Yan, W. Lu, C.K. Li, Y.H.A. Yang, A nonalkoxide sol-gel method for
the preparation of magnetite (Fe) nanoparticles, Curr. Nanosci. 7 (2011) 381-88.
https://doi.org/10.2174/157341311795542426

[83] Y. Zhang, C.P. Chay, Y.J. Luo, L. Wang, G.P. Li, Synthesis structure and
electromagnetic properties Fe_3O_4 aerogels by sol-gel method, J. Mater. Sci. Eng. B.
188 (2014) 13-19. https://doi.org/10.1016/j.mseb.2014.06.002

[84] O.M. Lemine, K. Omri, B. Zhang, L. El Mir, M. Sajieddine, A. Alyamani, M.
Bououdina, Sol-gel synthesis of 8 nm magnetite (Fe_3O_4) nanoparticles and their
magnetic properties, SuperlatticMicrost. 52 (2012) 793-799.
https://doi.org/10.1016/j.spmi.2012.07.009

[85] A. Hasanpour, M. Niyaifar, M.E. Asan, J. Amighian, Synthesis and characterization
of Fe_3O_4 and ZnO nanocomposites by the sol-gel method, J. Magn . Magn. Mater. 334
(2013) 41-44. https://doi.org/10.1016/j.jmmm.2013.01.016

[86] H.T. Cui, Y. Liu, W.Z. Ren, Structure switch between alpha- Fe_2O_3, gamma- Fe_2O_3
and Fe_3O_4 during the large scale and low temperature sol-gel synthesis of nearly
monodispersed iron oxide nanoparticles, Adv. Powder Technol. 24 (2013) 93-97.
https://doi.org/10.1016/j.apt.2012.03.001

[87] S. Sertel, T. Eichhorn, P.K. Plinkert, T. Efferth, Chemical composition and
antiproliferative activity of essential oil from the leaves of a medicinal herb,
Levisticum officinale, against UMSCC1 head and neck squamous carcinoma cells,
Anticancer. Res. 31 (2011) 185-1891.

[88] W.W. Wang, Y.J. Zhu, Microwave-assisted synthesis of magnetite nano sheets in mixed solvents of ethylene glycol and water, Curr. Nano. Sci. 3 (2007) 171-176. https://doi.org/10.2174/157341307780619233

[89] Y.B. Khollam, S.R. Dhage, H.S. Potdar, S.B. Deshpande, P.P. Bakare, S.D. Kulkarni, S.K. Date, Microwave hydrothermal preparation of submicron-sized spherical magnetite (Fe$_3$O$_4$) powders, Mater. Lett. 56 (2002) 571-577. https://doi.org/10.1016/S0167-577X(02)00554-2

[90] H.Y. Hu, H. Yang, P. Huang, D.X. Cui, Y.Q. Peng, J.C. Zhang, F.Y. Lu, J. Lian, D.L. Shi, Unique role of ionic liquid in microwave-assisted synthesis of monodisperse magnetite nanoparticles, Chem. Commun. 46 (2010) 3866-3868. https://doi.org/10.1039/b927321b

[91] W.C. Xiao, H.C. Gu, D. Li, D.D. Chen, X.Y. Deng, Z. Jiao, J. Lin, Microwave-assisted synthesis of magnetite nanoparticles for MR blood pool contrast agents, J. Magn. Magn. Mater. 324 (2012) 488-494. https://doi.org/10.1016/j.jmmm.2011.08.029

[92] S. Komarneni, W.W. Hu, Y.D. Noh, A. Van Orden, S.H. Feng, C.Z. Wei, H. Pang, F. Gao, Q.Y. Lu, H. Katsuki, Magnetite syntheses from room temperature to 150 degrees C with and without microwaves, Ceram Int. 38 (2012) 2563-2568. https://doi.org/10.1016/j.ceramint.2011.11.027

[93] A. Nan, R. Turcu, I. Crăciunescu, O. Pana, H. Scharf, J. Liebscher, Microwave-Assisted Graft Polymerization of epsilon-caprolactone on to magnetite, J. Poly. Sci. Part A: Poly. Chem. 47 (2009) 5397-5404. https://doi.org/10.1002/pola.23589

[94] M. Gobe, K. Konno, K. Kandori, A. Kitahara, Preparation and characterization of monodisperse magnetite sols in W/O microemulsion, J. Coll. Inter. Sci. 93 (1983) 293-295. https://doi.org/10.1016/0021-9797(83)90411-3

[95] T. Lu, J.H. Wang, J. Yin, A.Q. Wang, X.D. Wang, T. Zhang, Surfactant effects on the microstructures of Fe$_3$O$_4$ nanoparticles synthesized by microemulsion method, J. Coll. Surf. A. 436 (2013) 675-683. https://doi.org/10.1016/j.colsurfa.2013.08.004

[96] M.T. Reetz, W. Helbig, Size selective synthesis of nanostructured transition metal clusters, J. Am. Chem. Soc. 116 (1994) 7401-7402. https://doi.org/10.1021/ja00095a051

[97] C. Pascal, J.L. Pascal, F. Favier, M.A.E. Moubtassin, C. Payen, Electrochemical synthesis for the control of γ- Fe$_2$O$_3$ nanoparticle size, Morphology, Microstructure and Magnetic Behaviour, Chem. Mater. 11 (1999) 141-147. https://doi.org/10.1021/cm980742f

[98] B.S. Vasile, O.R. Vasile, C. Ghitulica, E. Andronescu, R. Dobranis, E. Dinu, R. Trusca, Yttria, Totally stabilized zirconia nanoparticles obtained through the pyrosol method, Phys. Status Solid. A. 207 (2010) 2499-2504. https://doi.org/10.1002/pssa.200925623

[99] W.S.A. Thongsuwan, P. Singjai, Preparation of iron oxide nanoparticles by a pyrosol technique, Key Eng. Mater. 353-358 (2007) 2175-2178 https://doi.org/10.4028/www.scientific.net/KEM.353-358.2175

[100] A. Ito, K. Ino, K. Shimizu, H. Honda, M. Kamihira, Fabrication of 3D tissue-like structure using magnetite nanoparticles and magnetic force, IEEE International Symposium on Micro-Nano Mechatronics and Human Sci. 1 (2006) 256-261. https://doi.org/10.1109/MHS.2006.320291

[101] T.K. Indira. P.K. Lakshmi, Magnetic Nanoparticles - A Review, Int. J. Pharm. 3 (2010) 1035-1042. https://doi.org/10.37285/ijpsn.2010.3.3.1

[102] S. Mann, R.B. Frankel, R.P. Blakemore, Structure, morphology and crystal growth of bacterial magnetite, Nature. 310 (1984) 405-407. https://doi.org/10.1038/310405a0

[103] D.A. Bazylinski, R.B. Frankel, H.W. Jannasch, Anaerobic magnetite production by a marine, magnetotactic bacterium, Nature. 334 (1988) 518-519. https://doi.org/10.1038/334518a0

[104] S. Mann, N.H.C. Sparks, R.B. Frankel, D.A. Bazylinski, H.W. Jannasch, Biomineralization of ferrimagnetic greigite (Fe_3O_4) and iron pyrite (FeS2) in a magnetotactic bacterium, Nature. 343 (1990) 258-260. https://doi.org/10.1038/343258a0

[105] H. Lee, A.M. Purdon, V. Chu, R.M. Westervelt, Controlled assembly of magnetic nanoparticles from magnetotactic bacteria using micro electromagnets arrays, Nano. Lett. 4 (2004) 995-998. https://doi.org/10.1021/nl049562x

[106] C. Lang, D. Schuler, Biogenic nanoparticles: production, characterization, and application of bacterial magnetosomes, J. Phys. Cond. Matt. 18 (2006) 2815-2828. https://doi.org/10.1088/0953-8984/18/38/S19

[107] J.H. Fendler, Nanoparticles and nanostructured films: preparation, characterization and applications, John Wiley & Sons, 1998. https://doi.org/10.1002/9783527612079

[108] C. Yee, G. Kataby, A. Ulman, T. Prozorov, H. White, A. King, M. Rafailovich, J. Sokolov, A. Gedanken, Self-assembled monolayers of alkanesulfonic and phosphonicacids on amorphous iron oxide nanoparticles, J. Lang. 15 (1999) 7111-7115. https://doi.org/10.1021/la990663y

[109] Y. Sahoo, H. Pizem, T. Fried, D. Golodnitsky, L. Burstein, C.N. Sukenik, G. Markovich, Alkyl phosphonate/phosphate coating on magnetite nanoparticles: a comparison with fatty acids, J. Lang. 17 (2001) 7907-7911. https://doi.org/10.1021/la010703+

[110] D. Portet, B. Denizot, E. Rump, J. Lejeune, P. Jallet, Nonpolymeric coatings of iron oxide colloids for biological use as magnetic resonance imaging contrast agents, J. Coll. Int. Sci. 238 (2001) 37-42. https://doi.org/10.1006/jcis.2001.7500

[111] B. Denizot, G. Tanguy, F. Hindre, E. Rump, J. Jeune, P. Jallet, Phosphorylcholine coating of iron oxide nanoparticles, J. Coll. Int. Sci. 209 (1999) 66-71. https://doi.org/10.1006/jcis.1998.5850

[112] C.C. Berry, S. Wells, S. Charles, A.S.G. Curtis, Dextran and albumin derivatised iron oxide nanoparticles: influence on fibroblasts in vitro, J. Bioma. 24 (2003) 4551-4557. https://doi.org/10.1016/S0142-9612(03)00237-0

[113] J. Ugelstad, P. Stenstad, L. Kilaas, W.S. Prestvik, R. Herje, A. Bererge, E. Hornes, Monodisperse magneticpolymer particles. New biochemical and biomedical applications, Blood. Purif. 11 (1993) 349-369. https://doi.org/10.1159/000170129

[114] H. Kawaguchi, K. Fujimoto, Y. Nakazawa, M. Sakagawsa, Y. Ariyoshi, M. Shidara, H. Okazaki, Y. Ebisawa, Modification and functionalization of hydrogel microspheres, J. Coll. Surf. A. Phy. Eng. Asp. 109 (1996) 147-154. https://doi.org/10.1016/0927-7757(95)03482-X

[115] F. Sauzedde, A. Elaïssari, C. Pichot, Hydrophilic magnetic polymer latexes 1. Adsorption of magneticiron oxide nanoparticles onto various cationic latexes, J. Coll. Poly. Sci. 277 (1999) 846-855. https://doi.org/10.1007/s003960050461

[116] K. Furusawa, K. Nagashima, C. Anzai, Synthetic process to control the total size and component distribution of multilayer magnetic composite particles, J. Coll. Poly. Sci. 272 (1994) 1104-1110. https://doi.org/10.1007/BF00652379

[117] J. Lee, T. Isobe, M. Senna, Preparation of ultrafine Fe particles by precipitation in the presence of PVA at high pH, J. Coll. Int. Sci. 177 (1996) 490-494. https://doi.org/10.1006/jcis.1996.0062

[118] G. Kataby, A. Ulman, R. Prozorov, A. Gedanken, Coating of amorphous iron nanoparticles by long-chain alcohols, J. Lang. 14 (1998) 1512-1515. https://doi.org/10.1021/la970978i

[119] A.K. Gupta, A.S.G. Curtis, Lactoferrin and ceruloplasmin derivatized superparamagneticiron oxide nanoparticles for targeting cell surface receptors, J. Biomat. 25 (2004) 3029-3040. https://doi.org/10.1016/j.biomaterials.2003.09.095

[120] A.T. Florence, The oral absorption of micro- and nanoparticulates: neither exceptional nor unusual, J. Pharm. Res. 14 (1997) 259-266. https://doi.org/10.1023/A:1012029517394

[121] M. Chen, S. Yamamuro, D. Farrell, S.A. Majetich, Gold-coated iron nanoparticles for biomedical applications, J. Appl. Phys. 93 (2003) 7551-7553. https://doi.org/10.1063/1.1555312

[122] J. Lin, W. Zhou, A. Kumbhar, J. Fang, E.E. Carpenter, C.J. Connor, Gold-coated iron (Fe@Au) nanoparticles: synthesis, characterization and magnetic field-induced self-assembly, J. Sol. Stat. Chem. 159 (2001) 26-31. https://doi.org/10.1006/jssc.2001.9117

Materials Research Forum LLC
https://doi.org/10.21741/9781644902332-1

[123] E.E. Carpenter, Iron nanoparticles as potential magnetic carriers, J. Magn. Magn. Mat. 225 (2001) 17-20. https://doi.org/10.1016/S0304-8853(00)01222-1

[124] W.L. Zhou, E.E. Carpenter, J. Lin, A. Kumbhar, J. Sims, C.J. Connor, Nanostructures of gold coated iron core-shell nanoparticles and the nanobands assembled under magnetic field, Eur. Phys. J.D. 16 (2001) 289-292. https://doi.org/10.1007/s100530170112

[125] P. Mulvaney, L.M. Liz-Marzan, M. Giersig, T. Ung, Silica encapsulation of quantum dots and metal clusters, J. Mater. Chem. 10 (2000) 1259-1270. https://doi.org/10.1039/b000136h

[126] P. Tartaj, T. Gonzalez-Carreno, C.J. Serna, Synthesis of nanomagnets dispersed in colloidal silica cages with applications in chemical separation, J. Lang. 18 (2002) 4556-4558. https://doi.org/10.1021/la025566a

[127] P. Tartaj, T. Gonzalez-Carreno, C.J. Serna, Single-step nanoengineering of silica coated maghemite hollow spheres with tunable magnetic properties, J. Adv. Mat. 13 (2001) 1620-1624. https://doi.org/10.1002/1521-4095(200111)13:21<1620::AID-ADMA1620>3.0.CO;2-Z

[128] A.M. Morawski, P.M. Winter, K.C. Crowder, S.D. Caruthers, R.W. Fuhrhop, M.J. Scott, J.D. Robertson, D.R. Abendschein, G.M. Lanza, S.A. Wickline, Targeted nanoparticles for quantitative imaging of sparse molecular epitopes with MRI, J. Magn. Res. Med. 51 (2004) 480-486. https://doi.org/10.1002/mrm.20010

[129] P.G. Shepherd, J. Popplewell, S.W. Charles, A method of producing ferrofluid with gadolinium particles, J. Phy. D. App. Phy. 3 (1970) 1985-1986. https://doi.org/10.1088/0022-3727/3/12/430

[130] S.L. Tie, Y.Q. Lin, H.C. Lee, Y.S. Bae, C. Lee, Amino acid-coated nano-sized magnetite particles prepared by two-step transformation, Colloids. Surf. A. 273 (2006) 75-83. https://doi.org/10.1016/j.colsurfa.2005.08.027

[131] D. Evoy, M. Lieberman, T.J. Fahey, M.J. Daly, Immunonutrition: the role of arginine, Nutrition. 14 (1998) 611-617. https://doi.org/10.1016/S0899-9007(98)00005-7

[132] R.C. Winterhalder, F.R. Hirsch, G. K. Kotantoulas, W. A. Franklin, & P. A. Bunn Jr., Chemoprevention of lung cancer--from biology to clinical reality. Ann Oncol. 15 (2004) 185-196. https://doi.org/10.1093/annonc/mdh051

[133] M. Zafarullaha, W.Q. Lia, J. Sylvstera, M. Ahmad, Cell Molecular mechanisms of N-acetylcysteine actions, Mol. Lif Sci. 60 (2003) 6-20. https://doi.org/10.1007/s000180300001

[134] X. Hong, W. Guo, H. Yuan, J. Li, Y. Liu, L. Ma, Y. Bai, T. Li, Periodate oxidation of nanoscaled magnetic dextran composites, Journal of Magnetism and Magnetic Materials, 269 (2004) 95-100. https://doi.org/10.1016/S0304-8853(03)00566-3

[135] M. Mahmoudi, A. Simchi, M. Imani, A.S. Milani, P. Stroeve, Optimal design and characterization of superparamagnetic iron oxide nanoparticles coated with polyvinyl alcohol for targeted delivery and imaging, J. Phys. Chem. B. 112 (2008) 14470-14481. https://doi.org/10.1021/jp803016n

[136] T.K. Jain, S.P. Foy, B. Erokwu, S. Dimitrijevic, C.A. Flask, V. Labhasetwar, Magnetic resonance imaging of multifunctional pluronic stabilized iron-oxide nanoparticles in tumor-bearing mice, Biomat. 30 (2009) 6748-6756. https://doi.org/10.1016/j.biomaterials.2009.08.042

[137] D. Arndt, V. Zielasek, W. Dreher, M. Bäumer, Ethylene diamine-assisted synthesis of iron oxide nanoparticles in high-boiling polyolys, Journal of Colloid and Interface Science, 417 (2014) 188-198. https://doi.org/10.1016/j.jcis.2013.11.023

[138] H.J. Chung, H. Lee, K.H. Bae, Y. Lee, J. Park, S.W. Cho, J.Y. Hwang, H. Park, R. Langer, D. Anderson, T.G. Park, Facile synthetic route for surface-functionalized magnetic nanoparticles: Cell labeling and magnetic resonance imaging studies, ACS Nano, 5 (2011) 4329-4336. https://doi.org/10.1021/nn201198f

[139] M. Babic, D. Horák, M. Trchová, Poly(L-lysine)-modified iron oxide nanoparticles for stem cell labeling, Bioconjugate Chemistry. 19 (2008) 740-750. https://doi.org/10.1021/bc700410z

[140] C. Schweiger, C. Pietzonka, J. Heverhagen, T. Kissel, Novel magnetic iron oxide nanoparticles coated with poly(ethylene imine)-g-poly(ethylene glycol) for potential biomedical application: synthesis, stability, cytotoxicity and MR imaging, Int. J. Pharm. 408 (2011)130-137. https://doi.org/10.1016/j.ijpharm.2010.12.046

[141] H.S. Lee, K.E. Hee, H. Shao, K.B. Kook, Synthesis of SPIO-chitosan microspheres for MRI-detectable embolotherapy, J. Magn. Magn. Mater. 293 (2005) 102-105. https://doi.org/10.1016/j.jmmm.2005.01.049

[142] L. Lartigue, P. Hugounenq, D. Alloyeau, S.P. Clarke, M. Lévy, J.C. Bacri, Cooperative organization in iron oxide multi-core nanoparticles potentiates their efficiency as heating mediators and MRI contrast agents, ACS Nano. 6 (2012)10935-10949. https://doi.org/10.1021/nn304477s

[143] H.T.R. Wiogo, M. Lim, V. Bulmus, L. Gutiérrez, R.C. Woodward, R. Amal, Insight into serum protein interactions with functionalized magnetic nanoparticles in biological media, Langmuir. 28 (2012) 4346-4356. https://doi.org/10.1021/la204740t

[144] Bar-Shir, L. Avram, S. Yariv-Shoushan, D. Anaby, S. Cohen, N. Segev-Amzaleg, Alginate-coated magnetic nanoparticles for noninvasive MRI of extracellular calcium, NMR Biomed. 27 (2014) 774-783. https://doi.org/10.1002/nbm.3117

[145] F. Hu, K.W. Macrenaris, E.A. Waters, E.A. Schultz-Sikma, A.L. Eckermann, T.J. Meade, Highly dispersible, superparamagnetic magnetite nanoflowers for magnetic resonance imaging, Chem. Commun. 46 (2010) 73-75. https://doi.org/10.1039/B916562B

[146] B. Sivakumar, R.G. Aswathy, Y. Nagaoka, M. Suzuki, T. Fukuda, Y. Yoshida, Multifunctional carboxymethyl cellulose-based magnetic nanovector as a theragnostic system for folate receptor targeted chemotherapy, imaging, and hyperthermia against cancer, Langmuir. 29 (2013) 3453-3466. https://doi.org/10.1021/la305048m

[147] G. Wang, X. Zhang, A. Skallberg, Y. Liu, Z. Hu, X. Mei, One-step synthesis of water dispersible ultra-small Fe_3O_4 nanoparticles as contrast agents for T1 and T2 magnetic resonance imaging, Nanoscale. 6 (2014) 2953-2963. https://doi.org/10.1039/c3nr05550g

[148] E. Illes, M. Szekeres, E. Kupcsik, E. Kupsick, I.Y. Toth, K. Farkas, A. Jedlovszky-Hajdu, PEGylation of surfacted magnetite core-shell nanoparticles for biomedical application, Colloids Surf. A. 460 (2014) 429-440. https://doi.org/10.1016/j.colsurfa.2014.01.043

[149] Z. Chen, B. Li, J. Zhang, L. Qin, D. Zhou, Y. Han, Z. Du, Z. Guo, Y. Song, R. Yang, Quorum sensing affects virulence associated proteins F1, LcrV, KatYand pH6 etc. of Yersinia pestis as revealed by protein microarray-based antibody profiling, Microbes Infect. 8 (2006) 2501-2508. https://doi.org/10.1016/j.micinf.2006.06.007

[150] K.J. Widder, A.E. Senyei, D.F. Ranney, Magnetically responsive microspheres and other carriers for the biophysical targeting of antitumour agents, J. Adv. Pharmacol. Chemother. 16 (1979) 213-271. https://doi.org/10.1016/S1054-3589(08)60246-X

[151] G. Unsoy, U. Gunduz, O. Oprea, D. Ficai, M. Sonmez, M. Radulescu, M. Alexie, A. Ficai, Magnetite: from synthesis to applications. Curr Top Med Chem., 15(2015) 1622-1640. https://doi.org/10.2174/1568026615666150414153928

[152] P. Theamdee, R. Traiphol, B. Rutnakornpituk, U. Wichai, M. Rutnakornpituk, Surface modification of magnetite nanoparticle with azobenzene-containing water dispersible polymer, J. Nano. Part. Res. 13 (2011) 4463-4477. https://doi.org/10.1007/s11051-011-0399-7

[153] D. Dorniani, A.U. Kura, S.H. Hussein-Al-Ali, M.Z. Bin Hussein, S. Fakurazi, A.H. Shaari, Z. Ahmad, Release Behavior and Toxicity Profiles towards Leukemia (WEHI-3B) Cell Lines of 6-Mercaptopurine-PEG-Coated Magnetite Nanoparticles Delivery System, Sci. World J. 2014 (2014) 1-11. https://doi.org/10.1155/2014/972501

[154] A.F. Wang, W.X. Qi, N. Wang, J.Y. Zhao, F. Muhammad, K. Cai, H. Ren, F.X. Sun, L. Chen, Y.J. Guo, M.Y. Guo, G.S. Zhu, A smart nanoporoustheranostic platform for simultaneous enhanced MRI and drug delivery, Micropor. Mesopor. Mat. 180 (2013) 1-7. https://doi.org/10.1016/j.micromeso.2013.06.015

[155] N.K. Verma, K. Crosbie-Staunton, A. Satti, S. Gallagher, K.B. Ryan, T. Doody, C. McAtamney, R. MacLoughlin, P. Galvin, C.S. Burke, Y. Volkov, Y.K. Gun'ko, Magnetic core-shell nanoparticles for drug delivery by nebulization, J. Nanobiotechnol. 11 (2013) 1-12. https://doi.org/10.1186/1477-3155-11-1

[156] Q. Yuan, R. Venkatasubramanian, S. Hein, R.D.K. Misra, A stimulus-responsive magnetic nanoparticle drug carrier: Magnetite encapsulated by chitosan-grafted-copolymer, Acta. Biomater. 4 (2008) 1024-1037. https://doi.org/10.1016/j.actbio.2008.02.002

[157] G. Unsoy, S. Yalcin, R. Khodadust, P. Mutlu, O. Onguru, U. Gunduz, Chitosan magnetic nanoparticles for pH responsive Bortezomib release in cancer therapy, Biomed. Pharma. 68 (2014) 641-648. https://doi.org/10.1016/j.biopha.2014.04.003

[158] G. Unsoy, R. Khodadust, S. Yalcin, P. Mutlu, U. Gunduz, Synthesis of Doxorubicin loaded magnetic chitosan nanoparticles for pH responsive targeted drug delivery, Eur. J Pharm. Sci. 62 (2014) 234-250. https://doi.org/10.1016/j.ejps.2014.05.021

[159] M. Konishi, Y. Tabata, M. Kariya, H. Hosseinkhani, A. Suzuki, K. Fukuhara, M. Mandai, A. Takakura, S. Fujii, In vivo anti-tumor effect of dual release of cisplatin and adriamycin from biodegradable gelatin hydrogel, J. Cont. Rel. 103 (2005) 7-19. https://doi.org/10.1016/j.jconrel.2004.11.014

[160] J.H. Kim, Y.S. Kim, K. Park, S. Lee, H.Y. Nam, K.H. Min, H.G. Jo, J.H. Park, K. Choi, S.Y. Jeong, R.W. Park, I.S. Kim, K. Kim, I.C. Kwon, Antitumor efficacy of cisplatin-loaded glycol chitosan nanoparticles in tumor-bearing mice, J. Cont. Rel. 127 (2008) 41-49. https://doi.org/10.1016/j.jconrel.2007.12.014

[161] X.Y. Zhang, L. Xue, J. Wang, Q. Liu, J.Y. Liu, Z. Gao, W.L. Yang, Effects of surface modification on the properties of magnetic nanoparticles/PLA composite drug carriers and in vitro controlled release study, Colloid Surface A. 431 (2013) 80-86. https://doi.org/10.1016/j.colsurfa.2013.04.021

[162] S. Sieben, C. Bergemann, A. Lukbbe, B. Brockmann, D. Rescheleit, Comparison of different particles and methods for magnetic isolation of circulating tumor cells, J. Magn. Magn. Mat. 225 (2001) 175-179. https://doi.org/10.1016/S0304-8853(00)01248-8

[163] Y.Q. Zhang, L.L. Li, F. Tang, J. Ren, Controlled Drug Delivery System Based on Magnetic Hollow Spheres/Polyelectrolyte Multilayer Core-Shell Structure, J. Nanosci. Nanotech. 6 (2006) 3210-3214. https://doi.org/10.1166/jnn.2006.469

[164] L. Li, D. Chen, Y. Zhang, Z. Deng, X. Ren, X. Meng, F. Tang, J. Ren, L. Zhang, Magnetic and fluorescent multifunctional chitosan nanoparticles as a smart drug delivery system, Nanotech. 18 (2007) 102-108. https://doi.org/10.1088/0957-4484/18/40/405102

[165] S.H. Hu, C.H. Tsai, C.F. Liao, D.M. Liu, S.Y. Chen, Controlled rupture of magnetic polyelectrolyte microcapsules for drug delivery, Langmuir, 24 (2008) 11811-11818. https://doi.org/10.1021/la801138e

[166] M.C. Urbina, S. Zinoveva, T. Miller, C.M. Sabliov, W.T. Monroe, C.S.S.R. Kumar, Investigation of Magnetic Nanoparticle–Polymer Composites for Multiple-controlled

Drug Delivery, J. Phy. Chem. C. 112 (2008) 11102-11108. https://doi.org/10.1021/jp711517d

[167] Q. He, Y. Tian, Y. Cui, H. Möhwald, J. Li, Layer-by-layer assembly of magnetic polypeptide nanotubes as a DNA carrier, J. Mat. Chem. 18 (2008) 748-754. https://doi.org/10.1039/b715770c

[168] P.V. Finotelli, D. Da Silva, M. Sola-Penna, A.M. Rossi, M. Farina, L.R. Andrade, A.Y. Takeuchi, M.H. Rocha-Leao, Microcapsules of alginate/chitosan containing magnetic nanoparticles for controlled release of insulin, J. Colloids and Surfaces B: Biointerfaces. 81 (2010) 206-211. https://doi.org/10.1016/j.colsurfb.2010.07.008

[169] S.S. Banerjee, D.H. Chen, Cyclodextrin-conjugated nanocarrier for magnetically guided delivery of hydrophobic drugs, J. Nano. Res. 11 (2009) 2071-2078. https://doi.org/10.1007/s11051-008-9572-z

[170] B. Chen, H. Zhang, C. Zhai, N. Du, C. Sun, J. Xue, D. Yang, H. Huang, B. Zhang, Q. Xie, Y. Wu, Carbon nanotube-based magnetic-fluorescent nanohybrids as highly efficient contrast agents for multimodal cellular imaging, J. Mat. Chem. 20 (2010) 9895-9902. https://doi.org/10.1039/c0jm00594k

[171] B. Koppolu, M. Rahimi, S. Nattama, A. Wadajkar, K.T. Nguyen, Development of multiple-layer polymeric particles for targeted and controlled drug delivery, Nanomedicine: Nanotech, Bio, and Med. 6 (2010) 355-361. https://doi.org/10.1016/j.nano.2009.07.008

[172] S. Purushotham, R.V. Ramanujan, Thermoresponsive magnetic composite nanomaterials for multimodal cancer therapy, Acta Biomaterialia. 6 (2010) 502-510. https://doi.org/10.1016/j.actbio.2009.07.004

[173] P.Y. Chen, H.L. Liu, M.Y. Hua, H.W. Yang, C.Y. Huang, P.C. Chu, L.A. Lyu, I.C. Tseng, L.Y. Feng, H.C. Tsai, S.M. Chen, Y.J. Lu, J.J. Wang, T.C. Yen, Y.H. Ma, T. Wu, J.P. Chen, J.I. Chuang, J.W. Shin, C. Hsueh, K.C. Wei, Novel magnetic/ultrasound focusing system enhances nanoparticle drug delivery for glioma treatment, Neuro-Oncology. 12 (2010) 1050-1060. https://doi.org/10.1093/neuonc/noq054

[174] K. Katagiri, M. Nakamura, K. Koumoto, Preparation of hybrid hollow capsules formed with Fe$_3$O$_4$ and polyelectrolytes via the layer-by-layer assembly and the aqueous solution process, American Chem. Soc. 2 (2010) 768-773. https://doi.org/10.1016/j.jcis.2009.09.014

[175] A. Shkilnyy, E. Munnier, K. Hervé, M. Soucé, R. Benoit, S. Cohen-Jonathan, P. Limelette, M. Saboungi, P. Dubois, I. Chourpa, Synthesis and Evaluation of Novel Biocompatible Superparamagnetic Iron Oxide Nanoparticles as Magnetic Anticancer Drug Carrier and Fluorescence Active Label, Phy. Chem. C. 114 (2010) 5850-5858. https://doi.org/10.1021/jp9112188

[176] S. Shen, J. Ren, J. Chen, X. Lu, C. Deng, X. J. Jiang, Development of magnetic multiwalled carbon nanotubes combined with near-infrared radiation-assisted desorption for the determination of tissue distribution of doxorubicin liposome injects in rats, Chromatography A. 1218 (2011) 4619-4626. https://doi.org/10.1016/j.chroma.2011.05.060

[177] M.M. Yallapu, S.F. Othman, E.T. Curtis, B.K. Gupta, M. Jaggi, S.C. Chauhan, Synthesis of pH responsive hydrogel-silver nanocomposite for use as biomaterials, Biomat. 32 (2011) 1890-1905. https://doi.org/10.1016/j.biomaterials.2010.11.028

[178] H.W. Yang, M.Y. Hua, H.L. Liu, C.Y. Huang, K.C. Wei, Potential of magnetic nanoparticles for targeted drug delivery, Nanotech Sci. and App. 5 (2012) 73-86. https://doi.org/10.2147/NSA.S35506

[179] A.L. Glover, J.B. Bennett, J.S. Pritchett, S.M. Nikles, D.E. Nikles, J.A. Nikles, C.S. Brazel, Magnetic heating of iron oxide nanoparticles and magnetic micelles for cancer therapy, IEEE Trans. Magn. 49 (2013) 231-235. https://doi.org/10.1109/TMAG.2012.2222359

[180] M. Kawashita, M. Tanaka, T. Kokubo, Y. Inoue, T. Yao, S. Hamada, T. Shinjo, Preparation of ferrimagnetic magnetite microspheres for in situ hyperthermic treatment of cancer, J. Bio. Mat. 26 (2005) 2231-2238. https://doi.org/10.1016/j.biomaterials.2004.07.014

[181] D.H. Kim, S.H. Lee, K.N. Kim, K.M. Kim, I.B. Shim, Y.K. Lee, Cytotoxicity of ferrite particles by MTT and agar diffusion methods for hyperthermic application, J. Mang. Mang. Mater. 293 (2005) 287-292. https://doi.org/10.1016/j.jmmm.2005.02.078

[182] C.C. Berry, A.S.G. Curtis, Functionalisation of magnetic nanoparticles for applications in biomedicine, J. Phys. D. 36 (2003) R198-R206. https://doi.org/10.1088/0022-3727/36/13/203

[183] S.V.S. Mornet, F. Grasset, E. Duguet, Magnetic nanoparticle design for medical diagnosis and therapy, J. Mater. Chem. 14 (2004) 2116-2175. https://doi.org/10.1039/b402025a

[184] E. Andronescu, M. Ficai, G. Voicu, D. Ficai, M. Maganu, A. Ficai, Synthesis and characterization of collagen/hydroxyapatite: magnetite composite material for bone cancer treatment, J. Mater.Sci. Mater. M. 21 (2010) 2237-2242. https://doi.org/10.1007/s10856-010-4076-7

[185] D.H. Kim, S.H. Lee, K.N. Kim, K.M. Kim, I.B. Shim, Y.K. Lee, Cytotoxicity of ferrite particles by MTT and agar diffusion methods for hyperthermic application, J. Mang. Mang. Mater, 293 (2005) 287-292. https://doi.org/10.1016/j.jmmm.2005.02.078

[186] F.Q. Hu, K.W. MacRenaris, E.A. Waters, E.A. Schultz-Sikma, A.L. Eckermann, T.J. Meade, Highly dispersible, superparamagnetic magnetite nanoflowers for magnetic resonance imaging, Chem. Commun. 46 (2010) 73-75. https://doi.org/10.1039/B916562B

[187] L.H. Shen, J.F. Bao, D. Wang, Y.X. Wang, Z.W. Chen, L. Ren, X. Zhou, X.B Ke, M. Chen, A.Q. Yang, One-step synthesis of monodisperse, water-soluble ultra-small Fe3O4 nanoparticles for potential bio-application, Nanoscale. 5 (2013) 2133-2141. https://doi.org/10.1039/c2nr33840h

[188] U.O. Hafeli, G.J. Pauer, In vitro and in vivo toxicity of magnetic microspheres, J. Mang. Mang. Mater. 194 (1999) 76-82. https://doi.org/10.1016/S0304-8853(98)00560-5

[189] J.M. Qu, G. Liu, Y.M. Wang, R.Y. Hong, Preparation of Fe3O4 chitosan nanoparticles used for hyperthermia, Adv. Powder. Technol. 21 (2010) 461-467. https://doi.org/10.1016/j.apt.2010.01.008

[190] Z. Liu, M. Li, X.J. Yang, M.L. Yin, J.S. Ren, X.G. Qu, The use of multifunctional magnetic mesoporous core/shell heteronanostructures in a biomolecule separation system, Biomater. 32 (2011) 4683-4690. https://doi.org/10.1016/j.biomaterials.2011.03.038

[191] B. Rittich, A. Spanova, SPE and purification of DNA using magnetic particles, J. Sep. Sci. 36 (2013) 2472-2485. https://doi.org/10.1002/jssc.201300331

[192] H.L. Hsu, R. Selvin, J.W. Cao, L.S. Roselin, M. Bououdina, Facile Synthesis of Magnetically Separable Nanozeolites for Bio-Applications, Sci. Adv. Mater. 3 (2011) 939-943. https://doi.org/10.1166/sam.2011.1221

[193] K. Aguilar-Arteaga, J.A. Rodriguez, E. Barrado, Magnetic solids in analytical chemistry: A review, Anal. Chim. Acta, 674 (2010) 157-165. https://doi.org/10.1016/j.aca.2010.06.043

[194] Z. Liu, M. Li, F. Pu, J.S. Ren, X.J. Yang, X.G. Qu, Hierarchical magnetic core-shell nanoarchitectures: non-linker reagent synthetic route and applications in a biomolecule separation system, J. Mater. Chem. 22 (2012) 2935-2942. https://doi.org/10.1039/C1JM14088D

[195] S. Panseri, C. Cunha, T. Dalessandro, M. Sandri, G. Giavaresi, M. Marcacci, C.T. Hung, A. Tampieri, Intrinsically superparamagnetic Fe-hydroxyapatite nanoparticles positively influence osteoblast-like cell behavior, J. Nanobiotechnol. 10 (2012) 1-10. https://doi.org/10.1186/1477-3155-10-32

[196] T. Kito, R. Shibata, M.H.I. Suzuki, T. Himeno, Y. Kataoka, T. Murohara, iPS cell sheets created by a novel magnetite tissue engineering method for reparative angiogenesis, Scientific Report. 3 (2013) 1418. https://doi.org/10.1038/srep01418

[197] M.E.A. Konczol, Cytotoxicity and genotoxicity of size-fractionated iron oxide (magnetite) in A549 human lung epithelial cells: role of ROS, JNK, and NF-kappa B, Chem. Res. Toxicol. 24 (2011) 1460-1475. https://doi.org/10.1021/tx200051s

[198] W. Kai, X. Xiaojun, P. Ximing, H. Zhenqing, Z. Qiqing, Cytotoxic effects and the mechanism of three types of magnetic nanoparticles on human hepatoma BEL-7402 cells, Nanoscale Res. Lett. 6 (2011) 480. https://doi.org/10.1186/1556-276X-6-480

[199] M. Ishii, R. Shibata, Y. Numaguchi, T. Kito, H. Suzuki, K. Shimizu, A. Ito, H. Honda, T. Murohara, Enhanced angiogenesis by transplantation of mesenchymal stem cell sheet created by a novel magnetic tissue engineering method, Arterioscler. Thromb.Vasc. Biol. 31 (2011) 2210-2215. https://doi.org/10.1161/ATVBAHA.111.231100

[200] A. Ito, K. Ino, M. Hayashida, T. Kobayashi, H. Matsunuma, H. Kagami, M. Ueda, H. Honda, Novel methodology for fabrication of tissue-engineered tubular constructs using magnetite nanoparticles and magnetic force, Tissue Eng. 11 (2005) 1553-1561. https://doi.org/10.1089/ten.2005.11.1553

[201] A. Ito, M. Hayashida, H. Honda, K. Hata, H. Kagami, M. Ueda, T. Kobayashi, Construction and harvest of multilayered keratinocyte sheets using magnetite nanoparticles and magnetic force, Tissue Eng. 10 (2004) 873-880. https://doi.org/10.1089/1076327041348446

[202] K.C. Barick, S. Singh, D. Bahadur, M.A. Lawande, D.P. Patkar, P.A. Hassan, Carboxyl decorated Fe_3O_4 nanoparticles for MRI diagnosis and localized hyperthermia, J. Colloid Interf. Sci. 418 (2014) 120-125. https://doi.org/10.1016/j.jcis.2013.11.076

[203] E.H. Kim, H.S. Lee, B.K. Kwak, B.K. Kim, Synthesis of ferrofluid with magnetic nanoparticles by sonochemical method for MRI contrast agent, J. Mang. Mang. Mater. 289 (2005) 328-330. https://doi.org/10.1016/j.jmmm.2004.11.093

[204] A.S. Teja, P.Y. Koh, Synthesis, properties, and applications of magnetic iron oxide nanoparticles, Prog. Cryst. Growth Charact. Mater. 55 (2009) 22-45. https://doi.org/10.1016/j.pcrysgrow.2008.08.003

[205] M. Faraji, Y. Yamini, E. Tahmasebi, A. Saleh, F. Nourmohammadian, Cetyltrimethyl ammonium Bromide-Coated Magnetite Nanoparticles as Highly Efficient Adsorbent for Rapid Removal of Reactive Dyes from the Textile Companies Wastewaters, J. Iran Chem. Soc. 7 (2010) S130-S144. https://doi.org/10.1007/BF03246192

Materials Research Forum LLC
https://doi.org/10.21741/9781644902332-2

Chapter 2

Magnetic Spinel Ferrite Nanoparticles: From Synthesis to Biomedical Applications

M.V. Nikolic

Institute for Multidisciplinary Research, University of Belgrade, Belgrade, Serbia

mariavesna@imsi.rs

Abstract

Spinel ferrites are a widely investigated and applied class of materials with a cubic spinel lattice structure. Their unique multifunctional properties (magnetic characteristics, tunable shape/size, large number of active surface sites, high values of specific surface area, good chemical stability, and possibilities for enhancing properties through surface modification) influenced by the synthesis procedure make them attractive for biomedical applications as magnetic nanoparticles in drug delivery to a set target, magnetic hyperthermia, tissue engineering, magnetic extraction of biological components and magnetic diagnostics. In this review, we give an overview of up-to-date synthesis procedures for obtaining magnetic spinel nanoparticles and nanocomposites with optimal properties for biomedical applications.

Keywords

Magnetic Nanoparticles, Spinel Ferrites, Nanocomposites, Synthesis, Biomedical Applications

Contents

1.   Introduction

Magnetic nanoparticles (MNPs) have been an important material in science and technology ever since the pioneering work of Neel [1, 2]. Magnetic nanoparticles with a size range 1 to 100 nm exhibit size-dependent magnetic properties and a high surface area-to-volume ratio (SAVR). Magnetic nanoparticles include metal nanoparticles, for instance iron, chromium, nickel or their compounds, such as oxides, ferrites or other alloys [3]. Application of magnetic nanoparticles has increased greatly, especially in the field of biomedical applications, focusing on hyperthermia, targeted drug delivery, bone tissue repair, bioseparation, diagnostics [4-10], as shown in Fig. 1. To be able to fulfill the requirements of biomedical applications, MNPs need to be monodisperse with a uniform composition, show good biocompatibility and low toxicity, and enable tunable magnetic properties that can meet the requirements of the chosen application. Magnetic properties, such as coercivity and anisotropy need to be strong to enable control handling an externally applied magnetic field. MNPs also need to have a high SAVR and feasibility for easy surface functionalization. Surface functionalization is applied to prevent agglomeration and aggregation, improve chemical stability and biocompatibility [11].

Spinel ferrites are a group of mixed metal oxides that have a characteristic general formula MFe_2O_4, in which M can be Co, Ni, Zn, Mn or other metal cations. They possess good chemically and thermally stable magnetic characteristics. Spinel ferrites have tunable magnetic properties primarily influenced by the material's crystal lattice structure [12]. Nanostructured spinel ferrites have a small grain size and high value of the specific surface area. The properties of nanostructured spinel ferrites are significantly different compared to bulk spinel ferrite materials [13]. Spinel ferrite nanostructures are simple to prepare, they are low cost, chemically stable, and as a magnetic material easily recovered and reused. The application field of spinel ferrites is large and includes, besides a wide range of biomedical applications [13, 14], gas sensing [15, 16, 17], EMI shielding [18, 19], photocatalysis [20, 21], and others, such as the recently proposed MNP assisted RNA extraction protocol applied in prospective detection of the COVID-19 virus [22]. The synthesis procedure can have a considerable influence on the resulting multifunctional properties of these versatile materials. In this review, we will focus on synthesis procedures of spinel ferrites and their nanocomposites from the viewpoint of obtaining properties best suited for biomedical applications, such as hyperthermia, drug delivery to a set target, tissue engineering, photothermal therapy, magnetic diagnostics, and magnetic extraction of biological components including future perspectives, needs, and trends.

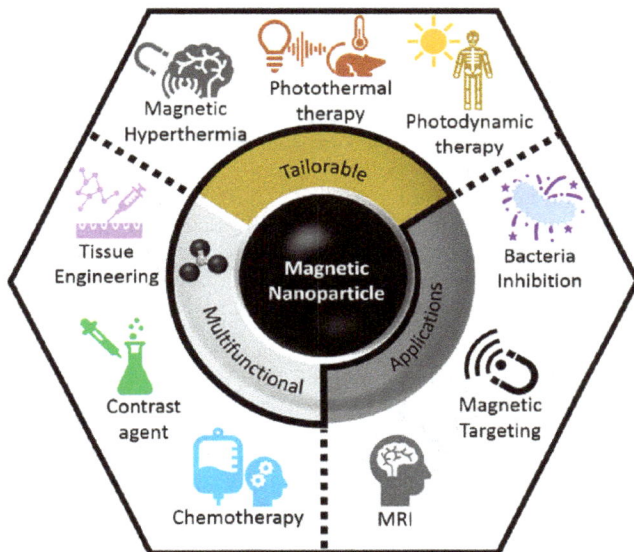

Fig. 1. Schematic illustration of some common biomedical application examples for magnetic nanoparticles. Reprinted with permission from [1] Copyright 2021, American Chemical Society

2. Spinel ferrites: A short overview

Spinel ferrites have a cubic crystal lattice structure ($Fd\bar{3}m$) showing close packing of oxygen atoms that have tetrahedral and octahedral sublattices. In the general formula - MFe_2O_4 (M is Co, Ni, Zn, Mn, Mg, and others), M^{2+} and Fe^{3+} cations can occupy the two distinctive crystallographic sites with tetrahedral (A sites) and octahedral (B sites) oxygen coordination [21, 23], as shown in Fig. 2. In a normal spinel the 8 A sites are filled with M^{2+} cations, while the 16 B sites are filled with Fe^{3+} cations. In an inverse spinel the 8 A sites are completely filled with Fe^{3+} cations, while 16 B sites are randomly filled with Fe^{3+} and M^{2+} cations. Most spinels have a mixed or partially inverse cubic spinel structure. In this case A and B sites are mutually filled with M^{2+} and Fe^{3+} cations. An inversion parameter (i) is often used to depict the non-convergent cation disorder ranging over tetragonal and octahedral sites [23]. The spinel ferrite general formula would then be $[M_{1-i}^{2+}Fe_i^{3+}]^A[M_i^{2+}Fe_{2-i}^{3+}]^BO_4$, with $i = 0$ representing a normal spinel structure, $i = 1$, representing a wholly inverse spinel structure and $i = 0.666$ a fully disordered spinel. Cation composition and ordering in the spinel structure have a crucial influence on the physical properties. Thus, the remaining magnetic moment of spinel ferrites mostly depends on the cation switching between tetrahedral (A) and octahedral (B) sites and the cation composition [12, 24].

Magnetite (Fe_3O_4) is a ferrite with an inverse spinel structure and cation allocation defined as $[Fe^{3+}]^A[Fe^{2+}Fe^{3+}]^BO_4$. Magnetitte has been applied as a remedy for customary ailments since ancient Greek times [25]. $NiFe_2O_4$ and $CoFe_2O_4$ are examples of other inverse spinel structures, $ZnFe_2O_4$ is an example of a normal spinel structure, while $MnFe_2O_4$ is a typical mixed spinel structure, due to the multiple valence of Mn. According to the recent review by Kefeni et al. [12] Fe_3O_4, $CoFe_2O_4$, $MnFe_2O_4$, $NiFe_2O_4$, and $ZnFe_2O_4$ have mostly been the subject of research for biomedical applications. Spinel ferrites have a wide range of physical properties that can be personalized for specific applications applying different synthesis procedures combined with varying the cation composition, surface functionalization, or the formation of nanocomposites with other materials.

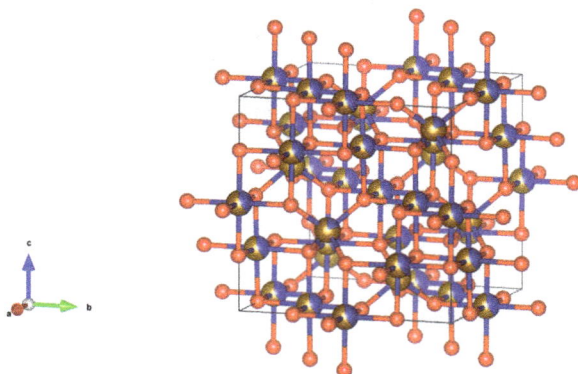

Fig. 2 Spinel ferrite structure, (CoFe₂O₄, Co – blue, Fe - brown, O – red), drawn using VESTA [26]

2.1   Doping and substitution with metal ions

Doping or cation substitution is widely applied especially to adjust the magnetic properties of spinel ferrites for application in biomedical and other diverse applications. Many metal ions have been applied and there is much research and literature data on this subject. For example, Saha et al [27] investigated substitution of Fe^{3+} cations with Zn^{2+} in $Zn_xFe_{3-x}O_4$, varying the Zn content (x) from 0 to 1. Hollow spherical nanoparticles with enhanced saturation magnetization (M_s) were achieved by Zn doping of Fe_3O_4, until $x = 0.2$. This can be attributed to substitution of Fe^{3+} cations by Zn^{2+} cations on tetrahedral sites disrupting termination of the magnetic moment of Fe^{3+} on tetrahedral and octahedral sites. Magnesium (Mg^{2+}) substitution of Fe^{3+} in $Mg_xFe_{3-x}O_4$ (in the case where x=0.1, 0.2, 0.4) led to smaller particle size, but increased particle agglomeration and a substantial saturation magnetization rise of 40% for $x = 0.1$ [24].

In a recent review, Sharifianjazi et al. listed 27 dopants for $CoFe_2O_4$ [28]. Zinc substituted $CoFe_2O_4$ nanoparticles ($Zn_{0.5}Co_{0.5}Fe_2O_4$) showed a high M_s value, low coercivity, and residual magnetization [29]. Cobalt ferrite has also been doped with rare earth elements, for example, Tm had a significant influence on magnetic properties [30]. When the amount of dopant was between 0.04 and 0.08 the superparamagnetic nature of cobalt ferrite at room temperature changed to ferromagnetic. In cobalt ferrite the value of magnetization was enhanced, while the coercivity value was reduced, following the growth in Ce and Dy content [31].

Substitution of Zn^{2+} with Ca^{2+} in $Ca_xZn_{1-x}Fe_2O_4$ ($x = 0.1$, 0.3 and 0.5) increased the saturation magnetization at 15 K and enhanced the heating efficiency needed for application in magnetic hyperthermia [32]. Superparamagnetic behavior was noted in pure $ZnFe_2O_4$ and Mg doped $ZnFe_2O_4$ (Mg substitutes Zn), while Ga doping (Ga substitutes Fe) resulted in a paramagnetic state [33]. Co-doping with both dopants resulted in intermediate behavior. In another study, Bini et al [34] investigated the influence of Ca and Gd doping on the superparamagnetic effect of $ZnFe_2O_4$ and determined that a more effective role was played by Ca ions replacing Zn, than Gd replacing Fe.

Mixed metal ions spinel ferrites, such as Mn-Zn [18, 19, 35], Mg-Co [20], Ni-Zn [36, 37], Mg-Zn [38, 39], Ni-Mg [40], Co-Mn [41], Co-Zn [42, 43] with varying metal ion content have been extensively investigated. Thus, enhanced magnetization, leading to an increased heating capacity suitable for magnetic hyperthermia application was achieved for the $Co_{0.5}Zn_{0.5}Fe_2O_4$ composition [42]. Higher Zn concentrations in nickel-zinc spinel nanoferrites led to a decrease in Curie temperature and spontaneous magnetization, improving the potential for application in self-controlled magnetic hyperthermia [37].

Even more complex spinel ferrites, combining three metal ions besides iron enable further tailoring of properties needed in diverse biomedical applications. Thus, copper substitution of zinc up to $x = 0.3$ in $Cu_xZn_{0.5-x}Mg_{0.5}Fe_2O_4$, lead to reduction of the lattice constant, and growth of saturation magnetization, accompanied by improved heat release and cell compatibility [44]. This composition was also assessed as a potential material for bone tissue regeneration and showed good capacity for the formation of apatite, *in vitro* degradation, antibacterial activity, and cell compatibility [45]. Manganese substitution of zinc in Mg-Zn ferrites enhances the superexchange relations involving tetrahedral (A) and octahedral (B) sites of the cubic lattice spinel ferrite structure [46]. Yang *et al.* [47] explored the influence of substituting Zn with Mg in Zn-Co ferrites on the Curie temperature (T_c), magnetic properties, and heating efficiency, showing that a higher in Mg^{2+} content results in T_c growth due to enhancement of the A-B super-exchange relations. Substitution of zinc with magnesium also increased the specific absorption rate (SAR), achieving a stable temperature of 44.7 °C for $x = 0.3$ in $Mg_xZn_{0.8-x}Co_{0.2}Fe_2O_4$. Cobalt substitution of nickel in Ni-Zn ferrites enables control of coercitivity and magnetic anisotropy [48].

2.2   Synthesis methods

A great amount of methods have been applied for production of spinel ferrites [13]. They include solid-state reaction [49, 50], mechanosynthesis [51], co-precipitation [24, 35, 46], microemulsion [52, 53], hydrothermal [47, 50], sol-gel [29, 54, 55] and combustion methods [20, 56, 57].

Milling of the starting powder combined with thermal treatment (mechanosynthesis + solid-state reaction method) is the simplest way to synthesize spinel ferrites [49]. The resulting morphology, electric, dielectric and magnetic properties depend on the applied milling and thermal treatment regime [51, 58]. This method is simple, but does not result in nanoparticles with magnetic properties suited for biomedical implementation.

One of the requirements of magnetic nanoparticles for biomedical application is to obtain stable, biocompatible nanoparticles of uniform size, suitable magnetic properties, and surface chemistry [5].

Co-precipitation is a chemical method applied for fabricating magnetic nanoparticles, including spinel ferrites [3] and especially magnetite [35]. An aqueous solution of ferric and ferrous salts was used to obtaining magnetic nanoparticles by slow addition of a precipitating agent. The precipitating agent was an aqueous solution of ammonia, sodium hydroxide, or sodium carbonate. In the case of magnetite precipitation was complete for pH values ranging from 8 to 14 [35]. Thus, Sharma et al [46] used a 97% precipitating solution of NaOH with an aqueous solution of magnesium(II) chloride hexahydrate, zinc chloride, manganese(II) chloride tetrahydrate, and iron chloride tetrahydrate mixed in the 1:2 M ratio in double distilled water to obtain powder precipitates that were dried at 373 K, followed by calcination at 1173 K to obtain manganese substituted magnesium-zinc ferrite nanoparticles suitable for hyperthermia applications. A reduction in particle size, but higher agglomeration was noted in magnesium doped magnetite obtained using the co-precipitation method when sodium hydroxide was used for precipitation [24]. The type of metal precursor, precipitating agent, pH value of the solution, the ratio between Fe^{2+} and Fe^{3+} cations, reaction time, temperature have a considerable effect on the resulting size and morphology of obtained nanoparticles. An ammonia solution was used as the precipitating agent to obtain spinel ferrite nanoparticles using the co-precipitation method, where the morphology depended on the reaction atmosphere [59]. The presence of a stabilizing agent also has an influence, thus, Mello et al [35] used polyethylene glycol to obtain Zn and Mn co-doped magnetite nanoparticles with an octahedral morphology and average particle size of 13 nm (Fig. 3).

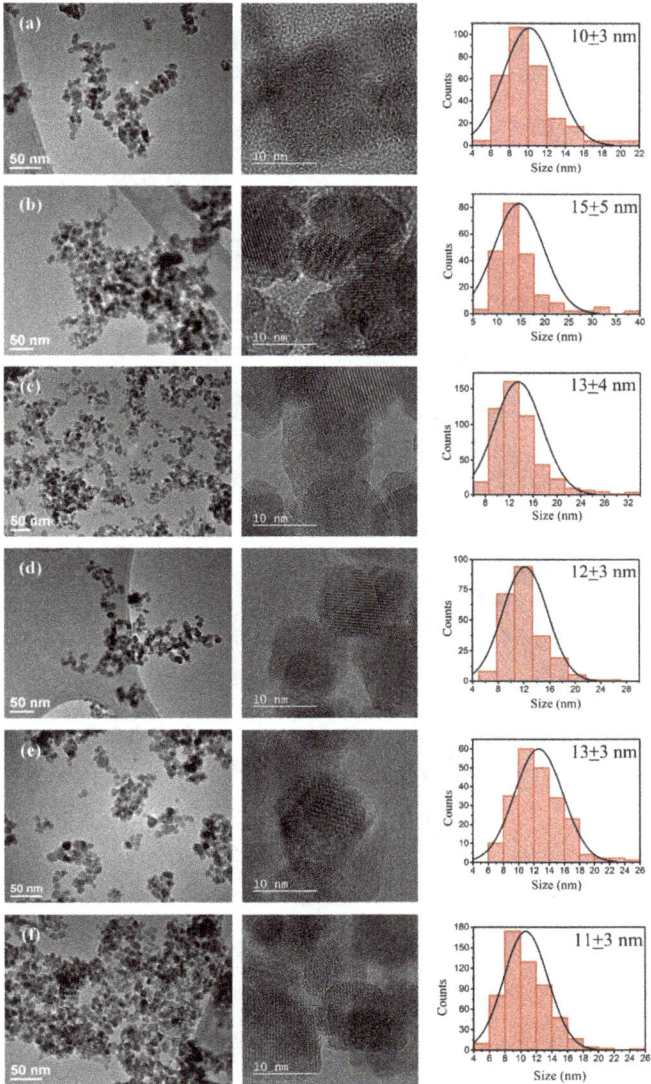

Fig. 3 TEM and HRTEM images of magnetite nanoparticles co-doped with Zn and Mn. Reprinted with permission from [35], Copyright 2019, Elsevier.

Microemulsions are characterized by colloidal dispersions which are thermodynamically stable and consist of a polar solvent, usually water, and a non-polar solvent that coexist as a single phase with distinct hydrophilic areas [60]. The globular type of microemulsion in the form of oil-in-water or water-in-oil has been utilized in the production of spinel ferrite nanoparticles [52]. In this case, each droplet is regarded as an individual nanoreactor for nucleation, growth, and crystallization of nanoparticles. The change in the oil-phase concentration and precipitating agent in the microemulsion enabled tailoring the average particle size of Mn-Zn nanoparticles between 2.4 and 9.4 nm (Fig. 4). The advantage of this method is that there is no need for additional calcination, but one limitation is that the organic solvents used are not environmentally friendly.

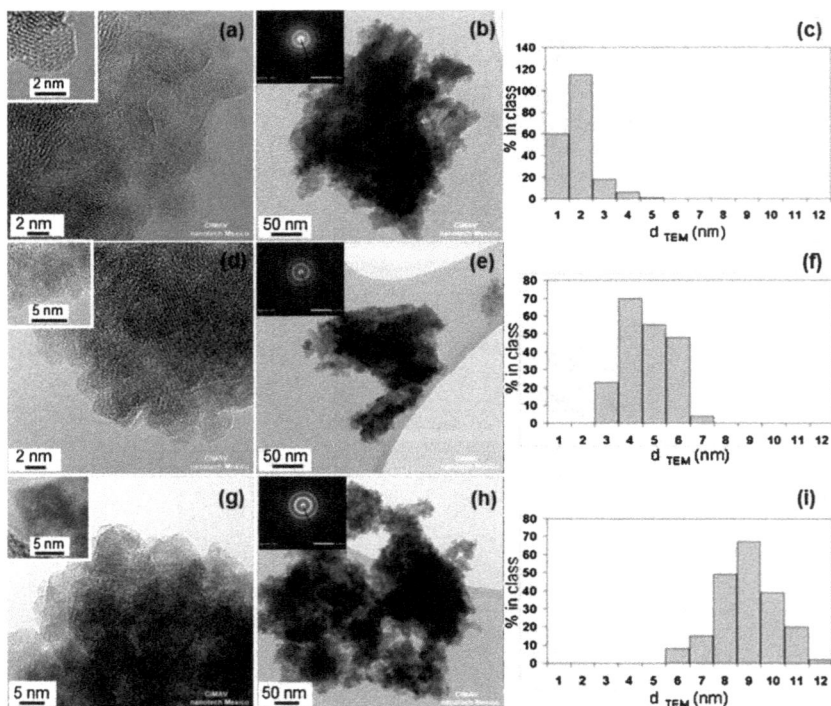

Fig. 4 Transmission electron microscopy images of Mn-Zn ferrite nanoparticles. Reprinted with permission from [52], Copyright 2014, Elsevier.

The hydrothermal synthesis method involves synthesis with help of chemical reactions in the presence of an aqueous solution at temperatures above water boiling value and under high pressures. The advantages of the hydrothermal method include good control of nucleation, improved dispersion and shape control, combined with a reduction in the operating temperature [61]. Hydrothermal treatment combined with precipitation has been applied to obtain zinc ferrite [50]. Jovanovic et al [61] determined that the pH value of the used solution had a dominant impact on the formation of cobalt ferrite nanoparticles, their growth, crystallization and magnetic properties.

Solvothermal synthesis involves chemical reactions that take place in a solvent (non-aqueous solution) at temperatures above the boiling point and high pressure. This method is simple and enables good control of particle shape, morphology, and crystallinity. It has been applied to obtain different magnetic spinel ferrites [62]. The link between cation distribution in spinel ferrites and the synthesis circumstances has been illustrated using polyols as metal ion solvents and complexing agents. Change in the polyol mediated annealing temperature has been revealed to have a substantial effect on the structure and cation distribution of nickel ferrite NPs and subsequent properties [63].

The sol-gel method combined with calcination is often used to obtain various spinel ferrites. It generally includes mixing of metal precursor solutions with gelation and chelating agent to form a xyrogel that is calcined to form the spinel ferrite. Thus, citric acid was used to synthesize zinc substituted cobalt ferrite nanoparticles [29], while urea was used to obtain cerium doped copper ferrite nanoparticles [54].

Synthesis using the sol-gel auto-combustion method is also commonly used to obtain spinel ferrites, such as manganese ferrite [57], or Mg-Co mixed ferrites [20]. The gelation and chelating agent in the formed gel when heated induces a thermal reaction acting as fuel, resulting in fine particles and a narrow particle size distribution [56]. Glycine, urea, or citric acid is commonly used as fuel. The sol-gel auto-combustion method is simple and compared to others more eco-friendly. Thus, Omelyanchik et al [64] used citric acid as the gelling and chelation agent and fuel to obtain biocompatible zinc-doped cobalt ferrite with high and flexible magnetic properties.

"Green synthesis" uses "green materials" such as plant, fruit, and vegetable extracts in the sol-gel process. Plants contain a variety of compounds which render reduction of metal ions and nanoparticles formation [65]. These include polyphenols, carotenoids, caffeine, tea, citrus (lemon, orange) peel, and other plant extracts. The utilization of plant extracts is inexpensive, non-toxic, and easily available. Cobalt ferrite with different grain size has been synthesized using orange fruit residue and three procedures: by extracting orange albedo and pectin, by extracting only pectin or utilizing the raw orange fruit residue bagasse [66]. Olive leaf extract containing phenolic compounds (oleuropein, phenolic acids, phenolic alcohols and flavonoids) was used as a chelating agent in "green" sol-gel synthesis of cobalt ferrite NPs resulting in a grain size in from 15 to 30 nm [67]. The amount of plant extract used influenced the obtained saturation magnetization and coercivity values. Magnesium ferrite NPs with an average grain size from 18 to 65 nm were synthesized using a *Solanum Lycopersicum* extract as the chelating agent [68]. Vitamin C present in the tomato extracts owes an important role in the chelating process in spinel ferrite formation.

Cobalt-zinc ferrite NPs were synthesized using honey as the "green material" and their possible application in magnetic hyperthermia therapy was investigated [65]. $Co_xZn_{1-x}Fe_2O_4$ NPs with a zinc content of $x = 0.6$ showed specific and intrinsic loss power higher than commercial Fe_3O_4.

2.3 Surface functionalization, modification, and coating

The purpose of functionalization of MNPs planned for biomedical use is to enhance biocompatibility, stability, and capability for encapsulating imaging agents or transport of therapeutic payloads [5]. MNPs for biomedical applications are often coated in order to achieve lower leachability and toxicity, reduce agglomeration, increase dispersibility and thermal stability [12]. Coating of MNPs is usually performed using synthetic or natural polymers with the aim to ensure colloidal stability through the prevention of aggregation, settling, and dipole-dipole reactions between particles [69]. Thus, polyethylene glycol (PEG) was used to coat nickel ferrite NPs obtained using the sol-gel combustion method to enhance biocompatibility and form a protective surface [70].

Surface modification of MNPs utilizing ligands, peptides, antibodies, genes, and drugs besides confirming colloidal stability can prevent agglomeration, sedimentation, and oxidation [5]. Oleic acid is non-toxic, hydrophilic, and helps improve biocompatibility. It has been used to coat cobalt ferrite [71, 72], magnesium ferrite [73], zinc ferrite [74] and magnetite [75]. Addition of oleic acid combined with solvothermal synthesis resulted in 6 nm well dispersed, non-agglomerated cobalt ferrite NPs [71]. The influence of different ligands, such as dextran, citric acid, or (3-aminopropyl) trierhoxylane on colloidal stability and biocompatibility of magnetite on colloidal stability and heating efficiency when applied for magnetic hyperthermia showed that coating magnetite NPs with (3-aminopropyl) trierhoxylane showed the highest specific absorption rate (SAR) [75]. A higher zeta potential reflecting longer balance in water, improved SAR and cell viability were obtained with a citric acid coating of sol-gel fabricated nickel-zinc ferrite NPs [76]. The addition of a combination of surfactants (oleic acid and tri-n-octylphosphine oxide) in a combined organic-hydrothermal synthesis procedure enabled formation of monodisperse $M_xFe_{3-x}O_4$ (M = Fe, Mg, Zn) spinel ferrite NPs using the hydrothermal method [77]. Malic acid was used as a surface passivating agent to synthesize magnetite NPs using the co-precipitation method that was highly biocompatible and dispersible in water as an anticancer drug carrier [78].

Coating spinel ferrite NPs can affect the magnetic moment [12] that could be due to spin pinning when coated ligands reduce the magnetic involvement of surface atoms.

2.4 Nanocomposites

Magnetic polymer nanocomposites combine good properties of the magnetic NPs with stability with high biocompatibility [79]. Magnetic polymer nanocomposites can be core-shell organic-inorganic nanocomposites, inorganic nanocomposites, and self-assembled colloidal nanocomposites. Silica is often used as a coating and encapsulating material to form a core-shell inorganic nanocomposite. Silica can significantly improve the colloidal

Materials Research Forum LLC
https://doi.org/10.21741/9781644902332-2

and chemical stability, facilitate surface functionalization and reduce toxicity [80]. Thus, magnetite NPs were obtained by co-precipitation, stabilized by coating agents, and encapsulated in hollow-shell mesoporous silica NPs ($Fe_3O_4@MSN$) with the aim of drug loading and drug delivery application, as shown in Fig. 5 [81].

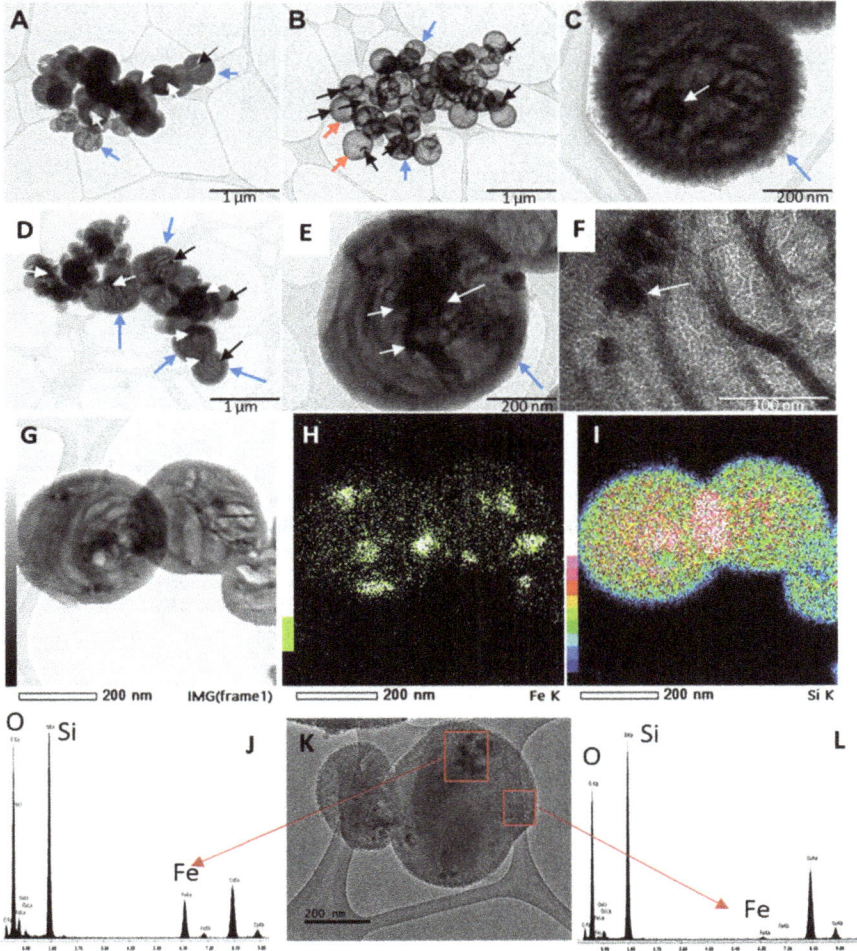

Fig. 5 TEM and STEM images of $Fe_3O_4@MSN$. Black and white arrows correspond to Fe_3O_4 Reprinted with permission from [81], Copyright 2021, MDPI.

Magnetite NPs synthesized using chemical coprecipitation were functionalized with curcumoid – extracted from the *Curcuma long* plant acting as a fluorescent biomarker and used to form a multicore nanocomposite encapsulated with silica as a prospective theranostic platform [80].

Recent research has focused on multifunctional nanocomposites combining a spinel ferrite, such as cobalt ferrite or copper ferrite with a bioceramic material such as magnesium silicate to form a multifunctional 3D nanocomposite scaffold or core-shell nanocomposites aimed for bone cancer therapy and bone tissue regeneration [82, 83, 84]. This type of nanocomposite has the potential to address issues of possible infection combined with bone cancer treatment, besides regeneration of the bone.

A core-shell nanocomposite consisting of magnetic $CoFe_2O_4$ NPs and hydroxyapatite to ensure high compatibility and drug conjugation was synthesized with the purpose of treating by combining drug release with magnetic hyperthermia [85].

3. Tailoring for biomedical applications

Each biomedical application of spinel ferrite NPs requires specific particle properties in view of the structure, morphology, and magnetic properties.

3.1 Hyperthermia

In 1979 Gordon *et al.* [86] suggested using locally induced heat energy created by treating colloidally suspended submicron particles (magnetite nanoparticles) injected into tumor cells with an external magnetic field. Magnetite remains the most commonly researched spinel ferrite NP that is often surface coated with an organic or inorganic coating agent for hyperthermia treatment [1]. Hyperthermia treatment is not new, and besides cancer treatment, it has been used in medicine to treat cardiac arrhythmia, varicose veins, endometrial bleeding, and excess fat. Magnetic hyperthermia treatment represents a form of localized cancer treatment that generates heat in magnetic nanoparticles injected into the tumor. When the temperature reaches around 43 °C the cancer cells die [9]. The higher metabolic rate of tumor cells makes them more susceptible to hyperthermic effects than is the case for celles that are healthy [12]. Hyperthermia treatment can cause cell death that is the result of various mechanisms as discussed comprehensively in the review by Chang et al [87]. Research has shown that is hyperthermia treatment of cancer with MNPs is more efficient when combined with other types of therapy, such as radiotherapy or chemotherapy. Heat-mediated drug delivery for cancer treatment has been intensively investigated [88].

The heat release of MNPs used for hyperthermia treatment depends on the particle dimensions, morphology, composition, applied surface modification, and fabrication method [25]. Thus, the heating efficiency was different for spherical, cubic and elongated magnetite (Fe_3O_4) NPs obtained using the solvothermal synthesis method [89]. The synthesized nanoparticle shape also influences the temperature distribution [90]. Magnetite nanorods were synthesized with the precursor of an iron oleate and the thermal decomposition method. To make them more suited for biological applications and hyperthermia they were treated with an amphilic surfactant – tetramethylammonium 11-aminoundecanoate (TMAAD) resulting in the transformation of the formed hydrophobic oleate coated iron oxide nanorods into hydrophilic, as shown in Fig. 6 [91]. *In vitro* evaluation of the hydrophilic rods in relation to MCF-7 breast cancer cell lines expressed anticancer activity with reduced hemolysis due to minimal erythrocyte damage.

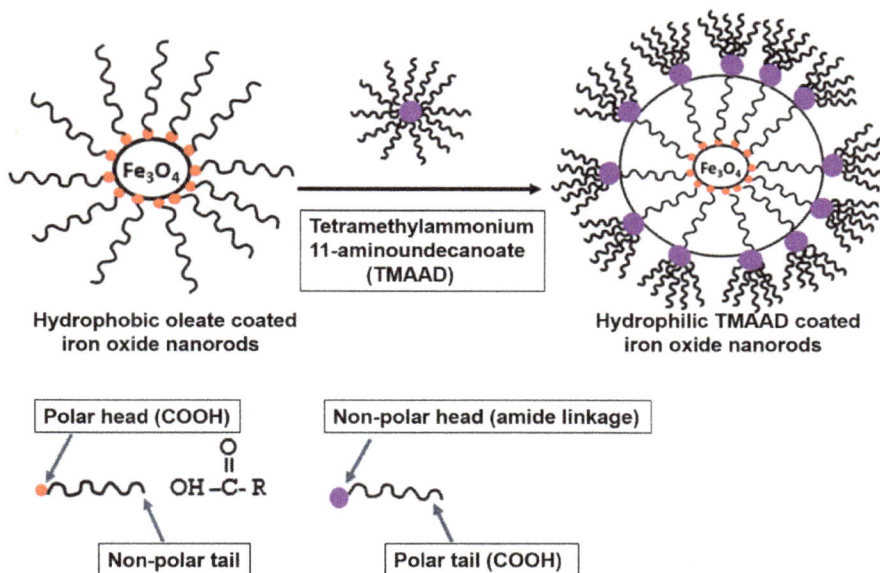

Fig. 6 Schematic representation of the development of hydrophobic oleate coated and hydrophilic TMAAD coated magnetite nanorods, Reprinted with permission from [91], Copyright 2021, American Chemical Society.

The solvothermal reflux technique was used to synthesize $CoFe_2O_4$ NPs with a narrow 10 nm average size distribution of [92]. The nearly monodisperse NPs showed increased magnetic moment leading to increased hyperthermal efficiency. Copper ferrite nanofibers were obtained using the electrospinning technique and subsequent calcination, showing promise for hyperthermia applications [93].

The frequency and amplitude square the applied external magnetic field also influence the heat release of MNPs. The Neél relaxation mechanism is responsible for heat release of NPs smaller than 20 nm, while for larger NPs it is the Brownian relaxation mechanism and hysteresis loss. Kouzoudis et al [94] investigated the dependence of hyperthermia power on frequency for clustered magnetite NPs (MIONs) in the frequency range 400-1100 KHz. No resonance was observed in this range, even though Neél relaxation times could occur in the applied frequency range. This was attributed to the morphology i.e. amplified interactions between particles of the concentrated magnetic clusters.

Research has focused on a variety of spinel ferrites, as the composition besides particle morphology has a significant influence on the heat release. In order to minimize the quantity of MNPs used for hyperthermia treatment, much research has been devoted to developing NPs with higher heating efficiency [87]. As mentioned above, doping/ substitution with metal ions is applied to tailor the magnetic characteristics of spinel ferrites. Magnesium doping of nearly monodisperse magnetite NPs obtained by the solvothermal reflux method lead to enhancement of the specific heat generation by modifying the magnetic susceptibility. Possible application of many mixed spinel ferrites in magnetic hyperthermia has been the subject of research. They include Co-Zn [32, 38, 42], Ni-Mg [40], Ni-Zn [36, 37, 76], Co-Mn [41] and Mn-Zn [35] spinel ferrites and also Cu doped Zn-Mg [44], Mn doped Mg-Zn [46], Mg doped Zn-Co [47].

3.2 Targeted drug delivery

In targeted drug distribution procedures MNPs are bound to the drug component enabling drug delivery directly into the center of the disease in a varying magnetic field [5]. Different parameters are varied and they include the pH value, light, temperature, electric current, and magnetic field, as illustrated in Fig. 7.

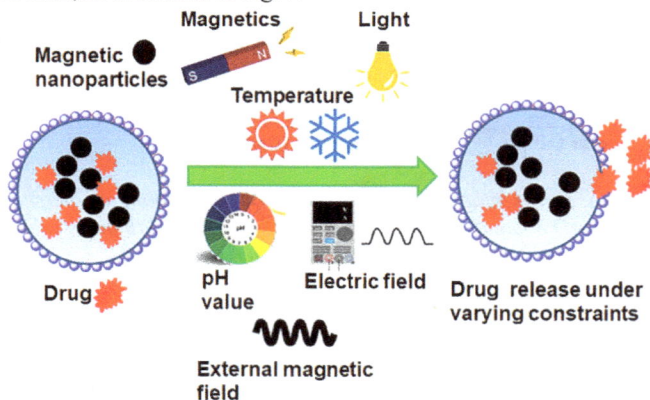

Fig. 7 Schematic representation of drug release using a MNP nanocomposite with applied external magnetic field under varying constraints, Reprinted with permission from [5], Copyright 2021, Americal Chemical Society.

Targeted drug delivery has been the subject of much research as it is simple, effective, and easy to prepare [12]. In comparison with the conventional method of drug administration that causes toxicity in the entire body due to nonspecific cell and tissue distribution and reduced therapeutic effectiveness, targeted drug delivery enables direct delivery to the target. This is especially significant in the case of cytotoxic drugs, as they can be encapsulated. Thus, doxorubicin (DOX) targeted delivery was regulated by the pH of the cells around the tumor cells [96]. Manganese zinc ferrite NPs with an average particle size of 25 nm were fabricated using the co-precipitation approach, stabilized with chitosan, and coated with polylactic-co-glycolic acid, a biocompatible and biodegradable polymer. Chitosan was selected to enable pH-sensitive release of DOX encapsulating the nanoparticle composite. Further research by Montha et al [97] focused on combining chitosan with a thermo-responsive polymer (N-isopropylacrylamide) to enable chemo-hyperthermia treatment with DOX. Monodisperse cobalt ferrite NPs (on average 15 nm) were obtained by a modified thermal decomposition approach and used to develop a nanocomposite with dopamine and DOX enabling on demand magnetic response induced hyperthermia and release of doxurubicin [98]. In-vivo experiments showed a worth noting tumor regression with no obvious toxicity occurrence.

Cobalt ferrite nanoparticles fabricated using the solvothermal approach were coated with L-arginine to form mesoporous nanoclusters that showed low toxicity, good loading capacity, and pH response release of DOX [99].

3.3 Tissue engineering

Magnetic nanoparticles incorporated into magnetoactive structures inside scaffolds can be beneficial for tissue healing combined with external magnetic field stimulation [100]. The presence of magnetic nanoparticles can provide better conditions for bone regeneration. Thus, magnetic iron oxide NPs were synthesized by in-situ precipitation into a hydroxyapatite/chitosan bone scaffold and tested for potential tissue engineering applications [101]. Chitosan was selected as part of the scaffold as a biocompatible, biodegradable polymer that also has inherent healing properties. Hydroxyapatite is a biomaterial with great potential for bone tissue repair and replacement [102]. Increased biocompatibility of magnetite NPs was achieved by coating with hydroxyapatite using a biological synthesis route and *Enterobacter aerogenes* [103].

Recent work has focused on simultaneous bone cancer therapy and regeneration using core-shell nanocomposites of magnetic NPs such as $CoFe_2O_4$ or $CuFe_2O_4$ and Mg_2SiO_4 as the bioceramic material [82, 83, 84].

Introduction of a bioactive coating with magnetic response ability based on cobalt ferrite and PVDF and Terfenol-D (TrFE) forming a $CoFe_2O_4@P(VDF-TrFE)$ nanocomposite in the form of a coating enabled improved healing and reparation of injured bone tissue due to enhanced osteogenic differentiation [109].

The magnetoelectric effect provides coupling between magnetic and electrical properties in magnetoelectric materials [104]. Bioreactor systems using magnetic stimulation of magnetoelectric composite materials utilize magnetic NPs such as $CoFe_2O_4$ [105]. Cobalt ferrite NPs have been used in many different composite materials in the form of scaffolds [106, 107], membranes [108], or coatings [109]. Thus, poly(vinylidene fluoride) (PVDF) a piezoelectric polymer was combined with cobalt ferrite NPs as magnetostrictive components to form a nanocomposite for application as a scaffold in bone tissue engineering, as shown in Fig. 8 [107].

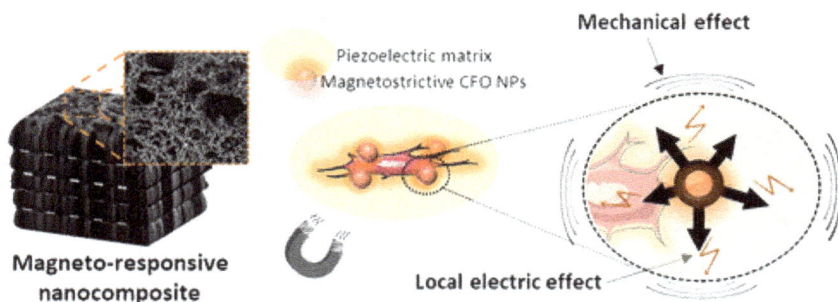

Fig. 8 Schematic representation of 3D scaffolds composed of PVDF and $CoFe_2O_4$ and their magnetoelectric effect in an external magnetic field, Reproduced with permission from [107], Copyright 2019, American Chemical Society

3.4 Photothermal therapy

Photothermal therapy (PTT) was developed as an extension of the photodynamic therapy (PDT) [110]. Both involve the utilization of photosensitizing agents (PA) that are introduced into the tumor tissue and excited with a specific band light. In PDT upon activation, singlet oxygen is generated by the PA that is acutely cytotoxic and leads to irreversible free radical damage to tissues in a radius of 20 nm. In PTT the PA upon activation converts the absorbed light into heat leading to ablation of malignant tissue due to local heating to around 43 °C, while the temperature of the surrounding healthy tissue remains at a normal level [111]. The basic requirements of PA agents for PTT are minimal toxicity, particle diameter between 30 and 200 nm, ability to absorb NIR radiation in the first biological window range (in the range 650-900 nm, with a peak usually around 800 nm), and high absorption cross-section to be able to maximize the conversion from light to heat. Different nanomaterials are used as PA agents for PTT and they include spinel ferrite MNPs. PA agents are transported to the tumor site in two ways. The first is intravenously by targeted delivery and the second is by direct injection into the tumor. Targeted delivery can be active and passive, where for the active form the PA agent needs to be functionalized with a peptide or antibody enabling specific recognition by the proteins overexpressed in the tumor cells. In the case of passive targeting, it is built on the effect of retention and permeability enhancement.

Different spinel ferrites have been investigated and applied in photothermal therapy, and they include magnetite (Fe_3O_4), zinc-ferrite ($ZnFe_2O_4$), manganese ferrite ($MnFe_2O_4$), and nickel ferrite ($NiFe_2O_4$) [112, 113]. As mentioned above functionalization of PA agents is significant and has been the subject of much research. Functionalization of manganese ferrite with a carboxylic acid functionalized IR806 dye enabled the formation of a nanocomposite with insignificant toxicity, effective photothermal damage mediated by NIR and ROS cytotoxicity, thus combining both photodynamic and photothermal therapy in one nanocomposite material [114]. Immunologically modified manganese ferrite nanoparticles synthesized by seed-mediated growth enable simultaneous treatment of cancer using immune- and photothermal therapy [115]. This type of treatment can cause cancer tumor ablation and stimulation of the host immune reaction studied in vivo and in vitro in the case of breast cancer, as depicted in Fig. 9. The R837 adjuvant was used as it is a Toll-like receptor-7 agonist capable of significant stimulation of DC cell maturation and some host immunosuppression, combined with ovalbumine that is a strong immunogenic antigen.

Besides tumor treatment ferrite nanoparticles have been used for photothermal therapy of other medical problems, such as treatment of artherosclerotic plaque. A theranostic platform aimed at multimodal image guidance of PTT for plaque neovascularizations was developed using manganese ferrite NPs and perfluorohexane made stable using polylactic-glycolic acid shells conjugated to the surface of an anti-vascular endothermal growth factor receptor antibody [116]. Hatamie et al [117] synthesized nanocomposites combining WS_2, $CoFe_2O_4$, and cubic Fe_3O_4 in a nanocomposite aimed for combined photothermal therapy and MRI targeting for ophthalmic treatment and research into corneal cell migration.

Much research has focused on nanocomposites formed by combining spinel ferrites with other materials. One such combination is with carbon or graphene [113, 118, 119]. Thus a nickel ferrite/C nanocomposite was investigated as a possible novel PA agent in melanoma cancer therapy by phototherapy [113]. In the case of a zinc-ferrite/reduced graphene (rGO) nanocomposite the cancer ablating ability was related to both local heating of cancer cells and the protein composition and formation of corona content under irradiation of continuous laser [119].

These nanocomposites can have multiple roles, such as a $MnFe_2O_4/C$ nanocomposite synthesized by Gorgizadeh et al [120] and investigated as a potential theranostic vehicle for MRI, photothermal and sonodynamic therapy. $Fe_3O_4@Au$ NPs, particle size ranging from 10 to 60 nm were fabricated using the hydrothermal method, used for the formation of a nanocomposite structure with rGO, and investigated as a combined phtothermal and radiotherapy agent for the squamous carcinoma KB cell line [121].

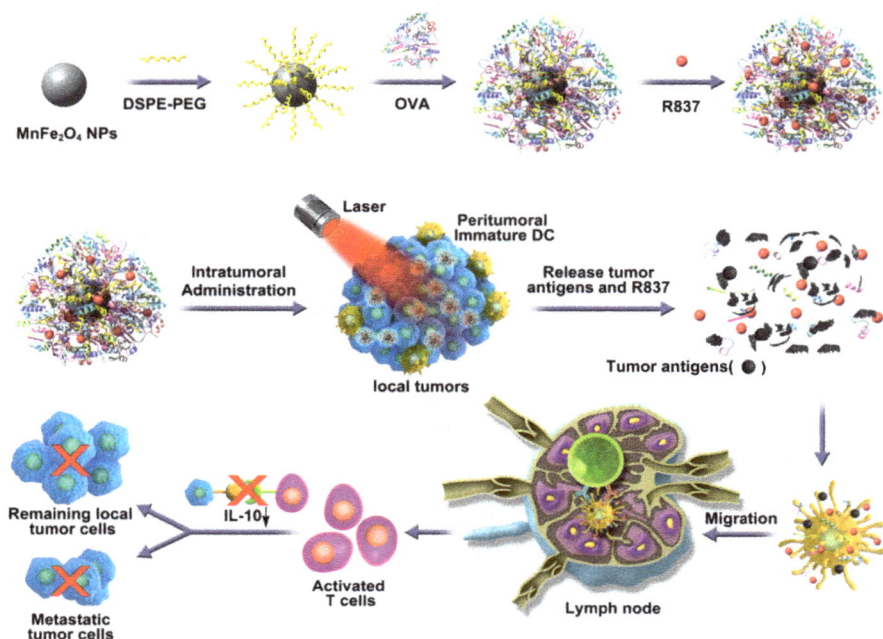

Fig. 9 Schematic representation of R837-OVA-PEG-MnFe$_2$O$_4$ NP synthesis and the procedure of laser irradiation triggered immune response, Reprinted with permission from [115], Copyright 2020, Elsevier.

Recent research on multifunctional magnetic nanoparticles has included the development of a nanocomposite of manganese ferrite combined with glucose oxidase and encapsulated with an erythrocyte (red blood cell) membrane enabling combined cancer diagnostics and treatment through MRI, photodynamic, photothermal and chemodynamic therapy, and starving of cancer cells [122].

A trimodal system was designed to combine magnetic hyperthermia with photothermal therapy, and luminescent nanothermometry by synthesizing multifunctional zinc substituted manganese ferrite NPs using the hydrothermal method, with citrate coating forming a core-shell structure when coated with silica with Nd^{3+} embedded in the amorphous silica layer, as shown in Fig. 10 [123].

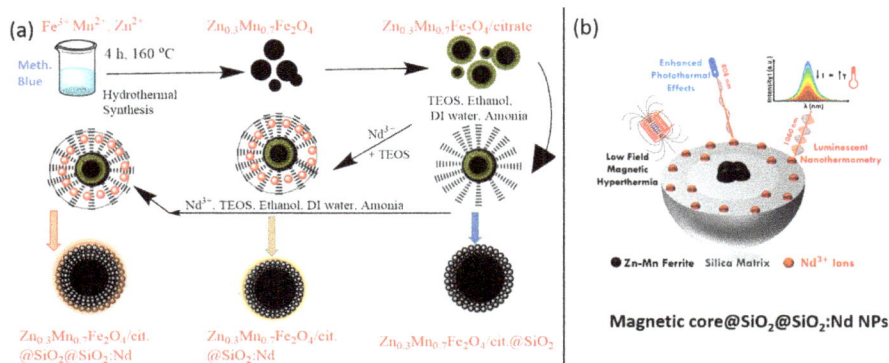

Fig. 10 Schematic diagram of a) synthesis of the $Zn_xMn_{1-x}Fe_2O_4@SiO_2"zNd^{3+}$ core-shell nanocomposite and b) combined therapy model, Reprinted with permission from [123], Copyright 2021, American Chemical Society.

3.5   Magnetic diagnostics

Magnetic resonance imaging (MRI) a standard clinical non-invasive, non-ionizing, and radiation-free technique applied for diagnosis of the tumor presence, location, and size in the human body has utilized magnetic nanoparticles as contrasting agents [12]. These include iron oxide Fe_3O_4 [124] and other spinel ferrites, such as $CoFe_2O_4$ [125]. The magnetic resonance imaging technique is based on a contrast created between healthy and sick tissue. It is supported using a contrasting agent whose purpose is to shorten the time of relaxation of water protons leading to improved tissue contrast [126]. Magnetic nanoparticles utilized as contrasting agents need to have good colloidal stability, low cytotoxicity and high biocompatibility. In this sense surface modification is essential.

Much research has centered on modification of magnetic nanoparticles utilized in MRI that includes different covalent and non-covalent modifications of Fe_3O_4. Non-covalent binding has included the application of diverse amphiphilic molecules. Derivatives of carboxylic, sulfonic, phosphoric, and more recently phosphonic acids were also used to obtain stable colloidal solutions. Thus, N-(phosphonomethyl) iminodiacetic acid (PMIDA) was recently investigated for surface modification of Fe_3O_4 NPs synthesized using the co-precipitation method [127]. In-vivo and in vitro investigations indicated that the resulting multifunctional NPs showed high selectivity in recognition of HER-2 receptors conveyed on MCF tumor cells. Recent research by Neto et al [128] has focused on methylene phosphonic acid – diemethylenetriaminepenta (DTPMP) as a capping agent for Fe_3O_4 and compared two synthesis methods: hydrothermal and sonochemical. The sonochemical method was faster resulting in MNPs with a better performance as a T_2 MRI agent.

Co-precipitation combined with hydrothermal treatment was used to obtain core-shell Fe_3O_4 –hydroxyapatite NPs that enabled the protection of the magnetic NPs from oxidation and improved MRI contrast ability in both the T_1 and T_2 weighing modes [129].

Recent research has centered on developing magnetic-fluorescent nanocomposites capable of bimodal imaging [130]. Iron oxide (Fe_3O_4) NPs were obtained by thermal decomposition and used to form a nanocomposite with CdTe quantum dots (QDs) coated with silica using a coordination-driven self-assembly method to obtain a magnetic-fluorescent nanocomposite enabling besides MRI imaging, multicolor fluorescent imaging, as shown in Fig. 11.

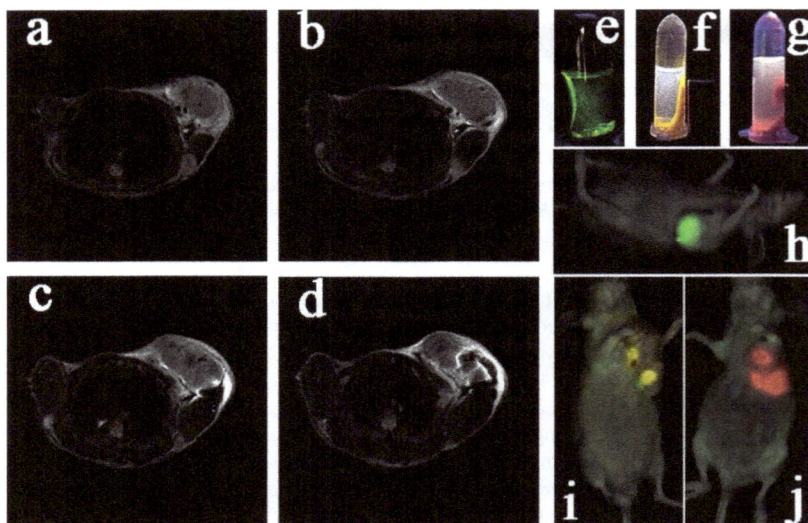

Fig. 11 In-vitro and in-vivo imaging using silica coated Fe_3O_4-QD nanocomposite, a-d MRI images, e-g digital photographs of the nanocomposite under an UV lamp, h-j in-vivo multicolor fluorescent imaging, Reprinted with permission from [130], Copyright 2020, American Chemical Society.

3.5 Magnetic separation of biological objects

Bioseparation using magnetic nanoparticles is a significant area of biomedicine enabling the separation of proteins, peptides, cells, enzymes, antibodies, and blood purification [5, 131, 132]. This process involves several steps that include specific production of magnetic nanoparticles, often in the form of magnetic nanocomposites, their modification, adsorption and separation, washing, elution, and recycling as shown in Fig. 12 [132]. Synthesis methods applied for obtaining MNPs suitable for application in bioseparation include co-precipitation, thermal decomposition, polyol, and solvothermal synthesis methods. Surface modification has been achieved by physical coating with a surfactant or polymer or chemical bonding often by conjunction with affinity ligands to separate biomolecules.

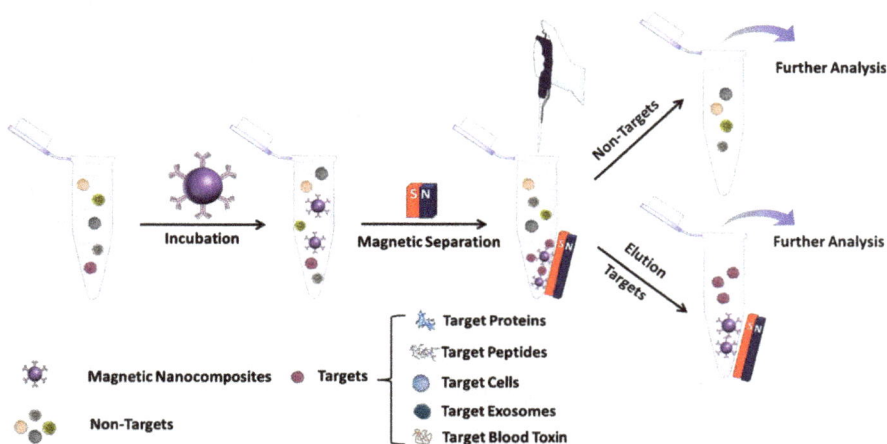

Fig. 12 Schematic illustration of the bioseparation process of magnetic nanocomposites, Reprinted with permission from [132], Copyright 2020, Elsevier.

Research has focused on core-shell magnetic nanoparticles, such as magnetite coated with silica or dopamine for magnetic bioseparation [133]. Magnetic beads formed from hollow shell structured $Fe_3O_4@SiO_2$ synthesized using the surfactant template approach were applied as adsorbents for the purification of plasmid DNA [134]. Further improvement of these magnetic beads was accomplished by control of the SiO_2 shell density with an intermediate surfactant [135]. Large $Fe_3O_4@SiO_2$ particles around 200 nm in size were functionalized with Au and poly(vinylpyrolidone) achieving properties suited for bioseparation and surface-enhanced Raman spectroscopy (SERS) sensing and tested successfully for targeting a glucagen-like peptide 1 [136]. Recent research has focused on magnetite (Fe_3O_4) –SiO_2 core shell nanoparticles with Au seeds as multifunctional NPs [137]. The synthesis procedure involved synthesis of superparamagnetic Fe_3O_4 through a modified polyol method, coated with SiO_2 to produce a core-shell structure, followed by seed assisted growth of Au seeds, functionalization, followed by a second growth of Au or Ag NPs. Enhanced SERS enhancement and a clean surface for direct functionalization were achieved enabling imaging and separation of a targeted cancer cell.

Recent research has analyzed development of microcryogels with embedded Fe_3O_4 magnetic nanoparticles for adsorption of bilirubin as an alternative for currently used methods [138].

4. Future perspectives, needs, and trends

In view of the expanding field of application of magnetic spinel ferrite nanoparticles in biomedicine, the main challenge remains to obtain nanoparticles or nanocomposites with spinel ferrite nanoparticles with a suitable morphology and magnetic properties through simple, environmentally friendly, scalable, and reproducible synthesis methods. Functionalization/coating of the synthesized MNPs needs to meet the requirements of the planned application and optimize the magnetic performance and biocompatibility. One future trend seems to be a synthesis of multifunctional magnetic nanoparticles often in the form of nanocomposites that include a magnetic spinel ferrite component to enable combined diagnostics and multiple treatment methods under the influence of an external magnetic field.

Acknowledgement

I would like to express my gratitude to Prof. Charanjeet Singh for useful advice while writing this paper. This research was funded by the Ministry for Education, Science and Technological Development of the Republic of Serbia under the contract 451-03-68/2022-14/200053.

References

[1] P. M. Martins, A. C. Lima, S. Ribeiro, S. Lanceros-Mendez, P. Martins, Magnetic nanoparticles for biomedical applications: from the soul of the earth to deep history of ourselves, ACS Appl. Bio. Mater. 4 (2021) 5839-5870. https://doi.org/10.1021/acsabm.1c00440

[2] L. Neel, Some theoretical aspects of rock-magnetism, Adv. Phys. 4 (1955) 191-243. https://doi.org/10.1080/00018735500101204

[3] T. I. Shabatina, O. I. Vernaya, V. P. Shabatin, M. Ya. Melnikov, Magnetic nanoparticles for biomedical purposes: Modern trends and prospects, Magnetochemistry 6 (2020) 30. https://doi.org/10.3390/magnetochemistry6030030

[4] S. H. Noh, S. H. Moon, T. H. Shin, Y. Lim, J. Cheon, Recent advances of magneto-thermal capabilities of nanoparticles: from design principles to biomedical applications, Nano Today 13 (2017) 61-76. https://doi.org/10.1016/j.nantod.2017.02.006

[5] S. Khizar, N. M. Ahmad, N. Line, N. Jaffrezic-Renault, A. Errachid-el-salhi, A. Elaissari, Magnetic nanoparticles: from synthesis to therapeutic applications, ACS Appl. Nano. Mater. 4 (2021) 4284-4306. https://doi.org/10.1021/acsanm.1c00852

[6] Y. Lin, K. Zhang, R. Zhang, Z. She, R. Tan, Y. Fan, X. Li, Magnetic nanoparticles applied in targeted therapy and magnetic resonance imaging: crucial preparation parameters, indispensable pre-treatments, updated research advancements and future perspectives, J. Mater. Chem. B 8 (2020) 5973. https://doi.org/10.1039/D0TB00552E

[7] D. Fan, Q. Wang, T. Zhu, H. Wang, B. Liu, Y. Wang, Z. Liu, X. Liu, D. Fan, X. Wang, Recent advances of magnetic nanomaterials in bone tissue repair, Front. Chem. 8 (2020) 745. https://doi.org/10.3389/fchem.2020.00745

[8] P. Dong, K. P. Rakesh, H. M. Manukumar, Y. H. E. Mohammed, C. S. Karthik, S. Sumathi, P. Mallu, H. L. Qin, Innovative nano-carriers in anticancer drug delivery - a comprehensive review, Bioorganic Chem. 85 (2019) 325-336. https://doi.org/10.1016/j.bioorg.2019.01.019

[9] K. Tofani, S. Tiari, Magnetic nanoparticle hyperthermia for cancer treatment: a review on nanoparticle types and thermal analyses, ASME of Medicinal Diagnostics 4(2021) 030801. https://doi.org/10.1115/1.4051293

[10] D. D. Stueberm J. Villanova, I. Aponte, Z. Xiao, V. L. Colvin, Magnetic nanoparticles in biology and medicine: past, present and future trends, Pharmaceutics 13 (2021 943. https://doi.org/10.3390/pharmaceutics13070943

[11] R. K. Reddy, P. A. Reddy, C. V. Reddy, N. P. Shetti, B. Babu, K. Ravindranadh, M. V. Shankar, M. C. Reddy, S. Soni, S. Naveen, Chapter 10 - Functionalized magnetic nanoparticles/biopolymer hybrids: Synthesis methods, properties and biomedical applications, in V. Gurtler, A. S. Ball, S. Soni, Methods in Microbiology. Volume 46, Academic Press, 2019, pp. 227-254. https://doi.org/10.1016/bs.mim.2019.04.005

[12] K. K. Kefeni, T. A. M. Msagati, T. T. I. Nkambule, B. B. Mamba, Spinel ferrite nanoparticles and nanocomposites for biomedical applications and their toxicity, Mater. Sci. Eng. C 107(2020) 110314. https://doi.org/10.1016/j.msec.2019.110314

[13] M. Amiri, M. Salavati-Niasari, A. Akbari, Magnetic nanocarriers: Evolution of spinel ferrites for medical applications, Adv. Colloid. Interface Sci. 265(2019) 29-44. https://doi.org/10.1016/j.cis.2019.01.003

[14] S. Y. Srinivasan, K. M. Paknikar, D. Bodas, V. Gajbhiye, Applications of cobalt ferrite nanoparticles in biomedical nanotechnology, Nanomedicine 13(2018) 1221-1238. https://doi.org/10.2217/nnm-2017-0379

[15] M. V. Nikolic, V. Milovanovic, Z. Z. Vasiljevic, Z. Stamenkovic, Semiconductor gas sensors: Materials, Technology, Design and Application, Sensors, 20 (2020) 6694. https://doi.org/10.3390/s20226694

[16] A. Sutka, K. A. Gross, Spinel ferrite semiconductor gas sensors, Sens. Actuators B 222 (2016) 95-105. https://doi.org/10.1016/j.snb.2015.08.027

[17] M. V. Nikolic, Z. Z. Vasiljevic, M. D. Lukovic, V. P. Pavlovic, J. B. Krstic, J. Vujancevic, N. Tadic, B. Vlahovic, V. B. Pavlovic, Investigation of $ZnFe_2O_4$ spinel

Materials Research Forum LLC
https://doi.org/10.21741/9781644902332-2

ferrite nanocrystalline screen-printed thick films for application in humidity sensing, Int J. Appl. Ceram. Technol. 16 (2019) 981-993. https://doi.org/10.1111/ijac.13190

[18] D. Petrovic, M. Lazic, O. S. Aleksic, M. V. Nikolic, V. Ibrahimovic, M. Pajnic, Mn-Zn ferrite line EMI suppressor for power switching noise in the impulse/high current bias regime, Turkish J. Elec. Eng. & Comp. Sci., 26 (2018) 2426-2436. https://doi.org/10.3906/elk-1710-52

[19] N. V. Blaz, M. D. Lukovic, M. V. Nikolic, O. S. Aleksic, Lj. D. Zivanov, Heterotube Mn-Zn ferrite bundle EMI suppressor with different magnetic coupling configurations, IEEE Trans. Magn. 50 (2014), 8000907. https://doi.org/10.1109/TMAG.2014.2310436

[20] M. P. Dojcinovic, Z. Z. Vasiljevic, V. P. Pavlovic, D. Barisic, D. Pajic, N. B. Tadic, M. V. Nikolic, Mixed Mg-Co spinel ferrites: structure, morphology, magnetic and photocatalytic properties, J. Alloys Compd. 855 (2021) 157429. https://doi.org/10.1016/j.jallcom.2020.157429

[21] K. K. Kefeni, B. B. Mamba, Photocatalytic application of spinel ferrite nanoparticles in wastewater treatment:review, SM&T 23(2020) e00140 https://doi.org/10.1016/j.susmat.2019.e00140

[22] S. B. Somvanshi, P. B. Kharat, T. S. Saraf, S. B. Somwanshi, S. B. Shejul, K. M. Jadhav, Multifunctional nano-magnetic particles assisted viral RNA extraction protocol for potential detection of COVID-19, Mater. Res. Innov. 25 (2021) 3. https://doi.org/10.1080/14328917.2020.1769350

[23] C. M. B. Henderson, J. M. Charnok, D. A. Plant, Cation occupancies in Mg, Co, Ni, Zn, Al ferrite spinels: a multi-element EXAFS study, J. Phys.: Condens. Matter. 19 (2007) 076214. https://doi.org/10.1088/0953-8984/19/7/076214

[24] V. Kusigerski, G. Illes, J. Blanusa, S. Gyergyek, M. Boskovic, M. Perovic, V. Spasojevic, Magnetic properties and heating efficacy of magnesium doped magnetite nanoparticles obtained by co-precipitation method, J. Magn. Magn. Mater. 475 (2019) 470-478. https://doi.org/10.1016/j.jmmm.2018.11.127

[25] Y. Wang, Y. Miao, M. Su, X. Chen, H. Zhang, Y. Zhang, W. Jiao, Y. He, J. Yi, X. Liu, H. Fan, Engineering ferrite nanoparticles with enhanced magnetic response for advanced biomedical applications, Mater. Today Adv. 8 (2020) 100119. https://doi.org/10.1016/j.mtadv.2020.100119

[26] K. Momma, F. Izumi, VESTA 3 for three dimensional visualization of crystal, volumetric and morphology data, J. Appl. Crystallogr. 44 (2011) 1271-1276. https://doi.org/10.1107/S0021889811038970

[27] P. Saha, R. Rakshut, K. Mandal, Enhanced magnetic properties of Zn doped Fe_3O_4 nano hollow spheres for better bio-medical applications, J. Magn. Magn. Mater. 475 (2019) 130-136. https://doi.org/10.1016/j.jmmm.2018.11.061

[28] F. Sharifiankjazi, M. Meradi, N. Parvin, A. Nemati, A. J. Rad, N. Sheysi, A. Abouuchenari, A. Mohammadi, S. Karbasi, Z. Ahmadi, et al Magnetic $CoFe_2O_4$ nanoparticles doped with metal ions: a review, Ceram. Int. 46 (2020) 18391-18412. https://doi.org/10.1016/j.ceramint.2020.04.202

[29] S. Iqbal, M. Fakhar-e-Alam, M. Atif, N. Amin, K. S. Alimgeer, A. Ali, Aqrab-ulAhmad, A. Harif, W. Aslam Farooq, Structural, morphological, antimicrobial and in vitro photodynamic therapeutic assessments of novel Zn^{2+} substituted cobalt ferrite nanoparticles, Results Phys. 15 (2019) 102529. https://doi.org/10.1016/j.rinp.2019.102529

[30] M. A. Almessiere, Y. Slimani, A. D. Korkmaz, S. Guner, M. Sertkal, S. E. Shirsath, A. Baykal, Structural, optical and magnetic properties of Tm^{3+} substituted cobalt spinel ferrites synthesized via sonochemical approach, Ultrason. Sonochem. 54 (2019) 1-10. https://doi.org/10.1016/j.ultsonch.2019.02.022

[31] M. Hashimi, M. Raghasundra, S. S. Meena, J. Shah, S. E. Shirsath, S. Kumar, D. Ravinder, P. Bhatt, Alimuddin, R. Kumar, R. K. Kotnala, Influence of rare earth doping (Ce and Dy) on electrical and magnetic properties of cobalt ferrites, J. Magn. Magn. Mater. 449 (2018) 319-327. https://doi.org/10.1016/j.jmmm.2017.10.023

[32] A. Manohar, V. Vijayakanth, K. H. Kim, Influence of Ca doping on $ZnFe_2O_4$ nanoparticles magnetic hyperthermia and cytotoxicity study, J. Alloys Compd. 886 (2021) 161276. https://doi.org/10.1016/j.jallcom.2021.161276

[33] G. Gazzola, M. Ambrosetti, M. C. Mozzati, B. Albini, P. Galinetto, M. Bini, Tuning the superparamagnetic effect in $ZnFe_2O_4$ nanoparticles with Mg, Ga doping, Mater. Chem. Phys. 273 (2021) 125069. https://doi.org/10.1016/j.matchemphys.2021.125069

[34] M. Bini, C. Tondo, D. Capsoni, M. C. Mazzati, B. Albini, P. Galinetto, Superparamagnetic $ZnFe_2O_4$ nanoparticles: the effect of Ca and Gd doping, Mater. Chem. Phys. 204 (2018) 72-82. https://doi.org/10.1016/j.matchemphys.2017.10.033

[35] L. B. de Mello, L. C. Varanda, F. A. Sigoli, I. O. Mazali, Co-precipitation synthesis of (Zn-Mn)-co-doped magnetite nanoparticles and their application in magnetic hyperthermia, J. Alloys Compd. 779 (2019) 698-705. https://doi.org/10.1016/j.jallcom.2018.11.280

[36] A. Manohar, K. Chintagumpala, K. H. Kim, Mixed Zn-Ni spinel ferrites: structure, magnetic hyperthermia and photocatalytic properties, Ceram. Int. 47 (2021) 7052-7061. https://doi.org/10.1016/j.ceramint.2020.11.056

[37] A. I. Tovstolytkin, M. M. Kulyk, V. M. Kalita, S. M. Ryabchenko, V. O. Zamorskyi, O. P. Fedorchuk, S. O. Solopan, A. G. Belous, Nickel-zinc spinel nanoferrites: magnetic characterization and prospects of the use in self-controlled magnetic hyperthermia, J. Magn. Magn. Mater. 473 (2019) 422-427. https://doi.org/10.1016/j.jmmm.2018.10.075

[38] T. Tatarchuk, M. Myslin, I. Mironyuk, M. Bououdina, A. T. Pedziwatr, R. Grgula, B. F. Bogacz, P. Kurzydio, Synthesis, morphology, crystallite size and adsorption properties of nanostructured Mg-Zn ferrites with enhanced porous structure, J. Alloys Compd. 819 (2020) 152945. https://doi.org/10.1016/j.jallcom.2019.152945

[39] A. Nigan, S. J. Pawar, Structural, magnetic and antimicrobial properties of zinc-doped magnesium ferrite for drug delivery applications, Ceram. Int. 46 (2020) 4058-4064. https://doi.org/10.1016/j.ceramint.2019.10.243

[40] A. Manohar, V. Vijayakanth, M. R. Pallavolu, K. H. Kim, Effects of Ni- substitution on structural, magnetic, hyperthermia, photocatalytic and cytotoxicity study of $MgFe_2O_4$ nanoparticles, J. Alloys Compd. 879 (2021) 160515. https://doi.org/10.1016/j.jallcom.2021.160515

[41] S. Jung, J. G. Jung, H. Choi, M. Kim, I. B. Shim, C. S. Kim, Magnetic, Moessbauer and hyperthermia properties of $Co_{1-x}Mn_xFe_2O_4$ nanoparticles, J. Radioanal. Nucl. Ch. 330 (2021) 433-437. https://doi.org/10.1007/s10967-021-07802-z

[42] C. Gomez-Polo, V. Recarte, L. Cervera, J. J. Beato-Lopez, J. Lopez-Garcia, J. A. Rodriguez-Velamazan, M. D. Ugarte, E. C. Mendonca, J. G. S. Duque, Tailoring the structural and magnetic properties of Co-Zn nanosized ferrites for hyperthermia applications, J. Magn. Magn. Mater. 465 (2018) 211-219. https://doi.org/10.1016/j.jmmm.2018.05.051

[43] R. Ramadan, M. K. Ahmed, V. Uskokovic, Magnetic, microstructural and photoactivated antibacterial features of nanostructured Co-Zn ferrites of different chemical and phase compositions, J. Alloys Compd. 856 (2021) 157013. https://doi.org/10.1016/j.jallcom.2020.157013

[44] M. Ansari, A. Bigham, S. A. Hassanzadeh-Tabrizi, H. A. Ahangar, Copper-substituted spinel Zn-Mg ferrite nanoparticles as potential heating agents for hyperthermia, J. Am. Ceram. Soc. 101 (2018) 3649-3661. https://doi.org/10.1111/jace.15510

[45] M. Ansari, A. Bigham, H. A. Ahangar, Super-paramagnetic nanostructured CuZnMg mixed spinel ferrite for bone tissue regeneration, Mater. Sci. Eng. C 105 (2019) 110084. https://doi.org/10.1016/j.msec.2019.110084

[46] R. Sharma, P. Thakur, M. Kumar, P. B. Barman, P. Sharma, V. Sharma, Enhancement in A-B super-exchange interaction with Mn^{2+} substitution in Mg-Zn ferrites as heating source in hyperthermic applications, Ceram. Int. 43 (2017) 13661-13669. https://doi.org/10.1016/j.ceramint.2017.07.076

[47] R. Yang, X. Yu, V. Li, C. Wang, C. Wu, W. Zhang, W. Guo, Effect of Mg doping on magnetic induction heating of Zn-Co ferrite nanoparticles, J. Alloys Compd. 851 (2021) 156907. https://doi.org/10.1016/j.jallcom.2020.156907

[48] J. N. P. K. Chintala, M. C. Varma, G. S. V. R. K. Choudary, K. H. Rao, Control of coercivity and magnetic anisotropy through cobalt substitution in Ni-Zn ferrite, J.

Supercond. Nov. Magn. 34 (2021) 2357-2370. https://doi.org/10.1007/s10948-021-05965-0

[49] N. J. Labus, Z. Z. Vasiljevic, O. S. Aleksic, M. D. Lukovic, S. Markovic, V. B. Pavlovic, S. V. Mentus, M. V. Nikolic, Characterization of $Mn_{0.63}Zn_{0.37}Fe_2O_4$ powders after intensive milling and subsequent thermal treatment, Sci. Sintering 49 (2017) 455-467. https://doi.org/10.2298/SOS1704455L

[50] X. Zhu, C. Cao, S. Su, A. Xia, H. Zhang, H. Li, Z. Liu, C. Jiu, A comparative study of spinel $ZnFe_2O_4$ ferrites obtained via a hydrothermal and a ceramic route: Structural and magnetic properties, Ceram. Int. 47 (2021) 15173-15179. https://doi.org/10.1016/j.ceramint.2021.02.077

[51] M. M. Milutinov, M. V. Nikolic, S. G. Lukovic, N. V. Blaz, N. J. Labus, O. S. Aleksic, Lj. D. Zivanov, Influence of starting powder milling on magnetic properties of Mn-Zn ferrite, Proc. Appl. Ceram. 11 (2017) 160-169. https://doi.org/10.2298/PAC1702160M

[52] K. Pemartin, C. Solans, J. Alvarez-Quintala, M. Sanchez-Domingues, Synthesis of Mn-Zn ferrite nanoparticles by the oil-in-water microemulsion reaction method, Colloids Surf. A: Physicochem. Eng. Asp. 451 (2014) 161-171. https://doi.org/10.1016/j.colsurfa.2014.03.036

[53] D. S. Matthew, R. S. Juang, An overview of the structure and magnetism of spinel ferrite nanoparticles and their synthesis in microemulsions, Chem. Eng. J. 129 (2007) 51-65. https://doi.org/10.1016/j.cej.2006.11.001

[54] E. Elayakumar, A. Mandikandan, A. Dinesh, K. Thanrasu, K. Kanimani Raja, R. Thilak Kumar, Y. Slimani, S. K. Jaganathan, A. Baykal, Enhanced magnetic property and antibacterial biomedical activity of Ce^{3+} doped $CuFe_2O_4$ spinel nanoparticles synthesized bu sol-gel method, J. Magn. Magn. Mater. 478 (2019) 140-147. https://doi.org/10.1016/j.jmmm.2019.01.108

[55] M. A. Almessiere, A. V. Trukhanov, F. A. Khan, Y. Slimani, N. Tashkandi, V. A. Turchenko, T. I. Zubar, D. I. Tishkevich, S. V. Trukhanov, L. V. Panina, A. Baykal, Correlation between microstructure parameters and anti-cancer activity of the $[Mn_{0.5}Zn_{0.5}](Eu_xNd_xFe_{2-2x})O_4$ nanoferrites produced by modified sol-gel and ultrasonic methods, Ceram. Int. 46 (2020) 7346-7354. https://doi.org/10.1016/j.ceramint.2019.11.230

[56] A. Sutka, G. Mezinskis, Sol-gel auto-combustion synthesis of spinel-type ferrite nanomaterials, Front. Mater. Sci. 6 (2012) 128-141. https://doi.org/10.1007/s11706-012-0167-3

[57] S. V. Bhandare, R. Kumar, A. V. Anupama, H. K. Choudhary, V. M. Jali, B. Sahoo, Annealing temperature dependent structural and magnetic properties of $MnFe_2O_4$ nanoparticles grown by sol-gel auto-combustion, J. Magn. Magn. Mater. 433 (2017) 29-34. https://doi.org/10.1016/j.jmmm.2017.02.040

[58] M. Milutinov, M. V. Nikolic, M. D. Lukovic, N. Blaz, N. Labus, Lj. D. Zivanov, O. S. Aleksic, Influence of starting powder milling on structural properties, complex impedance, electrical conductivity and permeability of Mn-Zn ferrite, J. Mater. Sci.: Mater. Electron., 27 (2016) 11856-11865. https://doi.org/10.1007/s10854-016-5328-1

[59] S. Slimani, C. Meneghini, A. Abdolrahimi, A. Talone, J. P. Murillo, G. Barucca, N. Yacacoub, P. Imperetori et al. Spinel iron oxide by the co-precipitation method: effect of the reaction atmosphere, Appl. Sci. 11 (2021) 5433. https://doi.org/10.3390/app11125433

[60] I. Danielsson, B. Lindman, The definition of microemulsions, Collods Surf. A 3 (1981) 391-392. https://doi.org/10.1016/0166-6622(81)80064-9

[61] S. Jovanovic, M. Spreitzer, M. Otonicar, J. H. Jeon, D. Suvorov, pH control of magnetic properties in precipitation-hydrothermal derived $CoFe_2O_4$, J. Alloys Compd. 589 (2014) 271-277. https://doi.org/10.1016/j.jallcom.2013.11.217

[62] M. Rafienia, A. Bigham, S. A. Hassanzadeh-Tabrizi, Solvothermal synthesis of magnetic spinel ferrites, J. Med. Signals Sens. 8 (2018) 108-118. https://doi.org/10.4103/jmss.JMSS_49_17

[63] T. Gaudisson, S. Nowak, Z. Nehme, N. Menguy, N. Yaacoub, J.-M. Greneche, S. Ammar, Polyol-made spinel ferrite nanoparticles - local structure and operating conditions: $NiFe_2O_4$ as a case study, Front. Mater. 8 (2021) 668994. https://doi.org/10.3389/fmats.2021.668994

[64] A. Omelyanchik, K. Levada, S. Pshenichnikov, M. Abdolrahim, M. Baricic, A. Kapitunova, A. Galieva, S. Sukhikh et al. Green synthesis of Co-Zn spinel ferrite nanoparticles: magnetic and intrinsic antimicrobial properties, Materials 13 (2020) 5014. https://doi.org/10.3390/ma13215014

[65] T. Tatarchuk, A. Shyichuk, Z. Sojka, J. Grybos, Mh. Nanshad, V. Kotsynbynski, M. Kowalska, S. Kwiatowska-Marks, N. Danylink, Green synthesis, structure, cations distribution and bonding characteristics of superparamagnetic cobalt-zinc ferrites nanoparticles for Pb(II) adsorption and magnetic hyperthermia applications J. Mol. Liq. 328 (2021) 115375. https://doi.org/10.1016/j.molliq.2021.115375

[66] E. P. Muniz, L. S. D. De Assuncao, L. M. deSouza, J. J. K. Ribeiro, W. P. Marquez, R. D. Perreira, P. S. S. Porto, J. R. C. Proreti, E. C. Passamani, On cobalt ferrite production by sol-gel from orange fruit residue by three related procedures and its application in oil removal, J. Cleaner Prod. 265 (2020) 121712. https://doi.org/10.1016/j.jclepro.2020.121712

[67] S. Banifatemi, E. Davar, B. Aghabaran, J. A. Segura, F. J. Alonso, S. M. Ghoreishi, Green synthesis of $CoFe_2O_4$ nanoparticles using olive leaf extract and characterization of their magnetic properties, Ceram. Int. 47 (2021) 19198-19204. https://doi.org/10.1016/j.ceramint.2021.03.267

[68] H. N. Chaudhari, P. N. Dhruv, C. Singh, S. S. Meena, S. Kavita, R. B. Jotania, Effect of heating temperature on structural, magnetic and dielectric properties of magnesium ferrites prepared in the presence of Solanum lycopersicum fruit extract, J. Mater. Sci.: Mater. Electron. 31 (2020) 18445-18463. https://doi.org/10.1007/s10854-020-04389-1

[68] V. Socoliuc, D. Peddis, V. I. Petrenko, M. V. Avdeev, D. Susan-Resiga, T. Szabo, R. Turcu, E. Tombacz, L. Vekas, Magnetic nanoparticle systems for nanomedicine - A materials science perspective, Magnetochemistry 6(2020) 2. https://doi.org/10.3390/magnetochemistry6010002

[70] M. R. Phadatare, V. M. Khot, A. B. Salukhe, N. D. Tharat, S. H. Pamar, Studies on polyethylene glycol coating on $NiFe_2O_4$ nanoparticles for biomedical applications, J. Magn. Magn. Mater. 324 (2012) 770-772. https://doi.org/10.1016/j.jmmm.2011.09.020

[71] S. Jovanovic, M. Spreitzer, M. Tramsek, Z. Trontelj, D. Suvorov, Effect of oleic acid concentration on the physicochemical properties of cobalt ferrite nanoparticles, J. Phys. Chem. C 118 (2014) 13844-13856. https://doi.org/10.1021/jp500578f

[72] P. B. Kharat, S. B. Somvanshi, P. P. Khirade, K. M. Jadhav, Induction heating analysis of surface functionalized nanoscale $CoFe_2O_4$ for magnetic fluid hyperthermia toward noninvasive cancer treatment, ACS Omega 5 (2020) 233778-23384. https://doi.org/10.1021/acsomega.0c03332

[73] S. B. Somvanshi, S. R. Patade, D. D. Andhare, S. A. Jadhav, M. V. Khedkar, P. B. Kharat, P. P. Khirade, K. M. Jadhav, Hyperthermic evaluation of olic acid coated nano-spinel magnesium ferrite: Enhancement via hydrophobic -to-hydrophylic surface transformation, J. Alloys. Compd. 835 (2020) 155422. https://doi.org/10.1016/j.jallcom.2020.155422

[74] S. B. Somvanshi, P. B. Kharat, M. V. Khedkar, K. M. Jadhav, Hydrophobic to hydrophylic surface transformation of nano-scale zinc-ferrite via oleic acid coating: Magnetic hyperthermia study towards biomedical applications, Ceram. Int. 46 (2020) 7642-7653 https://doi.org/10.1016/j.ceramint.2019.11.265

[75] M. Ognjanovic, D. M. Stankovic, Z. K. Jacimovic, M. Kosovic-Perutovic, B. Dojcinovic, The effect of surface-modifier of magnetite nanoparticles on electrochemical detection of dopamine and heating efficiency in magnetic hyperthermia, J. Alloys Compd. 884 (2021) 161075. https://doi.org/10.1016/j.jallcom.2021.161075

[76] P. V. Ramana, K. S. Rao, K. R. Kumar, G. Kapusetti, M. Choppadandi, J. N. Kiran, K. H. Rao, A study of uncoated and coated nickel-zinc ferrite nanoparticles for magnetic hyperthermia, Mater. Chem. Phys. 266 (2021) 124546. https://doi.org/10.1016/j.matchemphys.2021.124546

[77] H. Etemadi, P. G. Pleiger, Improvements in the organic-phase hydrothermal synthesis of monodisperse $MxFe3-x\neg O4$ (M = Fe, Mg, Zn) spinel nanoferrites for

magnetic fluid hyperthermia application, ACS Omega 5 (2020) 18091-18104.
https://doi.org/10.1021/acsomega.0c01641

[78] B. Dutta, S. Checker, K. C. Barick, H. G. Salunke, V. Gota, Malic acid grafted Fe_3O_4 nanoparticles for controlled drug delivery and efficient heating source for hyperthermia, J. Alloys Compd. 883 (2021) 160950.
https://doi.org/10.1016/j.jallcom.2021.160950

[79] S. Kalia, S. Kanoo, A. Kumar, Y. Haldorai, B. Kumari, R. Kumar, Magnetic polymer nanocomposites for environmental and biomedical applications, Colloid Polym. Sci. 292 (2014) 2025-2052. https://doi.org/10.1007/s00396-014-3357-y

[80] E. C. S. Santos, J. A. Cunha, M. G. Martins, B. M. Galeana-Villar, R. J. Carballo-Vivas, P. B. Leite, A. L. Rossi, F. Garcia, P. V. Finotelli, H. C. Farraz, Curcuminoids-conjugated multicore magnetic nanaoparticles: design and characterization of a potential theranostic platform, J. Alloys. Compd. 879 (2021) 160448.
https://doi.org/10.1016/j.jallcom.2021.160448

[81] M. Perez-Garnes, V. Morales, R. Sanz, R. A. Garcia-Munoz, Cytostatic and cytotoxic effects of hollow-shell mesoporous silica nanoparticles containing magnetic iron oxide, Nanomaterials, 11 (2021) 2455. https://doi.org/10.3390/nano11092455

[82] A. Bigham, A. H. Aghajanian, S. Behzadzadeh, Z. Sokhani, S. Shojaei, Y. Kaviani, S. A. Hassanzadeh-Tabrizi, Nanostructured magnetic Mg_2SiO_4-$CoFe_2O_4$ composite scaffold with multiple capabilities for bone tissue regeneration, Mater. Sci. Eng. C 99 (2019) 83-95. https://doi.org/10.1016/j.msec.2019.01.096

[83] A. Bigham, A. H. Aghajanian, S. Allahdaneh, S. A. Hassanzadeh-Tabrizi, Multifunctional mesoporous magnetic Mg_2SiO_4-$CuFe_2O_4$ core-shell nanocomposite for simultaneous bone cancer therapy and regeneration, Ceram. Int. 45 (2019) 19481-19488. https://doi.org/10.1016/j.ceramint.2019.06.205

[84] A. Bigham, A. H. Aghajanian, A. Sandi, M. Rafienia, Hierarchical porous Mg_2SiO_4-$CoFe_2O_4$ nanomagnetic scaffold for bone cancer therapy and regeneration: surface modification and in vitro studies, Mater. Sci. Eng. C 109 (2020) 110579.
https://doi.org/10.1016/j.msec.2019.110579

[85] S. A. Hassanzadeh-Tabrizi, N. Norbaksh, R. Pournajaf, M. Tayebi, Synthesis of mesoporous cobalt ferrite/hydroxyapatite core-shell nanocomposite for magnetic hyperthermia and drug release applications, Ceram. Int. 47 (2021) 18167-18176.
https://doi.org/10.1016/j.ceramint.2021.03.135

[86] R. T. Gordon, J. R. Hines, D. Gordon, Intracellular hyperthermia: a biophysical approach to cancer treatment via intracellular temperature and biophysical alterations, Med. Hypothesis 5 (1979) 83-102. https://doi.org/10.1016/0306-9877(79)90063-X

[87] D. Chang, M. Lim, J. A. C. M. Goos, R. Qiao, X. Y. Ng, F. M. Mansfeld, M. Jackson, T. P. Davis, M. Kavallaris, Front. Pharmacol. 9 (2018) 831.
https://doi.org/10.3389/fphar.2018.00831

[88] P. Das, M. Colombo, D. Prosperi, Recent advances in magnetic fluid hyperthermia for cancer therapy, Colloids Surf. B 174 (2019) 42-55. https://doi.org/10.1016/j.colsurfb.2018.10.051

[89] O. Polozhentsev, A. V. Soldatov, Efficiency of heating magnetite nanoparticles with different surface morphologies for the purpose of hyperthermia, J. Surf. Investig. 15 (2021) 799-805. https://doi.org/10.1134/S1027451021040364

[90] Y. Tang, R. C. C. Flesch, T. Jin, Y. Gao, M. Ho, Effect of nanoparticle shape on therapeutic temperature distribution during magnetic hyperthermia, J. Phys. D: Appl. Phys. 54 (2021) 165401. https://doi.org/10.1088/1361-6463/abdb0e

[91] A. J. Rajan, N. K. Sahu, Hydrophobic-tohydrophylic transition of Fe3O4 nanorods for magnetically induced hyperthermia, ACS Appl. Nano. Mater. 4 (2021) 4642-4653. https://doi.org/10.1021/acsanm.1c00274

[92] A. Manohar, D. D. Geleta, C. Krishnamoorthi, J. Lee, Synthesis, characterization and magnetic hyperthermia properties of nearly monodisperse CoFe2O4 nanoparticles, Ceram. Int. 46 (2020) 28035-28041. https://doi.org/10.1016/j.ceramint.2020.07.298

[93] S. Kumari, M. K. Maglam, L. K. Pradhan, L. Kumar, J. P. Borah, M. Kar, Modification in crystal structure of copper ferrite fiber by annealing and its hyperthermia application, Appl. Phys. A 127 (2021) 273. https://doi.org/10.1007/s00339-021-04429-5

[94] D. Kouzoudis, G. Samourgkanidis, A. Kolokithos-Ntoukas, G. Zoppellaro, K. Spiliotopoulos, Magnetic hyperthermia in the 400-1100 kHz frequency range using MIONs of condensed colloidal nanocrystal clusters, Front. Mater. 8 (2021) 131. https://doi.org/10.3389/fmats.2021.638019

[95] A. Manohar, C. Krishnamoorthi, Synthesis and magnetic hyperthermia studies on high susceptible Fe1-x–MgxFe2O4 superparamegnetic nanospheres, J. Magn. Magn. Mater. 443 (2017) 267-274. https://doi.org/10.1016/j.jmmm.2017.07.065

[96] W. Montha, W. Maneeprakorn, N. Buatong, I. M. Tang, W. Pon-On, Synthesis of doxorubicin-PLGA loaded chitosan stabilized (Mn,Zn)Fe2O4 nanoparticle bilogocal activity and pH response drug release, Mater. Sci. Eng. C 59 (2016) 235-240. https://doi.org/10.1016/j.msec.2015.09.098

[97] W. Montha, W. Maneeprakorn, I. M. Tang, W. Ponon, Hyperthermia evaluation and drug/propetin controlled release during alternating magnetic field stimuli-reponse Mn-Zn particles, RSC Adv. 10 (2020) 40206. https://doi.org/10.1039/D0RA08602A

[98] W. Jia, Y. Qi, Z. Hu, Z. Xlong, Z. Luo, Z. Xiang, J. Hu, W. Lu, Facile fabrication of monodisperse CoFe2O4 nanocrystals@dopamine@DOX hybrids for magnetic response on-demand cancer theranostic applications, Adv. Comp. Hybrid Mater. 4 (2021) 989-1001. https://doi.org/10.1007/s42114-021-00276-3

[99] Z. Shi, Y. Zeng, X. Chen, F. Zhou, L. Zheng, G. Wang, J. Gao, Y. Ma, L. Zheng, B. Fu, R. Yu, Mesoporous superparamegnetic cobalt ferrite nanoclusters: synthesis, characterization and application in drug delivery, J. Magn. Magn. Mater. 498 (2020) 166222. https://doi.org/10.1016/j.jmmm.2019.166222

[100] R. Eivazzadeh-Keihan, E. B. Noruzi, K. K. Chenab, A. Jafari, F. Radinekiyan, S. M. Hashemi, F. Ahmadpour, A. Behboudi, S. M. Hashemi, F. Ahmadpour, J. Mosafer, A. Mokhtarzadeh, A. Maleki, M. R. Hamblin, Metal-based nanoparticles for bone tissue engieering, J. Tissue Eng. Regen. Med. 14 (2020) 1687-1714. https://doi.org/10.1002/term.3131

[101] F. Heidari, M. E. Bahrololoom, D. Vashaee, L. Tayebi, In situ preparation of iron oxide nanoparticles in natural hydroxyapatite chitosan matrix for bone tissue engineering application, Ceram. Int. 41 (2015) 3094-3100. https://doi.org/10.1016/j.ceramint.2014.10.153

[102] B. Ghiasi, Y. Sefidbakht, S. Mozaffari-Jorin, B. Gharehcheloo, M. Mehrarya, A. Khodadadi, M. Rezai, S. Omid, R. Siadat, V. Uskokovic, Hydroxyapatite as a biomaterial - a gift that keeps on giving, Drug. Dev. Ind. Pharm. 46 (2020) 1035-1062. https://doi.org/10.1080/03639045.2020.1776321

[103] E. Ahmadzadeh, F. T. Rowsham, M. Hosseini, A biological method for in-situ synthesis of hydroxyapatite-coated magnetite nanoparticles using Enterobacter aerogenes: Characterization and acute toxicity assessments, Mater. Sci. Eng. C 73 (2017) 220-224. https://doi.org/10.1016/j.msec.2016.12.012

[104] N. Castro, M. M. Fernandes, C. Ribeiro, V. Correira, R. Minguez, S. Lanceros-Mendez, Magnetic bioreactor for magneto-, mechano- and electroactive tissue engineering, Sensors 20 (2020) 3340. https://doi.org/10.3390/s20123340

[105] S. Kopyl, R. Surmenev, M. Surmeneva, Y. Fetisov, H. Kholkin, Magnetoelectric effect: principles and applications in biology and medicine - a reviw, Materials Today Bio 12 (2021) 100149. https://doi.org/10.1016/j.mtbio.2021.100149

[106] E. O. Carvalho, C. Ribeiro, D. M. Correira, G. Botello, S. Lanceros-Mendez, Biodegradeable hydrogels loaded with magnetically responsive microspheres as 2D and 3D scaffolds, Nanomaterials 10 (2020) 2421. https://doi.org/10.3390/nano10122421

[107] M. M. Fernandez, D. M. Correia, C. Ribeiro, N. Castro, V. Correira, S. Lanceros-Mendez, Bioinsipred three-dimensional magnetoactive scaffolds for bone tissue engineering, ACS Appl. Mater. Interfaces 11 (2019) 45265-45275. https://doi.org/10.1021/acsami.9b14001

[108] A. Reizabal, R. Brito-Pereirra, M. M. Fernandes, N. Castro, V. Correira, C. Ribeiro, C. M. Costa, L. Perez, J. L. Vilas, S. Lanceros-Mendez, Silk fibroin magnetoactive nanocomposite films and membranes for dynamic bone tissue

engineering strategies, Materialia 12 (2020) 100709.
https://doi.org/10.1016/j.mtla.2020.100709

[109] B. Tang, X. Shen, Y. Yang, Z. Xu, J. Yi, Y. Yao, M. Cao, Y. Zhang, H. Xia, Enhanced cellular osteogenic differentiation on $CoFe_2O_4$/P(VDF-TrFE) nanocomposite coatings under static magnetic field, Colloids Surf. B 198 (2021) 111473. https://doi.org/10.1016/j.colsurfb.2020.111473

[110] J. R. Melamed, R. S. Edelstein, E. S. Day, Elucidating the fundamental mechanisms of cell death triggered by photothermal therapy, ACS Nano 9 (2015) 6-11. https://doi.org/10.1021/acsnano.5b00021

[111] J. Esterrich, M. A. Busquets, Iron oxide nanoparticles in photothermal therapy, Molecules 23 (2018) 1567. https://doi.org/10.3390/molecules23071567

[112] K. Wang, P. Yang, R. Guo, X. Yao, W. Yang, Photothermal performance of MFe_2O_4 nanoparticles, Chinese Chem. Lett. 30 (2019) 2013-2016. https://doi.org/10.1016/j.cclet.2019.04.005

[113] M. Gorgizadeh, N. Azarpira, N. Sattarahmady, In vitro and in vivo tumor annihilation by near-infrared photothermal effect, Colloids Surf. B 170 (2018) 393-400. https://doi.org/10.1016/j.colsurfb.2018.06.034

[114] K. Deng, Y. Chen, C. Li, X. Deng, Z. Hou, Z. Cheng, Y. Han, B. Xing, J. Lin, 808 nm light reposnsive nanotheranostic agents based near-infrared dye functionalized manganese ferrite for magnetic targeted and imaging-guided photodynamic/photothermal therapy, J. Mater. Chem. B. 5 (2017) 1803. https://doi.org/10.1039/C6TB03233H

[115] B. Zhou, Q. Wu, M. Wang, A. Hoover, X. Wang, F. Zhou, R. A. Towner, N. Smith, D. Aunders, J. Song, J. Qu, W. R. Chen, Immunologically modified $MnFe_2O_4$ nanoparticles to synergize photothermal therapy and immunotherapy for cancer treatment, Chem. Eng. J. 396 (2020) 125239. https://doi.org/10.1016/j.cej.2020.125239

[116] Z. Yang, J. Yao, J. Wang, C. Zhang, Y. Cao, L. Hao, C. Yang, C. Wu, J. Zhang, Z. Wang, H. Ran, Y. Tian, Ferrite-encapsulated nanoparticles with stable photothermal performance for multimodal imaging-guided atherosclerotic plaque neovascularization, Biomater. Sci. 9 (2021) 5652. https://doi.org/10.1039/D1BM00343G

[117] S. Hatamie, P. J. Shih, B. W. Chen, I. J. Wang, T. H. Young, D. J. Yao, Synergic effect of novel WS2 carriers holding spherical cobalt ferrite@cubic Fe_3O_4 (WS2-$CoFe_2O_4$@c-Fe_3O_4) nanocomposites in magnetic resonance imaging and photothermal therapy for ocular treatments and investigation of corneal endothermal cell migration, Nanomaterials 10 (2020) 2555. https://doi.org/10.3390/nano10122555

[118] O. Akhavan, A. Meidanchi, E. Ghaderi, S. Khoei, Zinc-ferrite spinel graphene in magneto-photothermal therapy of cancer, J. Mater. Chem. B 2 (2014) 3306. https://doi.org/10.1039/c3tb21834a

[119] M. J. Hajipour, O. Akhavan, A. Meidanchi, S. Laurent, M. Mahmoudi, Hyperthermia-induced protein corona improves the therapeutic effects of zinc ferrite spinel graphene sheets against cancer, RSC Adv. 4 (2014) 62557-62565. https://doi.org/10.1039/C4RA10862K

[120] M. Gorgizadeh, M. Behzadpour, F. salehi, F. Daneshvar, R. Dehdari Vais, R. Nazari-Vanani, N. Azarpira, M. Lotfi, N. Sattarahmady, A MnFe$_2$O$_4$/C nanocomposite as a novel thernaostic agent in MRI, sonodynamic therapy and photothermal therapy of a melanomic cancer model, J. Alloys Compd. 816 (2020) 152597. https://doi.org/10.1016/j.jallcom.2019.152597

[121] T. S. Ardakani, A. Meidanchi, A. Shokri, A. Shakeri-Zadeh, Fe$_3$O$_4$@Au reduced graphene oxide nanostructures: combinatorial effects of radiotherapy and photothermal therapy on oral squamous carcinoma KB cell line, Ceram. Int. 46 (2020) 28676-28685. https://doi.org/10.1016/j.ceramint.2020.08.027

[122] R. Sun, Y. H. Ge, H. Liu, P. He, W. Song, X. Zhang, Erythrocyte membrane-encapsulated glucose oxidase and manganese/ferrite nanocomposite as a bimemtric "All in One" nanoplatform for cancer therapy, ACS Appl. Bio. Mater. 4 (2021) 701-710. https://doi.org/10.1021/acsabm.0c01226

[123] M. Vincius-Araujo, N. Shrivastava, A. A. Sousa-Junior, S. A. Mendanha, R. C. De Santana, A. F. Bakuzis, ZnxMn$_{1-x}$–Fe$_2$O$_4$@SiO$_2$:Nd^{3+} core-shell nanoparticles for low-field magnetic hyperthermia and enhanced photothermal therapy with the potential for nanothermometry, ACS Appl. Nano. Mater. 4 (2021) 2190-2210. https://doi.org/10.1021/acsanm.1c00027

[124] Y. Li, H. Zhang, Fe$_3$O$_4$-based nanotheranostics for magnetic resonance imaging - synergized multifunctional cancer management, Nanomedicine 14 (2019) 1493-1512. https://doi.org/10.2217/nnm-2018-0346

[125] S. Y. Srinivasan, K. M. Paknikar, D. Bodas, V. Gajbhiye, Applications of cobalt ferrite nanoparticles in biomedical nanotechnology, Nanomedicine 13 92018) 1221-1238 https://doi.org/10.2217/nnm-2017-0379

[126] J. Huang, X. Zhong, L. Wang, L. Yang, H. Mao, Improving the magnetic responance imaging contrast and detection method with engineered magnetic nanoparticles, Theranostics 2 (2012) 86-102 https://doi.org/10.7150/thno.4006

[127] A. M. Demin, A. G. Pershina, A. S. Minin, A. V. Mekhaev, V. V. Ivanov, S. P. Lezhava, A. A. Zakharova, I. V. Byzov, M. A. Uimin, V. P. Krasnov, L. M. Ogorodova, PMIDA-modified Fe$_3$O$_4$ magnetic nanoparticles: synthesis and application for liver MRI, Langmuir 34 (2018) 3449-3458. https://doi.org/10.1021/acs.langmuir.7b04023

[128] D. M. A. Neto, L. S. Da Costa, F. L. De Menezes, L. M. U. D. Fechine, R. M. Freire, J. C. Denardin, M. Banober-Lopez, I. F. Vasconcelos, T. S. Ribeiro, L. K. A. M. Leal, J. A. C. De Souza, J. Gallo, P. B. A. Fechine, A novel aminophosphate-coated magnetic nanoparticle as MRI contrast agent, Appl. Surf. Sci. 543 (2021) 148824. https://doi.org/10.1016/j.apsusc.2020.148824

[129] V. Zheltova, A. Vlasova, N. Bobrysheva, I. Abdullin, V. Semenov, M. Osmolowsky, M. Vozhesenskiy, O. Osmolovskaya, Fe_3O_4@HAp core shell nanoparticles as MRI contrast agent: synthesis, characterization and theoretical and experimental study of shell impact on magnetic properties, Appl. Surf. Sci. 532 (2020) 147352. https://doi.org/10.1016/j.apsusc.2020.147352

[130] J. Li, J. Zhang, Z. Guo, H. Jiang, H. Zhang, X. Wang, Self-assembly fabrication of honeycomb-like magnetic-fluorescent Fe_3O_4-QDs nanocomposites for bimodal imaging, Langmuir 36 (2020) 14471-14477. https://doi.org/10.1021/acs.langmuir.0c00077

[131] H. Fatima, K. S. Kim, Magnetic nanoparticles for bioseparation, Korean J. Chem. Eng. 34 (2017) 589-599. https://doi.org/10.1007/s11814-016-0349-2

[132] Q. Yang, Y. Dong, Y. Qiu, X. Yang, H. Cao, Y. Wu, Design of functional nanocomposites for bioseparation, Colloids Surf. B 191 (2020) 111014. https://doi.org/10.1016/j.colsurfb.2020.111014

[133] E. Sahin, E. Turan, H. Tumturk, G. Demirel, Core-shell magnetic nanoparticles: a comparative study based on silica and polydopamine coating for magnetic bio-separation platforms, Analysts 137 (2012) 5654. https://doi.org/10.1039/c2an36211b

[134] G. S. An, D. H. Choi, J. U. Hur, A. H. Oh, H. H. Choi, S. C. Choi, Y. S. Oh, Y. G. Jung, Hollow structured Fe_3O_4@SiO_2 nanoparticles novel synthesis and enhanced adsorbents for purification of plasmid DNA, Ceram. Int. 44 (2018) 18791-18795. https://doi.org/10.1016/j.ceramint.2018.07.111

[135] J. S. Han, G. S. An, Preparation of dual-layered core shell Fe_3O_4@SiO_2 nanoparticles and their properties of plasmid DNA purification, Nanomaterials 11(2021) 3422. https://doi.org/10.3390/nano11123422

[136] S. A. Adams, J. L. Hauser, A. L. C. Allen, K. P. Lindquist, A. P. Ramirez, S. Oliver, J. Z. Zhang, Fe_3O_4@SiO_2 nanoparticles functionalized with gold and poly(vinylpyrolidone) for bio-separation and sensing applications, ACS Appl. Nano Mater. 1 92018) 1406-1412 https://doi.org/10.1021/acsanm.8b00225

[137] M. S. Kim, B. C. Park, Y. J. Kim, J. H. Lee, T. M. Koo, M. J. Ko, Y. K. Kim, Design of magnetic-plasmonic nanoparticle assemblies via interface engineering of plasmonic shells for targeted cancer cell imaging and separation, Small 16 (2020) 2001103. https://doi.org/10.1002/smll.202001103

[138] K. Cetin, Magnetic nanoparticles embedded microcryogels for bilirubin removal, Process Biochem. 112 (2022) 203-208. https://doi.org/10.1016/j.procbio.2021.12.004

Magnetic Nanoparticles for Biomedical Applications   Materials Research Forum LLC
Materials Research Foundations 143 (2023) 76-101   https://doi.org/10.21741/9781644902332-3

Chapter 3

Applications of Magnetic Oxide Nanoparticles in Hyperthermia

Robert C. Pullar

Department of Molecular Sciences and Nanosystems (DSMN), Ca' Foscari University of
Venice, Scientific Campus, Via Torino 155, 30172 Venezia Mestre (VE), Italy

robertcarlye.pullar@unive.it, http://orcid.org/0000-0001-6844-4482

Abstract

Magnetic oxide nanoparticles (NPs) are probably the most common nanomaterials in
everyday biomedicine, and have been in use since the 1990's. They are usually magnetic
iron oxide NPs, made of magnetite (Fe_3O_4) or maghemite (γ-Fe_2O_3), or a mixture of the
two. Both of these have the spinel structure, and other spinel ferrites such as $ZnFe_2O_4$,
$CoFe_2O_4$ and $NiFe_2O_4$ are also used. For applications in magnetic hyperthermia these NPs
must be below the magnetic domain size, making them superparamagnetic, which means
that their magnetisation can be "switched on" by the application of an external magnetic
field. Magnetic hyperthermia treatment is a form of thermotherapy which is used to kill
tumour cells with thermal energy (heat) in a very localised manner, by causing magnetic
oxide NPs to heat up near tumour cells. Under an applied AC magnetic field the magnetic
spin of the NPs switches rapidly in direction, transforming the magnetic energy into
thermal energy. Temperatures of 41-46 °C are sufficient, this localised heating elevating
the temperature of tumour cells, inhibiting growth, killing them, or inducing tumour cell
apoptosis. Magnetic NPs were first used in tumour thermotherapy in 1996, and since then
there has been a great deal of research in this field. The treatment can be applied alone, or
used in combination with other therapies such as surgery, radiotherapy and chemotherapy,
and it has shown excellent synergistic effects in combination with anticancer drugs
(chemotherapeutics).

Keywords

Magnetic Oxide Nano Particles, Spinel Structure, Cobalt Ferrite, Hyperthemia Application

Contents

1.   Introduction

Magnetic resonance imaging (MRI) was developed in the 1970's [1,2], Mansfield and Lauterbur won the 2003 Nobel Prize in Physiology or Medicine for this, and magnetic contrast agents were used to image various tissues, including those based on magnetic iron oxides (combination of MRI and hyperthermia will be looked at in section 5.4). The concept of using magnetic micro- and nanoparticles for drug delivery was also proposed in the late 1970s, and magnetic oxides are probably the most used oxide materials in common / everyday biomedicine and have been in full use since the 1990's, almost always used as magnetic nanoparticles (NPs).

These are usually magnetic iron oxide (magnetite Fe_3O_4 or maghemite γ-Fe_2O_3), both of which have the spinel structure, or other spinel ferrites, e.g. $ZnFe_2O_4$, $CoFe_2O_4$, $NiFe_2O_4$. They are often coated with a silica or polymer layer (core@shell NPs), to improve biocompatibility or allow biofunctionalisation. Since the beginning of the millennium, there has been an explosion in interest in such NPs for biomedical applications, which include separation, imaging, therapy, analysis and drug delivery.

These magnetic oxides are spinels, with the formula AB_2O_4 (A =2+, B = 3+). In normal spinels, the M^{2+} ions are in the tetrahedral A sites, and the M^{3+} ions are in octahedral the B sites (Fig. 1a). However, spinel ferrites are inverse spinels, with Fe^{3+} in tetrahedral sites, and Fe^{3+} and M^{2+} in octahedral sites (octahedral total charge = +2.5), where $M^{2+} = Co^{2+}$, Ni^{2+}, Zn^{2+} in ferrites or Fe^{2+} in magnetite, all of which are slightly larger ions than Fe^{3+} (Fig. 1b-d).

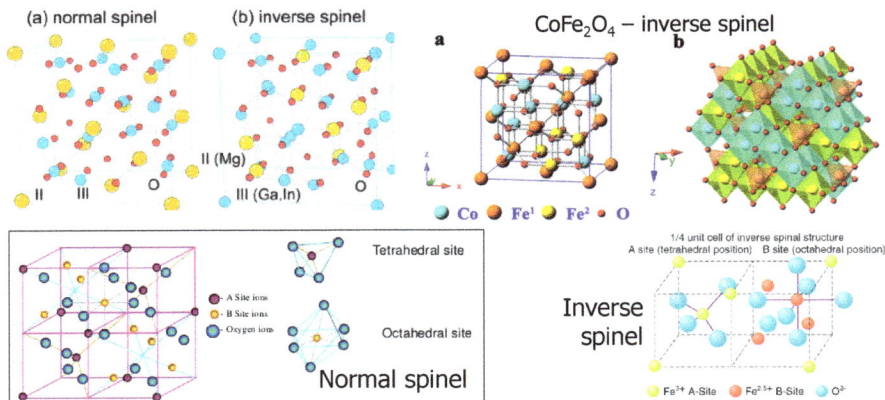

Fig. 1. The spinel AB_2O_4 structure, showing a) a normal spinel, and b-d) the inverse spinel structure found in spinel ferrites [3,4,5]. Reprinted with permission from [3], copyright 2006 by the American Physical Society; Reproduced from Ref [4] with permission from the Royal Society of Chemistry.

Magnetite is Fe_3O_4 and maghemite is γ-Fe_2O_3, but both are spinels, despite the nominal Fe_2O_3 structure of maghemite. This is because γ-Fe_2O_3 can be thought of as an iron-deficient Fe_3O_4 with Fe^{3+} ions only, and with some Fe^{3+} vacancies in the structure (Fig. 2). It is actually $Fe_{2.67}O_4\square_{0.33}$, where $\square$ = Fe^{3+} vacancy. In maghemite, Fe^{3+} ions are in tetrahedral A sites, and Fe^{3+} ions and the vacancies $\square$ are in octahedral B sites. Magnetite and maghemite have virtually identical X-ray diffraction (XRD) patterns, and so are difficult to tell apart from this technique. However, they do have quite different magnetisations, and so can be identified from magnetisation measurements or their magnetic hysteresis loops. Magnetite has a saturation magnetisation (M_s) = $\sim$ 90 A M^2 kg^{-1}, whereas maghemite M_s = $\sim$ 75 A M^2 kg^{-1}.

Maghemite can be formed from the reduction of hematite, α-Fe_2O_3 (hematite is trigonal / rhombohedral, not a spinel). Magnetic oxide NPs are also synthesised by many common nanosynthesis methods, such as coprecipitation, combustion/citrate synthesis, sol-gel, hydrothermal, and reverse micelle processes in a hydrophobic solvent [7]. Magnetite can also be made from the oxidation of metallic iron NPs. The range of properties of the various iron oxide based NPs investigated for magnetic hyperthermia, and their exact synthesis methods, can be found in Ref [7] by Hedayatnasab et al. They are almost always under 50 nm in diameter, and usually are less than 15 nm.

Materials Research Forum LLC

https://doi.org/10.21741/9781644902332-3

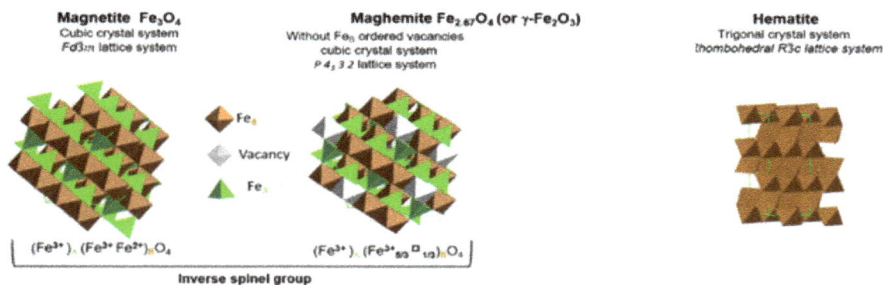

Fig. 2. The structures of magnetite, maghemite and hematite (a-Fe$_2$O$_3$) iron oxides [6]. Reproduced from Ref [6] with permission from the Royal Society of Chemistry.

These NPs are often coated with silica (e.g., the Stöber process), and they can also be coated with polymer layers, e.g. poly(methyl methacrylate) (PMMA) or polylactic acid (PLA). These silica or polymer layers can easily be functionalised with other organic / biological molecules for drug delivery, gene therapy, interaction with biomarkers / target molecules, etc. (Fig. 3).

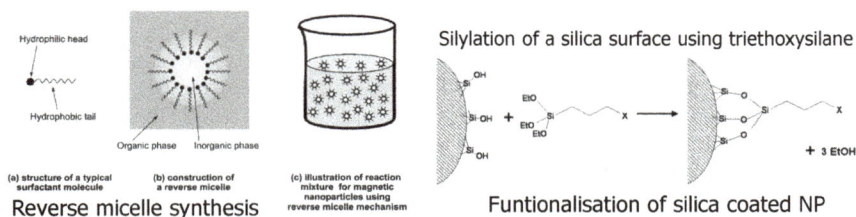

Fig. 3. Examples of processes for coating of magnetic NPs with polymers via a reverse micelle synthesis, or with silica using a Stöber-type synthesis.

2.   Magnetic nanoparticles

There are several properties of magnetic NPs in which are of particular interest in biomedicine:

1. Large surface area to volume ratio

2. The possibility of access in many kinds of biological tissues

3. Biocompatibility and relatively low toxicity in the human body

4. Less sensitivity to oxidation

5. More stability in magnetic response, and superparamagnetism

6. Ease of synthesis process and surface treatment (low cost)

7. Stable when dispersed in aqueous solutions or body fluids

8. Long blood circulation time (more time for specific localisation)

9. Low agglomeration probability after applying magnetic fields, so their movement in blood and subsequent extraction is possible – this is also good for microfluidics (lab-on-a-chip)

Magnetic oxide NPs are usually used as superparamagnetic NPs. As the NP size decreases, the number of magnetic domains possible within a particle decreases (as the magnetic domain size is constant). When an NP reaches the domain size, it reaches the maximum magnetisation (M, or M_s) and coercivity (H_c). Coercivity is due to domain wall movement in a magnetic field H, in which domains aligned with the field grow, as the others shrink and the domain wall moves, until the domains aligned with the field dominate. The coercivity is the size of magnetic field required to reverse this, randomising the orientation of the magnetic domains to leave a net magnetisation of zero, as the domains in opposing directions cancel each other out. In an NP below this magnetic domain size such wall movement is impossible, as the NP consists of just one domain. The individual particles can align in a magnetic field (if they are free to move/rotate), leading to a magnetisation with an applied field, but they will return to random orientation in the absence of a field, with no net magnetisation for all of the NPs considered as a bulk. This means that when below the domain size, Hc rapidly reduces to zero, known as the superparamagnetic state. Such NPs will align in magnetic field, creating a net magnetisation, but will have zero net magnetisation, and hence zero H_c, and no remnant magnetisation (M_r) in the absence of an external magnetic field (Fig. 4). This means that magnetisation can effectively be "switched" on and off by an applied magnetic field, H.

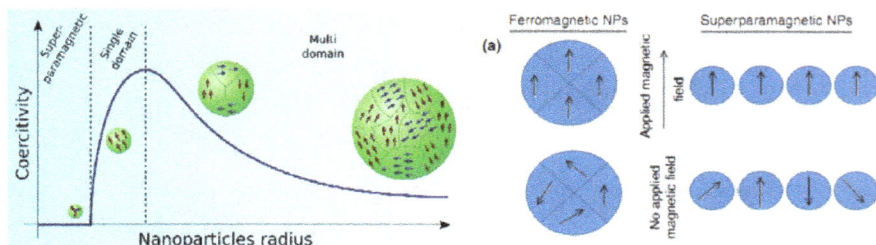

Fig. 4. Effect of nanoparticle size on coercivity, and superparamagnetism below the magnetic domain size [8]. Reproduced with permission from Ref [8].

Therefore, there exists a size limit for superparamagnetic oxide NPs. Typically an iron oxide or ferrite NP will become superparamagnetic just below the single domain size. The magnetic domain size is based on a theoretical calculation (that can also be measured), but it also depends on the morphology of the NP. For spherical magnetite / maghemite NPs, the limit for superparamagnetism is reported to be between 25-30 nm – they become superparamagnetic below this size. It was shown in 2001 that for $CoFe_2O_4$ this is very similar – it becomes superparamagnetic below 20 nm, as M_s also falls rapidly, and M_s was greatly reduced in 6 nm NPs (Fig. 5) [9]. This significant decrease in M_s is usually seen in very small NPs, below the critical domain size (at which M_s reached a maximum), as surface demagnetisation effects begin dominate, becoming greater in proportion as the NP size further decreases. Therefore, superparamagnetic NPs will have reduced magnetisation compared to their ferromagnetic analogues above the critical domain size, but have the advantage of being magnetically "switchable".

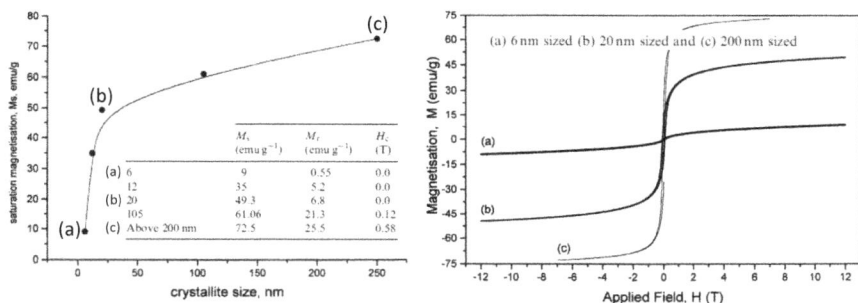

Fig. 5. The critical domain size for superparamagnetism in $CoFe_2O_4$ NPs, shown to be around 20 nm [9]. Reproduced from Ref. [9] with permission from Elsevier, Copyright 2001.

This absence of a net magnetisation in superparamagnetic NPs in the absence of an applied magnetic field is described as the magnetic relaxation process of NPs in the human body, magnetic relaxation meaning there is no overall net magnetisation of the individual NPs. They achieve a random orientation in body fluids via Brownian motion, the "random walk" of colloidal NPs in fluids (Fig. 6). Inside tissues and cells they can undergo Neel relaxation, the disordering of alignment of magnetic domains or moments between individual NPs (Fig. 6) [10].

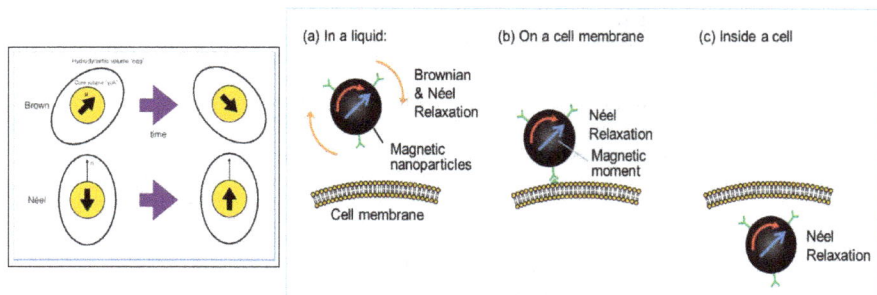

Fig. 6. Magnetic relaxation processes possible for NPs in the human body, via Brownian motion and Neel relaxation.

3.   Magnetic hyperthermia cancer treatment

Thermotherapy in general kills tumour cells with thermal energy. It can be applied alone, or used in combination with other therapies such as surgery, radiotherapy and chemotherapy, as exposing tumour cells to mildly elevated temperatures can sensitise them to chemotherapy and radiation, decreasing their viability [11]. It was fist demonstrated to be possibly effective as a cancer treatment as long ago as 1957, using a magnetic nanofluid based on 10-100 nm maghemite NPs, heated to 43-46 °C using a 1.2 MHz AC magnetic field against tumorous lymph nodes [12]. On *in-vitro* tests on rodent cells in 1991, it was shown that heating my magnetic hyperthermia could kill all types of cells, but that tumour cells had no increased sensitivity to the treatment compared to normal cells (unlike, for example, in radiation and chemotherapy). Therefore, for hyperthermia to be used as a cancer therapy, it must be applied specifically to the tumour cells in a localised manner [13]. When magnetic NPs were first used in tumour thermotherapy it was called magnetic fluid thermotherapy or MFH [14]. It is now more commonly known as magnetic hyperthermia treatment. It has been clinically trialled for the treatment of tumours such as glioblastoma multiforme [15], prostate cancer [16] and difficult-to-treat pancreatic cancers [17].

Magnetic hyperthermia treatment is not thermoablation, which uses higher temperatures up to 56 °C to crudely destroy cells by necrosis, coagulation and carbonisation, but exploits the localised heating of superparamagnetic NPs moving to align with an AC applied field, switching between opposing alignments (Fig. 7). Under this applied AC magnetic field, the magnetic spin switches from ↑ to ↓ rapidly, transforming the magnetic energy into thermal energy, along with thermal losses due to eddy currents [18]. This localised heating elevates the temperature of tumour cells, inhibiting growth, killing them, or inducing tumour cell apoptosis (Fig. 7). In a low-viscosity ferrofluid, the magnetic NPs can also physically rotate, which can mechanically damage cells, and also lead to frictional heating [19]. The localised heating is small (typically <45 °C) [20], the NPs can be guided by magnetic fields, monitored by MRI, and can also be used to release drugs when heated, or kill bacteria.

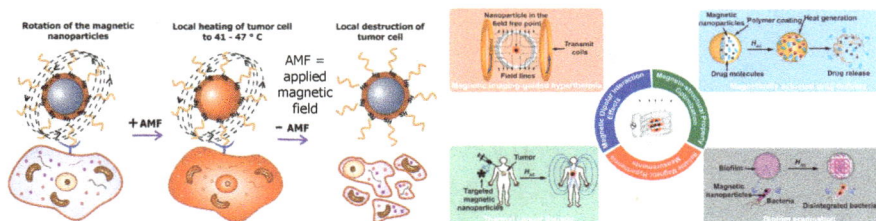

Fig. 7. Diagrams of the magnetic hyperthermia therapy process using an applied AC magnetic field (AMF), and other applications of magnetic NPs such as tracking using MRI, drug delivery and antimicrobial effects [21]. Reproduced with permission from Ref [21].

There are particular conditions that must be met for a successful magnetic hyperthermia treatment [22]. The frequency and strength of any externally applied AC magnetic field used to generate the heating is limited by the body's physiological responses to high frequency magnetic fields used. Such responses include the stimulation of peripheral and skeletal muscles, possible cardiac stimulation and arrhythmia, and any non-specific inductive heating of the surrounding tissue. Generally, the useable range of frequencies and magnetic field is considered to be between $f = 0.05$–1.2 MHz and $H <15$ kA m^{-1} (<200 Oe or <0.02 T).

More material can be localised in a tumour with direct injection of the NPs, and intravascular administration or antibody targeting can also be used. About 5–10 mg of 10 nm magnetic NPs concentrated in each cm^3 of tumour tissue is appropriate for magnetic hyperthermia in human patients [23], often used in the form of a ferrofluid. The NPs used must be biocompatible and non-toxic, so some ferrites such as $NiFe_2O_4$, containing toxic nickel, should be avoided. They are often also coated with a biocompatible layer which can be further functionalised with drugs, antibodies, immunotherapeutic agents, etc. (Fig. 8).

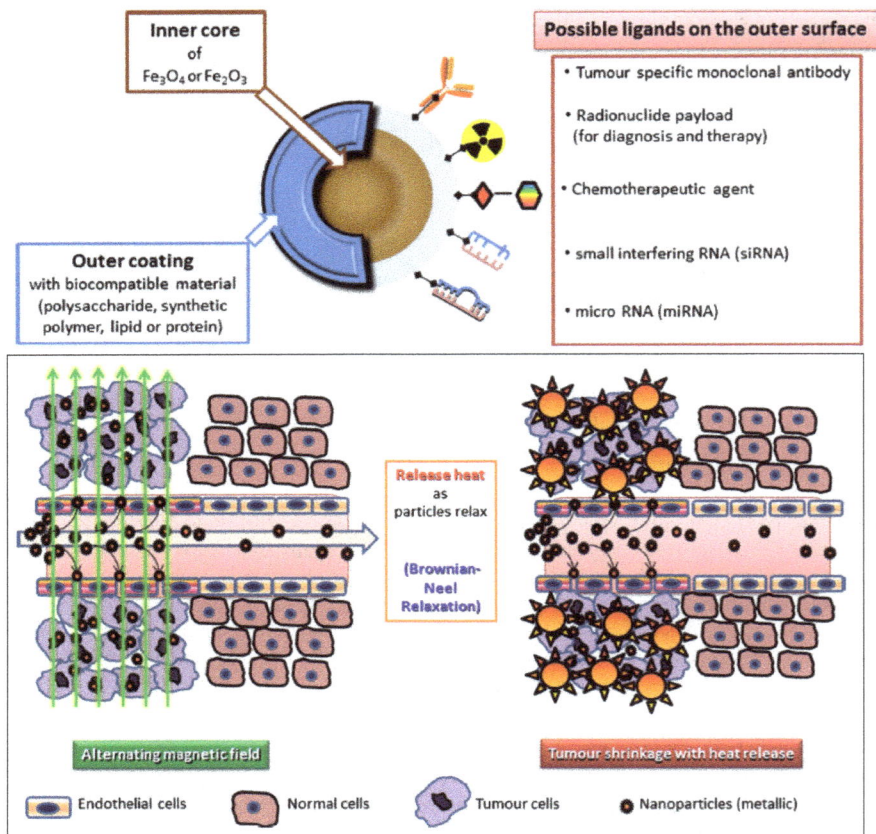

Fig. 8. Top: Schematic representation of a multifunctional magnetic NP for use in hyperthermia treatments. Bottom: Schematic representation of the mechanism of hyperthermia induced by magnetic NPs in the presence of an AC magnetic field, and its ability to kill local tumour cells without damaging neighbouring healthy tissue [23]. Reproduced from Ref. [23] with permission from Elsevier, Copyright 2016.

Localisation of NPs in the tumour allows localised heating that selectively kills tumour cells while keeping surrounding normal ones healthy, a major advantage of magnetic hyperthermia treatment (Fig. 8). This overcomes the main drawback of many current methods such as chemotherapy and radiotherapy, which show harmful secondary effects derived from a lack of treatment specificity. Low temperatures of 41-46 °C are enough to destroy the tumours, but for this the NPs must have sufficient magnetisation to generate a specific loss power (SLP, also known as specific absorption rate, or SAR) in the order of tens of W g^{-1}. It has been demonstrated that the highest SLP values are obtained for NPs that are superparamagnetic [24], below the single domain size, usually <20 nm, so a trade-off must be made between a loss in magnetisation as the particle size becomes small enough to achieve the superparamagnetic state. Indeed, 20 nm $CoFe_2O_4$ NPs, at the superparamagnetic size limit, were found to have the highest SLP [25]. Ideally, the NPs should also be as monodisperse as possible, and spherical or cuboid in shape. Although a coating layer is often useful, and sometimes essential for hydrophobic NPs, is should be as thin as possible, as SLP is reduced by this surface coating, as can be seen for chitosan-coated 20 nm magnetite NPs at the various applied magnetic fields in Fig. 9 [26].

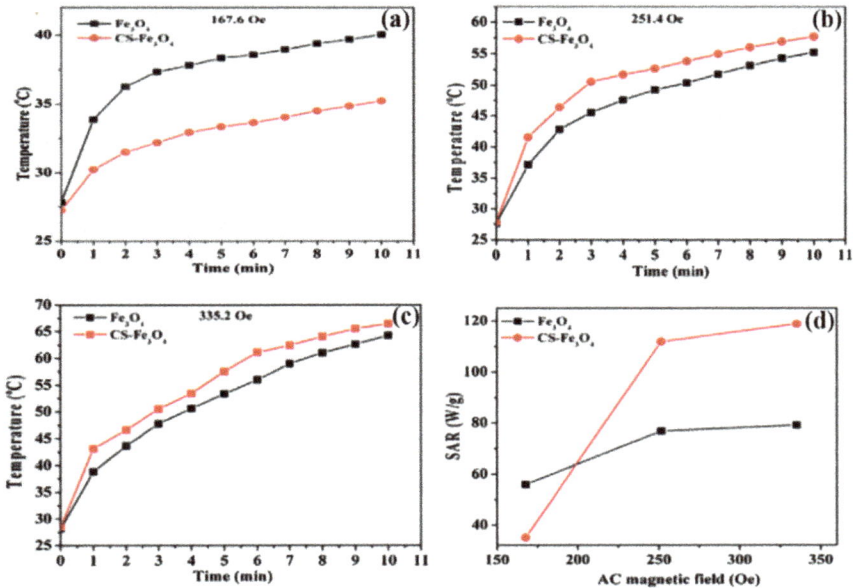

Fig. 9. Comparison of hyperthermia temperature obtained for chitosan (CS) coated magnetite NPs and uncoated NPs with applied fields of a) 168, b) 251 and c) 335 Oe, and d) the specific absorption rate at those fields [26]. Reproduced from Ref. [26] with permission from Elsevier, Copyright 2009.

Magnetic Nanoparticles for Biomedical Applications   Materials Research Forum LLC
Materials Research Foundations 143 (2023) 76-101   https://doi.org/10.21741/9781644902332-3

The tumour cells are killed by destroying vascular structure, reducing enzyme activity and inhibiting DNA and RNA synthesis. However, for effective hyperthermia the initial heating must be rapid. If cells are treated at >43 °C for a short time, the cells that survive are much more sensitive to any subsequent heat treatments, even at lower temperatures. This is referred to as step-down heating. Combined therapies with anticancer drugs can produce very good synergistic effects. For example, the therapeutic efficacy of paclitaxel (a common anti-tumour drug) can be increased by 10–100 fold when combined with hyperthermia treatment at 43 °C for 30 min, and the killing capacity of chemotherapeutics with low cytotoxicity at room temperature can be doubled after heating at temperatures of 41-45 °C.

4.   Examples of oxide NPs for magnetic hyperthermia

Since the invention of the technique in the late 1990s, there has been an ever increasing amount of research into magnetic NPs for use in magnetic hyperthermia treatments, with over 4500 articles up to 2021 (Fig. 10). Most of these are on magnetite and maghemite NPs, although there are also many studies on other ferrites, including non-spinel ferrites such as hexaferrites, and more exotic NPs. The hexaferrites investigated to date are based on strontium M-type ferrites ($SrFe_{12}O_{19}$, SrM), to avoid the problems of barium toxicity, and have included mixed spinel-hexaferrite phases [27,28] and pure [29] and substituted [30] SrM ferrites, although obtaining NPs small enough to be both biocompatible and superparamagnetic is problematic for the hexaferrites, due to their large domain sizes.

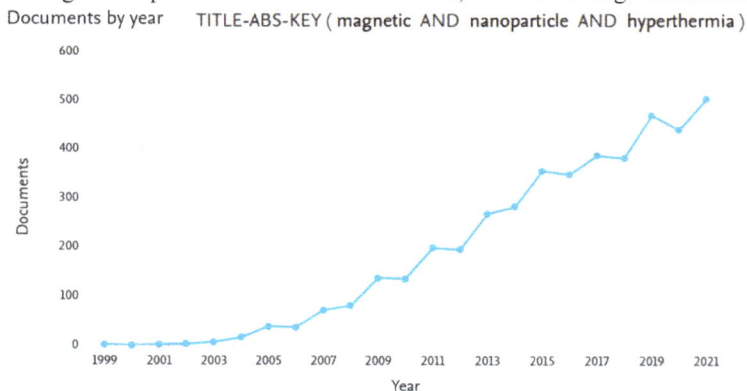

Fig. 10. Yearly articles published on magnetic NPs for hyperthermia treatment since 1999, with over 4500 articles in total up to 2021. (Search carried out on Scopus, search term: TITLE-ABS-KEY (magnetic AND nanoparticle AND hyperthermia)).

An example is Fe_3O_4 NPs made by hot injection in a non-polar organic solvent. These were not water soluble, so they were coated with Pluronic F127 surfactant in order to transfer them to a phosphate buffered saline solution. The NPs were 5-18 nm in diameter, and produced a SLP of up to 150 W g^{-1} with an applied field of only 14 kA m^{-1} (175 Oe, 0.0175 T) and an AC frequency of 400 kHz. The larger particles gave greater SLP values, up to a diameter of 16 nm, as the critical domain size was approached, as these would have greater magnetisation values, and at higher fields of 24 kA m^{-1} and SLP of around 450 W g^{-1} was achieved [31].

Once developed, such NPs have to be tested in-vivo, in living subjects, before progressing to medical trials. For example, $Fe_{1-x}Mn_xO$ NPs tested in-vivo in mouse tumour cells, showed excellent results 50 days after treatment compared to subjects with no treatment, with an almost complete suppression of the tumour volume compared to the control subject, accompanied by an increase in the subjects body weight of around 20% (compared to a small loss in weight for the control) (Fig. 11) [32].

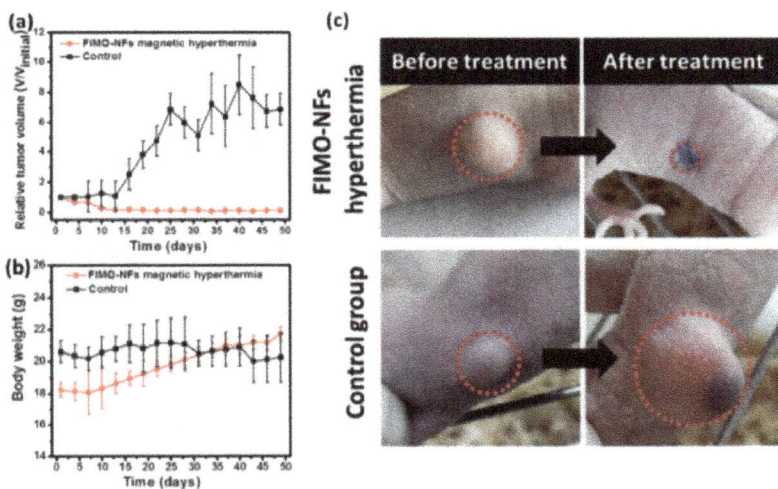

Fig. 11. $Fe_{1-x}Mn_xO$ NPs tested in-vivo for hyperthermia therapy in mouse tumour cells [32]. Reproduced from Ref. [32] with permission from John Wiley and Sons, copyright 2016.

5.   Combination therapies involving magnetic hyperthermia treatment

Many different combination therapies have been developed involving magnetic hyperthermia treatment, combined with drug delivery, new cell/bone growth in scaffolds, phototherapy and theranostics (the combination of magnetic hyperthermia treatment and chemotherapy drug delivery with MRI contrast agent capabilities in a single NP).

5.1 Drug delivary via magnetic hyperthermia

An increasingly investigated approach is to combine hyperthermia for both direct treatment and targeted drug delivery or gene therapy to treat tumours [33]. In 1998 it was found that the magnetic hyperthermia effect of magnetite NPs with cationic liposomes under an AC magnetic field could delay primary tumour growth, while also stimulating the immune response [34]. Since then, there has been much interest in combing magnetic hyperthermia therapy and immunotherapy or drug delivery.

In such therapies, the thermal energy of magnetic hyperthermia can also be used for targeted drug delivery (Fig. 12) to a specific site via:

a) Injection of the NPs

b) Application of magnetic field causing heat (hyperthermia treatment) and releasing the trapped drug

c) Removal of the applied magnetic field decreases temperature to body temperature, entrapping the residual drugd) Re-application of a magnetic field can heat again, releasing more residual drug and incurring further hyperthermia therapy

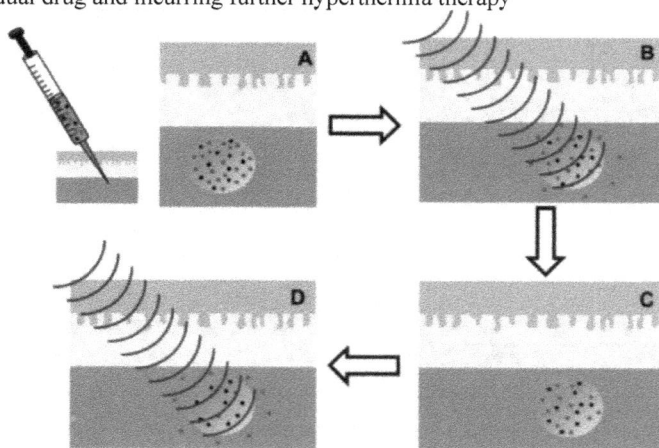

Fig. 12. Scheme of application and mechanism of remote-controlled drug with magnetic hyperthermia NPs. a) Injection of the NPs. b) Application of the magnetic field that causes heat and releases the trapped drug. c) Removal of the applied magnetic field that leads to the temperature decrease (until the initial body temperature), entrapping the residual drug. d) Application of a magnetic field heats to upper transition, releasing residual drug [35]. Reproduced from Ref. [35] with permission from John Wiley and Sons, copyright 2016.

Superparamagnetic $CoFe_2O_4@MnFe_2O_4$ NPs were used to induce immunogenic cell death by expressing tumour-associated antigens to inhibit metastatic tumours, along with hyperthermia therapy [36]. In a very interesting paper, ferrimagnetic vortex-domain iron oxide nanorings (FVIOs) used magnetic hyperthermia to incur an immune response to cause the immunogenic cell death of 4T1 tumour cells (Fig. 13) [37]. FVIOs coated with polyethylene glycol had higher magnetic-thermal conversion efficiency than traditional superparamagnetic iron oxides. The resulting magnetic hyperthermia exposed calreticulin proteins and released cytokines, causing the immune system to attack the cancer cells, via PD-L1 immune checkpoint blocking. $Zn_{0.2}Mn_{0.8}Fe_2O_4$ core@shell NPs were found to be able to generate sufficient heat under an AC magnetic field to prompt the expression of heat shock proteins that acted as immunogens [38], as well as acting as magnetic resonance imaging contrast agents (see section 5.4).

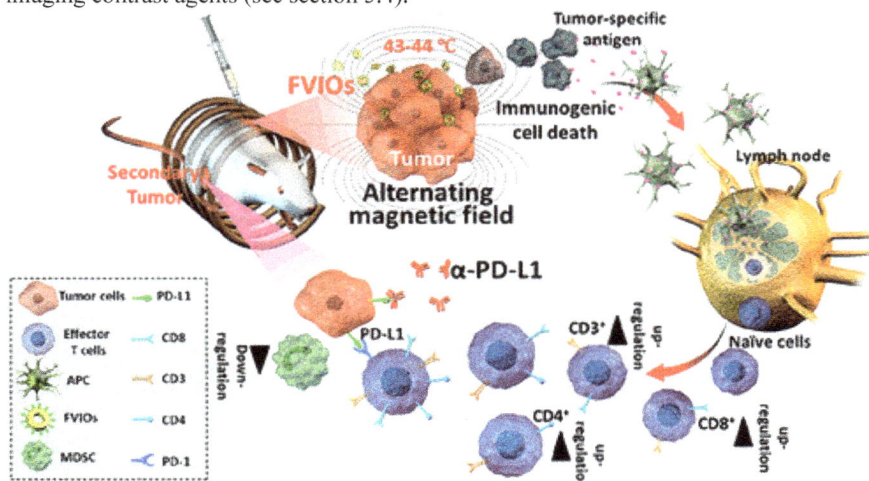

Fig. 13. Combination magnetic hyperthermia therapy and magnetically triggered drug delivery to treat tumours [37]. Reprinted with permission from Ref. [37], copyright 2019 American Chemical Society.

Magnetic heating also triggered the release of DOX on demand, being shown to be highly effective for in-vivo cancer therapy tested in mice. This was demonstrated with adamantane (Ad) grafted 6 nm $Zn_{0.4}Fe_{2.6}O_4$ superparamagnetic NPs (Ad-MNP), mixed with Ad-grafted polyamidoamine dendrimers (Ad-PAMAM) and coated with β-CD-grafted branched polyethylenimine (CD-PEI) or Ad-functionalized polyethylene glycol (Ad-PEG) via self-assembly, and functionalised with DOX (Fig. 14). This increased their size to between 70-160 nm, but the 160 nm NPs were poorly biocompatible, with a high retention of the NPs in the mouse body after 36 h, whereas the 70 and 100 nm NPs were biocompatible (100 nm is of the considered an approximate limit for good biocompatibility). Surprisingly, both the hyperthermia treatment (AMF in Fig. 14) and the DIX coated NPs without an applied magnetic field had almost no effect on the in-vivo growth of tumours on mice for 2 weeks (equal to the control), but the combined treatment of hyperthermia and DOX almost completely inhibited tumour growth, especially with a second NP injection and hyperthermia treatment on days 7-8 (Fig. 14) [39].

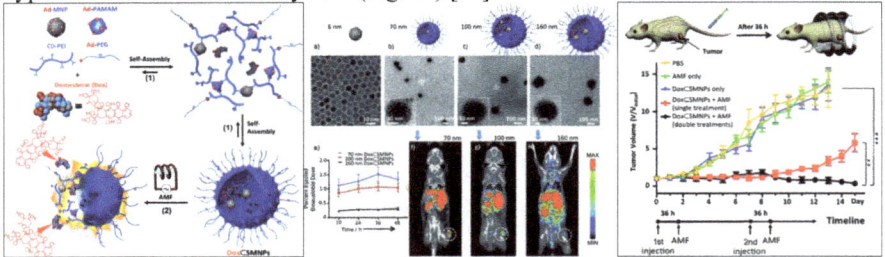

Fig. 14. Left: Synthesis composite NPs made from Ad-grafted $Zn_{0.4}Fe_{2.6}O_4$ NPs (Ad-MNP) mixed with Ad-grafted polyamidoamine dendrimers (Ad-PAMAM), coated with CD-PEI or Ad-PEG, and functionalised with DOX. Centre: TEM images of the composite NPs, and effects of composite NP size on biocompatibility (images of NP retention in mouse organs after 36 h). Right: Effects on tumour volume after two weeks in mice with no NPs (PBS), magnetic hyperthermia treatment only (AMF), DOX treatment only, and the combined hyperthermia-drug therapy (black and red lines) [39]. Reproduced and adapted from Ref. [39] with permission from John Wiley and Sons, copyright 2013.

Another interesting approach is to use magnetoelectric NPs, containing both magnetic and dielectric/ferroelectric components, for magnetic drug delivery. 30 nm $CoFe_2O_4@BaTiO_3$ magnetoelectric core-shell NPs were used to release drugs by applying a low alternating current magnetic field [40]. These NPs can couple external magnetic fields with the electric forces in the drug–carrier bonds, thus releasing the attached drug without exploiting any heat generated by the magnetic field (Fig. 15). This was extremely effective at low fields and frequencies of 66 Oe and 1 kHz. The NPs were studied for the field-triggered release after crossing the blood–brain barrier for the anti-AIDS retroviral drug AZTTP, but such magnetoelectric NPs could also be used for combined hyperthermia therapy and drug delivery.

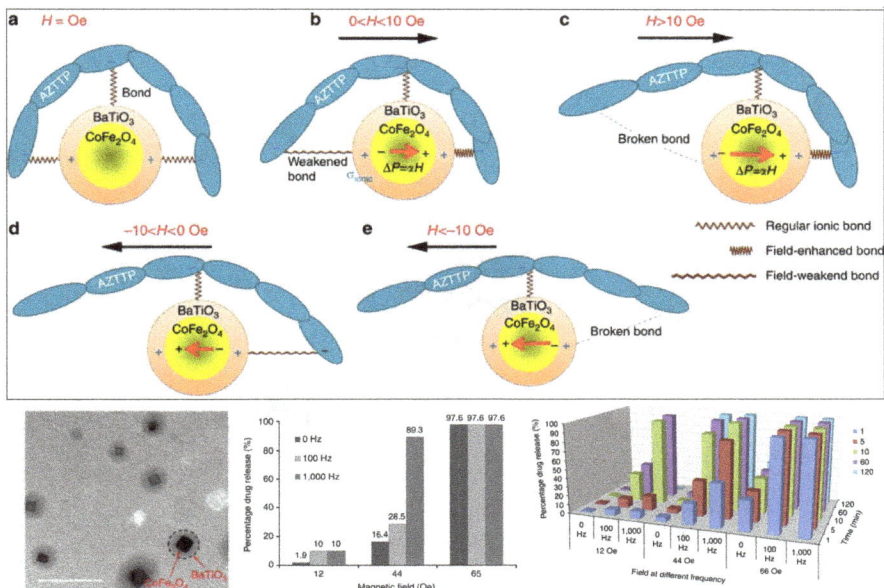

Fig. 15. 30 nm $CoFe_2O_4@BaTiO_3$ magnetoelectric NPs for magnetic field assisted AZTTP drug release. Top row: the magnetoelectric drug release mechanism under an AC magnetic field between 12-100 Oe. Bottom row: The $CoFe_2O_4@BaTiO_3$ magnetoelectric NPs (scale bar = 100 nm) and drug release capacity under various value of AC magnetic field strength and frequency field [40]. Adapted from Ref [40] with permission of Springer Nature, copyright 2013.

5.2   Combined hyperthermia & phototherapy NPs

This process combines magnetic hyperthermia and phototherapy treatments for tumours in one NP (Fig. 16). The phototherapy element is provided either by direct heating using lasers in photothermal therapy, or the release of reaction oxygen species (ROS, typically $O_2\bullet$ or $OH\bullet$ radicals) in photodynamic therapy, in which lasers create the ROS from a photocatalytic/photosensitive component, often a light activated polymer or biomolecule attached to the magnetic NPs. A magnetic field can also be applied to stimulate the hyperthermia treatment.

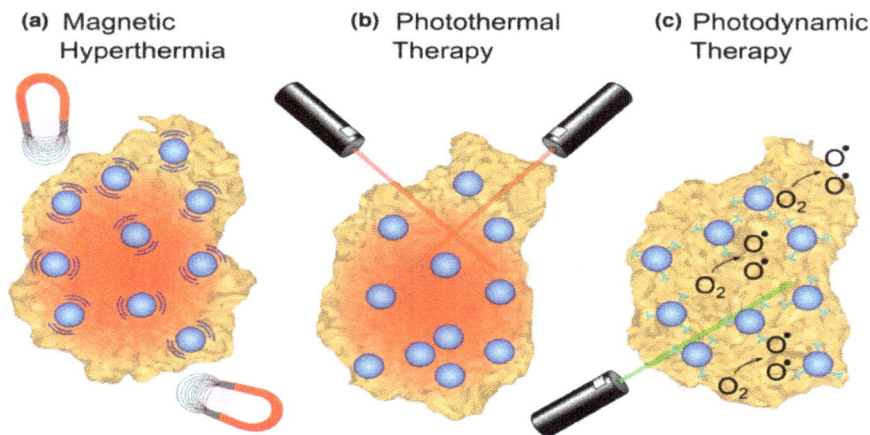

Fig. 16. Combined Hyperthermia & Phototherapy treatments for tumours [41].
Reproduced from Ref. [40] with permission from Elsevier, Copyright 2015.

Photothermal therapy kills cancer cells via heat energy that is converted from the light energy of a laser in an NP that possess a high photothermal conversion efficiency, such as gold or magnetite NPs [42]. It has also been shown that photothermal therapy can cause the immunogenic cell death of cancer by releasing tumour-associated antigens [43]. As magnetic iron oxide NPs have excellent photothermal performance, and have frequently been used for this therapy [44], the combination of phototherapy and magnetic hyperthermia is an obvious step. Magnetic iron oxide NPs have also been coated with a myeloid-derived suppressor cell membrane, with the added ability to target tumour cells and avoid immune clearance during such therapies [45]. Dual ferrite phase $CoFe_2O_4@MnFe_2O_4$ NPs polymerised pyrrole by Fe^{3+} in the presence of water and polyvinyl alcohol (PVA), the resulting nanocomposite combining the photothermal effect of pyrrole and the hyperthermia of the NPs. The magnetic NPs in the polymer matrix showed a decrease in M_s, but nevertheless they had a SAR of 930 W g^{-1}, with a relatively high applied field of 60.6 kA m^{-1} (765 Oe) at 200 kHz. Using an 808 nm laser together with this AC magnetic field, photothermal therapy killed 58% of 4T1 cancer cells, and magnetic hyperthermia killed 68%, with the combination of the two therapies exhibiting a synergistic effect [46].

Commonly used photosensitisers in photodynamic therapy include organic photosensitisers, inorganic photosensitisers and organic–inorganic hybrid materials [47]. Iron oxide NPs have been used for this, using combined/coated with an organic photosensitiser, or as a core@shell NP with a photosensitive gold core [48]. 20-30 nm magnetite NPs were coated with silica shells and the functionalised on the surface with 2,7,12,18-tetramethyl-3,8-di-(1-propoxyethyl)-13,17-bis-(3-hydroxypropyl) porphyrin (PHPP) as a photosensitiser [49], 50 nm $Fe_3O_4@SiO_2$ NPs were further coated with 3-Aminopropyl)triethoxysilane (APTES) and a conjugated chlorin photosensitiser pyropheophorbide-a (PPa) to form $Fe_3O_4@SiO_2@APTES@Glutaryl-PPa$ NPs which lowered the in vitro cell viability of human HeLa cervical cancer cells to 10.2% [50], and γ-$Fe_2O_3@Au$ NP-graphene nanocomposites were used as theranostic and photodynamic agents [51]. NPs have also been made with a magnetic iron oxide core and a lipid bilayer shell functionalised with an organic photosensitiser, and this was used in combined photodynamic therapy and magnetic hyperthermia under an AC magnetic field to complete cancer cell death in-vitro and total solid-tumour ablation in an in-vivo rodent model [52]. TiO_2 NPs are also well known to be photocatalysts able to produce ROS under UV light, have been used in phototherapy [53], and would be very suitable for combination with iron oxides in a core@shell NP for photodynamic therapy. The main limit of photodynamic therapy is that its use is generally restricted to superficial tumours, due to the limited penetration depth of the required radiation [54].

5.3   Hyperthermia induced in magnetic bone scaffolds

Bone scaffolds are typically 3D printed structures with a macroporosity on the order of hundreds of microns, implanted to replace voids in bone after surgery, on which new bone cells will grow and profligate, as the original scaffold is slowly resorbed into the body and replaced by new bone as it forms. Sucha scaffolds are often made of hydroxyapatite (HAp, $Ca_{10}(PO_4)_6(OH)_2$), the major constituent of bone, bioactive glasses, or biopolymers. Often scaffolds are used as implants to replace areas of bone removed during surgery for bone cancers, and if a magnetic nanoparticle can be included in the scaffold, it can also be used for hyperthermia treatment of the immediately surrounding tissue to ensure no cancerous cells remain. This combines magnetic stimuli that can favour cell adhesion and growth with hyperthermia for a combined bone cancer therapy consisting of local heating, and can also incorporate magnetically or thermally activated drug delivery. Various HAp-based scaffolds treated with magnetic NPs for such purposes have been studied, as well as dip coated biocompatible gelatin-based scaffolds in an aqueous solution of 10 nm Fe_3O_4 NPs made with a polyacrylic acid (PAA) surfactant [55]. The same magnetic 10 nm NPs were also incorporated into bioresorbable HAp NPs, and added to poly-ε-caprolactone (PCL) based polymeric matrices. These showed a temperature rise of up to 10 °C after heating for 300 s (293 kHz and 30 mT (300 Oe), when they contained at least 20 wt% of the FeHAp composite NPs. Other HAp-coated magnetite NPs nanoparticles had a high SLP of 85 W g^{-1} with an applied field of 180 Oe at a frequency of 409 kHz, killing all MG-63 osteosarcoma cells in 30 min due to damage to nucleotides, proteins, and membranes [56]. SrM hexagonal ferrites are also being investigated combined with HAp [57] and bioactive glasses [58] for use as hyperthermia materials in bone scaffolds and implants.

5.4   Theranostics-combined MRI, hyperthermia and chemotherapy NPs

The combination of magnetic hyperthermia and chemotherapy drug delivery with MRI contrast agent capabilities in one NP is called Theranostics (therapy + diagnostics) (Fig. 17). MRI (magnetic resonance imaging) is based on the nuclear magnetic resonance of protons, and can be used to image body tissue using large magnetic fields (typically H = 1-2 T, requiring a superconducting magnet). This aligns protons, and by measuring the RF radio waves emitted, various types / densities of tissue can be selectively imaged, as they respond differently to this, creating the image in 2D sections which can be combined into a 3D model. Many body tissues contain >70% water with different densities, and it can be hard to differentiate the contrast of them in MRI. For this, contrast agents are injected into the body to highlight different kinds of tissue in the image often based on Gd or Mn for T1 to combat loss of resonance intensity following pulse excitation, or on magnetic compounds for T2 to increase the width or broadness of resonance. Ferromagnetic and superparamagnetic NPs are excellent T2 contrast agents, and have a stronger magnetic response than the gadolinium based T1 contrast agents. Iron oxides based magnetic NPs have a long circulation time up of to 24 h in blood, are non-toxic (unlike Gd), have much lower magnetic field requirements than paramagnetic Gd, and greatly decrease the T2 relaxation time, increasing the relaxation rate to get a higher resolution image for certain types of tissue [59].

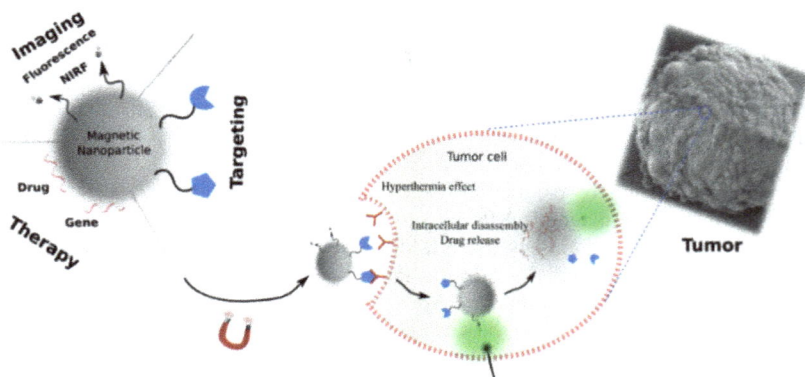

Fig.17. Theranostic multifunctional magnetic iron oxide NPs for tumour targeting, imaging (diagnosis) and therapy [60]. Reproduced from Ref. [39] with permission from John Wiley and Sons, copyright 2017.

Increased MRI contrast effects can be achieved by substituting of some Fe ions in iron oxide NPs with other magnetic or nonmagnetic atoms [61], for example, the substitution of Fe^{2+} ion with Zn^{2+} to produce $ZnFe_2O_4$ (or $Zn_xFe_{3-x}O_4$, $x \leq 1$) spinel ferrite, with increased magnetisation observed in 15 nm NPs [62]. The optimum Zn-to-Fe substitution ratio for enhancement of the T2 contrast effect was found to be $x = 0.4$ Zn^{2+}, to give the ferrite $Zn_{0.4}Fe_{2.6}O_4$, with a T2 contrast effect six times greater than that achieved by the commercially available iron oxide formulation Feridex. Another approach is to merge the T1 and T2 contrast effects in a dual-mode contrast agent, able to give more accurate MRI for all types of tissue, and better at distinguishing between normal and diseased areas [63,64]. Core@shell NPs were produced with a fine Gd-based T1 contrast agent shell for direct T1 contrast effects, surrounding a 15 nm superparamagnetic T2 contrast agent core of $MnFe_2O_4$, separated by a thin SiO_2 layer. Using this hybrid structure, the inherent ambiguity of a single mode contrast agent for different types of tissue was effectively eliminated, leading to a greater diagnostic accuracy.

An example of materials developed for such theranostic applications are γ-Fe_2O_3 (maghemite) NPs encapsulated within the membrane of a poly(trimethylene carbonate)-b-poly(l-glutamic acid) (PTMC-b-PGA) block copolymer as 100-400 nm vesicles (lipid membranes), using precipitation [65]. With 70 wt% 6 nm magnetic NPs in the vesicles, the common anticancer drug doxorubicin hydrochloride (DOX) was added, able to be released by application of the magnetic field and subsequent heating (Fig. 18). This enabled the NPs' position, and the affected tissue, to be monitored in the body by MRI, along with hyperthermia treatment and targeted drug delivery directly to the tumour cells.

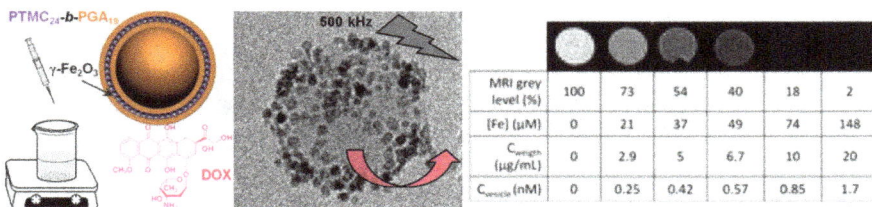

		100	73	54	40	18	2
MRI grey level (%)		100	73	54	40	18	2
[Fe] (µM)		0	21	37	49	74	148
C_{weight} (µg/mL)		0	2.9	5	6.7	10	20
$C_{vesicle}$ (nM)		0	0.25	0.42	0.57	0.85	1.7

Fig. 18. Theranostic NPs (γ-Fe_2O_3@PTMC-b-PGA functionalised with DOX) which can target tumours, treat them by hyperthermia and or delivered chemotherapy drugs, and image the results by enhanced contrast MRI [66]. Reproduced with permission from Ref [66].

Other imaging methods such as fluorescent tagging can also be incorporated for a truly multifunctional biomedical NP. Iron oxide NPs can be made to have all of these functionalities, and so are the most promising, although other spinel ferrite based materials have also been investigated. For example, hyperthermic DOX-encapsulated magnetic NPs (Dox@SMNPs) made from 6 nm $Zn_{0.4}Fe_{2.6}O_4$ NPs with polyamidoamine dendrimers (Ad-

PAMAM), polyethylenimine (CD-PEI), and polyethylene glycol (Ad-PEG) coating were made by self-assembly, and also had fluorescence imaging properties [66].

Conclusions and future outlook

Despite being a technique first demonstrated over 60 years ago, there seems to have been little interest in developing magnetic hyperthermia as a viable therapy for treating cancers until the 2000's, with a steady increase in investigations articles since then. This is at least partly due to the advances in nanoscience in the last two decades, as magnetic iron oxides need to be superparamagnetic for such applications, and also under 100 nm to be biocompatible to a sufficient degree.

Since then, much of the recent progress seems to be on combined therapies, most with simultaneous drug delivery or immunotherapy, but also with other treatments such as phototherapies, to create a truly multifunctional material offering multiple therapy options in a single treatment. Of particular interest seems to be theranostics, combining therapy with diagnostic and imaging capabilities, particularly as T2 MRI contrast agents in the case of magnetic iron oxide NPs.

The potential shown of magnetoelectric core@shell NPs has demonstrated their capability for drug delivery, and this is an ideal method to be combined with magnetic hyperthermia therapy, especially as there exist single phase magnetoelectric/multiferroic NPs with sufficiently high magnetisation, such as hexaferrites.

Another interesting therapy which seems to have potential and to date does not seem to have been combined with hyperthermia treatments is sonodynamic therapy. This is a non-invasive treatment that can induce the immunogenic cell death of tumours by generating ROS, similar to photodynamic therapy [67]. Ultrasound can penetrate deeper into tissue than a laser, so this method can induce the death of deep cancer cells, as long as the depth and range of sound waves is controlled sufficiently to avoid damage to any surrounding healthy cells. The sonodynamic therapy induced immune response is not sufficient alone to prevent tumour metastasis, and sonosensitive NPs have been combined with immunotherapy to eradicate tumours and also produce a sufficiently long-term immune memory effect to inhibited tumour recurrence and lung metastasis [68]. This would seem to be a potentially useful therapy to combine with magnetic hyperthermia in the future.

References

[1] P. Lauterbur, Image formation by induced local interactions: examples employing nuclear magnetic resonance, Nature 242 (1973) 190-191
https://doi.org/10.1038/242190a0

[2] R. Damadian, M. Goldsmith, L. Minkoff, NMR in cancer XVI: FONAR image of the live human body, Physiol. Chem. Physics 9 (1977) 97-100

[3] A. Seko, K. Yuge, F. Oba, A. Kuwabara, I. Tanaka, Phys. Rev. B 73 (2006) 184117
https://doi.org/10.1103/PhysRevB.73.184117

[4] X. Zeng, J. Zhang, S. Zhu, X. Deng, H. Ma, J. Zhang, Q. Zhang, P. Li, D. Xue, N. J. Mellors, X. Zhang, Y. Peng, Nanoscale 9 (2017) 7493-7500 https://doi.org/10.1039/C7NR02013A

[5] http://www.chemohollic.com/2016/07/spinels-normal-or-inverse.html

[6] Z. Liad, C. Chanéac, G. Berger, S. Delaunay, A. Graffd, G. Lefèvre, RSC Adv. 9 (2019) 33633-33642 https://doi.org/10.1039/C9RA03234G

[7] Z. Hedayatnasab, F. Abnisa , W. M. A. W. Daud, Mater. Design 123 (2017) 174-196 https://doi.org/10.1016/j.matdes.2017.03.036

[8] V. V. Mody, A. Singh, B. Wesley, Euro. J. Nanomed. 5 (2013) 11-21 https://doi.org/10.1515/ejnm-2012-0008

[9] M. Rajendran, R. C. Pullar, A. K. Bhattacharya, D. Das, S. N. Chintalapudi, C. K. Majumdar, J. Magn. Mag. Mater. 232 (2001) 71-83 https://doi.org/10.1016/S0304-8853(01)00151-2

[10] D. B. Reeves, J. B. Weaver, Crit. Rev. Biomed. Eng. 42 (2014) 85-93 https://doi.org/10.1615/CritRevBiomedEng.2014010845

[11] M. R. Horsman, J. Overgard, Clin. Oncol. 19 (2007) 418 https://doi.org/10.1016/j.clon.2007.03.015

[12] R. K. Gilchrist, R. Medal, W. D. Shorey, R. C. Hanselman, J. C. Parrott, C.B. Taylor, Ann. Surg. 146 (1957) 596-606 https://doi.org/10.1097/00000658-195710000-00007

[13] L. Roizin-Towle, J. P. Pirro, Int. J. Radiat. Oncol. Biol. Phys. 20 (1991) 751 https://doi.org/10.1016/0360-3016(91)90018-Y

[14] A. Jordan, P. Wust, R. Scholz, B. Tesche, H. Fähling, T. Mitrovics, T. Vogl, J. Cervós-Navarro, R. Felix, Int. J. Hyperthermia 12 (1996) 705 https://doi.org/10.3109/02656739609027678

[15] K. Mahmoudi, A. Bouras, D. Bozec, R. Ivkov, C. Hadjipanayis, Int. J. Hyperthermia -34 (2018) 1316 https://doi.org/10.1080/02656736.2018.1430867

[16] M. Johannsen, U. Gneveckow, L. Eckelt, A. Feussner, N. Waldöfner, R. Scholz, S. Deger, P. Wust, S. A. Loening, and A. Jordan, Int. J. Hyperthermia. 21 (2005) 637-47 https://doi.org/10.1080/02656730500158360

[17] F. Brero, M. Albino, A. Antoccia, P. Arosio, M. Avolio, F. Berardinelli, D. Bettega, P. Calzolari, M. Ciocca, M. Corti, A. Facoetti, S. Gallo, F. Groppi, A. Guerrini, C. Innocenti, C. Lenardi, S. Locarno, S. Manenti, R. Marchesini, M. Mariani, F. Orsini, E. Pignoli, C. Sangregorio, I. Veronese, A. Lascialfari, Nanomaterials 10 (2020) 1 https://doi.org/10.3390/nano10101919

[18] P. Moroz, S. K. Jones, B. N. Gray, Int. J. Hyperthermia 18 (2002) 267 https://doi.org/10.1080/02656730110108785

[19] A. J. Giustini, A. A. Petryk, S. M. Cassim, J. A. Tate, I. Baker, P. J. Hoopes, Nano LIFE 1 (2010) 17-32 https://doi.org/10.1142/S1793984410000067

[20] I. Hilger, Int. J. Hyperthermia 29 (2013) 828 https://doi.org/10.3109/02656736.2013.832815

[21] I. Belyanina, O. Kolovskaya, S. Zamay, A. Gargaun, T. Zamay, A. Kichkailo, Molecules 22 (2017) 975 https://doi.org/10.3390/molecules22060975

[22] K. Maier-Hauff, F. Ulrich, D. Nestler, H. Niehoff, P. Wust, B. Thiesen, H. Orawa, V. Budach, A. Jordan, J. Neuro-Oncology 103 (2011) 317 https://doi.org/10.1007/s11060-010-0389-0

[23] N. R. Datta, S. Krishnan, D. E. Speiser, E. Neufeld, N. Kuster, S. Bodis, H. Hofmann, Cancer Treatment Reviews 50 (2016) 217-227 https://doi.org/10.1016/j.ctrv.2016.09.016

[24] M. A. Gonzalez-Fernandez, T. E. Torres, M. Andrés-Vergés, R. Costo, P. de la Presa, C. J. Serna, M. P. Morales, C. Marquina, M. R. Ibarra, G. F. Goya, J. Solid State Chem. 182 (2009) 2779-2784 https://doi.org/10.1016/j.jssc.2009.07.047

[25] A. Sathya, P. Guardia, R. Brescia, N. Silvestri, G. Pugliese, S. Nitti, L. Manna, T. Pellegrino, Chem. Mater. 28 (2016) 1769-1780 https://doi.org/10.1021/acs.chemmater.5b04780

[26] P. B. Shete, R. M. Patil, N. D. Thorat, A. Prasad, R. S. Ningthoujam, S. J. Ghosh, S. H. Pawar, Appl. Surf. Sci. 288 (2014) 149-157 https://doi.org/10.1016/j.apsusc.2013.09.169

[27] P. Veverka, E. Pollert, K. Zaveta, S. Vasseur, E. Duguet, Nanotechnology 19 (2008) 215705 https://doi.org/10.1088/0957-4484/19/21/215705

[28] E. Pollert, P. Veverka, M. Veverka, O. Kaman, K. Zaveta, S. Vasseur, R. Epherre, G. Goglio, E. Duguet, Prog. Solid State Chem. 37 (2009) 1-14 https://doi.org/10.1016/j.progsolidstchem.2009.02.001

[29] A. U. Rashid, P. Southern, J. A. Darr, S. Awan, S. Manzoor, J. Magn. Mag. Mater. 344 (2013) 134-139 https://doi.org/10.1016/j.jmmm.2013.05.048

[30] M. Abdellahi, A. Najfinezhad, S. Saber-Samanadari, A. Khandan, H. Ghayoura, Chin. J. Phys. 56 (2018) 331-339 https://doi.org/10.1016/j.cjph.2017.11.016

[31] K. M. Krishnan, IEEE Trans Magn. 46 (2010) 2523-2558 https://doi.org/10.1109/TMAG.2010.2046907

[32] X. L. Liu, C. T. Ng, P. Chandrasekharan, H. T. Yang, L. Y. Zhao, E. Peng, Y. B. Lv, W. Xiao, J. Fang, J. B. Yi, H. Zhang, K. H. Chuang, B. H. Bay, J. Ding, H. M. Fan, Adv. Healthcare Mater. 5 (2016) 2092

[33] Y. Li, X. Liu, X. Zhang, W. Pan, N. Li, B. Tang, Chem. Commun. 57 (2021) 12087-12097 https://doi.org/10.1039/D1CC04604G

[34] M. Yanase, M. Shinkai, H. Honda, T. Wakabayashi, J. Yoshida, T. Kobayashi, Jpn. J. Cancer Res. 89 (1998) 775-782 https://doi.org/10.1111/j.1349-7006.1998.tb03283.x

[35] J. Huang, Y. Li, A. Orza, Q. Lu, P. Guo, L. Wang, L. Yang, H. Mao, Adv. Funct. Mater. 26 (2016) 3818 https://doi.org/10.1002/adfm.201504185

[36] J. Pan, P. Hu, Y. Guo, J. Hao, D. Ni, Y. Xu, Q. Bao, H. Yao, C. Wei, Q. Wu, J. Shi, ACS Nano 14 (2020) 1033-1044 https://doi.org/10.1021/acsnano.9b08550

[37] X. Liu, J. Zheng, W. Sun, X. Zhao, Y. Li, N. Gong, Y. Wang, X. Ma, T. Zhang, L. Y. Zhao, Y. Hou, Z. Wu, Y. Du, H. Fan, J. Tian, X. J. Liang, ACS Nano 13 (2019) 8811-8825 https://doi.org/10.1021/acsnano.9b01979

[38] A. Singh, V. Nandwana, J. S. Rink, S. R. Ryoo, T. H. Chen, S. D. Allen, E. A. Scott, L. I. Gordon, C. S. Thaxton, V. P. Dravid, ACS Nano 13 (2019) 10301-10311 https://doi.org/10.1021/acsnano.9b03727

[39] J.-H. Lee, K.-J. Chen, S.-H. Noh, M. A. Garcia, H. Wang, W.-Y. Lin, H. Jeong, B. J. Kong, D. B. Stout, J. Cheon, H.-R. Tseng, Angew. Chem. Int. Ed. 52 (2013) 4384-4388 https://doi.org/10.1002/anie.201207721

[40] M. Nair, R. Guduru, P. Liang, J. Hong, V. Sagar, S. Khizroev, Nature Commun. 4 (2013) 1707 https://doi.org/10.1038/ncomms2717

[41] R. A. Revia, M. Zhang, Mater. Today, 19 (2016) 157-168 https://doi.org/10.1016/j.mattod.2015.08.022

[42] C. L. Chen, L. R. Kuo, C. L. Chang, Y. K. Hwu, C. K. Huang, S. Y. Lee, K. Chen, S. J. Lin, J. D. Huang, Y. Y. Chen, Biomater. 31 (2010) 4104-4112 https://doi.org/10.1016/j.biomaterials.2010.01.140

[43] C. W. Ng, J. Li, K. Pu, Adv. Funct. Mater. 28 (2018) 1804688 https://doi.org/10.1002/adfm.201804688

[44] J. Estelrich, M. A. Busquets, Molecules 23 (2018) 1567 https://doi.org/10.3390/molecules23071567

[45] G.-T. Yu, L. Rao, H. Wu, L.-L. Yang, L.-L. Bu, W.-W. Deng, L. Wu, X. Nan, W.-F. Zhang, X.-Z. Zhao, W. Liu, Z.-J. Sun, Adv. Funct. Mater. 28 (2018) 1801389 https://doi.org/10.1002/adfm.201801389

[46] J. Wang, Z. G. Zhou, L. Wang, J. Wei, H. Yang, S. P. Yang, J. M. Zhao, RSC Adv. 5 (2015) 7349-7355 https://doi.org/10.1039/C4RA12733A

[47] D. K. Chatterjee, L. S. Fong, Y. Zhang, Adv. Drug Deliv. Rev. 60 (2008) 1627-37 https://doi.org/10.1016/j.addr.2008.08.003

[48] A. B. Seabra, Iron Oxide Magnetic Nanoparticles in Photodynamic Therapy: A Promising Approach Against Tumor Cells. In: M. Rai, R. Shegokar (eds), Metal Nanoparticles in Pharma., Springer, Cham (2017) https://doi.org/10.1007/978-3-319-63790-7_1

[49] Z-L. Chen, Y. Sun, P. Huang, X-X. Yang, X.-P. Zhou, Nanoscale Res. Lett. 4 (2009) 400-408 https://doi.org/10.1007/s11671-009-9254-5

[50] J. Cheng, G. Tan, W. Li, J. Li, Z. Wang, Y. Jin, RSC Adv. 6 (2016) 37610-37620 https://doi.org/10.1039/C6RA03128E

[51] H. Chen, F. Liu, Z. Lei, L. Ma, Z. Wang, RSC Adv. 5 (2015) 84980-84987 https://doi.org/10.1039/C5RA17143A

[52] R. Di Corato, G. Béalle, J. Kolosnjaj-Tabi, A. Espinosa, O. Clément, A. K. A. Silva, C. Ménager, C. Wilhelm, ACS Nano 9 (2015) 2904-2916 https://doi.org/10.1021/nn506949t

[53] M. Wang, Z. Hou, A. A. Al Kheraif, B. Xing, J. Lin, Adv. Healthcare Mater. 7 (2018) 1800351 https://doi.org/10.1002/adhm.201800351

[54] A. Quarta, C. Piccirillo, G. Mandriota, R. Di Corato, Materials 12 (2019) 139 https://doi.org/10.3390/ma12010139

[55] M. Bañobre-López, Y. Piñeiro-Redondo, M. Sandri, A. Tampieri, R. De Santis, V. A. Dediu, J. Rivas, IEEE Trans. Mag. 50 (2014) 5400507 https://doi.org/10.1109/TMAG.2014.2327245

[56] S. Mondal, P. Manivasagan, S. Bharathiraja, M. Santha Moorthy, V. T. Nguyen, H. H. Kim, Y. Nam, K. D. Lee, J. Oh, Nanomaterials 7 (2017) 426 https://doi.org/10.3390/nano7120426

[57] A. Najafinezhad, M. Abdellahi, S. Saber-Samandari, H. Ghayour, A. Khandan, J. Alloys Comp. 734 (2018) 290-300 https://doi.org/10.1016/j.jallcom.2017.10.138

[58] W. Leenakul, P. Intawin, J. Ruangsuriya, P. Jantaratana, K. Pengpat, Integrated Ferroelectrics 148 (2013) 81-89 https://doi.org/10.1080/10584587.2013.852034

[59] S. Khizar, N. M. Ahmad, N. Zine, N. Jaffrezic-Renault, A. Errachid-el-salhi, A. Elaissari, ACS Appl. Nano Mater. 4 (2021) 4284−4306 https://doi.org/10.1021/acsanm.1c00852

[60] V. F. Cardoso, A. Francesko, C. Ribeiro, M. Bañobre-López, P. Martins, S. Lanceros-Mendez, Adv. Healthcare Mater. 7 (2018) 1700845 https://doi.org/10.1002/adhm.201700845

[61] D. Yoo, J. H. Lee, T. H. Shin, J. Cheon, Acc. Chem. Res. 44 (2011) 863 https://doi.org/10.1021/ar200085c

[62] J. T. Jang, H. Nah, J. H. Lee, S. H. Moon, M. G. Kim, J. Cheon, Angew. Chem. Int. Ed. 48 (2009) 1234 https://doi.org/10.1002/anie.200805149

[63] J. S. Choi, J. H. Lee, T. H. Shin, H. T. Song, E. Y. Kim, J. Cheon, J. Am. Chem. Soc. 132 (2010) 11015 https://doi.org/10.1021/ja104503g

[64] Y. Li, K. Hu, B. Chen, Y. Liang, F. Fan, J. Sun, Y. Zhang, N. Gu, Colloids Surf. A 520 (2017) 348 https://doi.org/10.1016/j.colsurfa.2017.01.073

[65] C. Sanson, O. Diou, J. Thévenot, E. Ibarboure, A. Soum, A. Brûlet, S. Miraux, E. Thiaudière, S. Tan, A. Brisson, V. Dupuis, O. Sandre, S. Lecommandoux, ACS Nano 5 (2011) 1122-1140 https://doi.org/10.1021/nn102762f

[66] U. H. Sk, C. Kojima, Biomolecular Concepts 6 (2015) 205-217 https://doi.org/10.1515/bmc-2015-0012

[67] M. Xu, L. Zhou, L. Zheng, Q. Zhou, K. Liu, Y. Mao, S. Song, Cancer Lett. 497 (2021) 229-242 https://doi.org/10.1016/j.canlet.2020.10.037

[68] W. Yue, L. Chen, L. Yu, B. Zhou, H. Yin, W. Ren, C. Liu, L. Guo, Y. Zhang, L. Sun, K. Zhang, H. Xu, Y. Chen, Nat. Commun. 10 (2019) 2025 https://doi.org/10.1038/s41467-019-09760-3

Magnetic Nanoparticles for Biomedical Applications
Materials Research Foundations 143 (2023) 102-139

Materials Research Forum LLC
https://doi.org/10.21741/9781644902332-4

Chapter 4

Physical and Cellular Basis of Oncologic Magnetic Thermotherapy

Fiorela Ghilini[1], Mariana Tasso[1] and Marcela. B. Fernández van Raap[2]*

[1]Instituto de Investigaciones Fisicoquímicas Teóricas y Aplicadas (INIFTA), Departamento de Química, Facultad de Ciencias Exactas, Universidad Nacional de La Plata - CONICET, calle 64 y Diagonal 113, (1900) La Plata - Argentina.

[2]Instituto de Física La Plata (IFLP), Departamento de Física, Facultad de Ciencias Exactas, Universidad Nacional de La Plata - CONICET, Diagonal 113 y 63, (1900) La Plata - Argentina.

*raap@fisica.unlp.edu.ar

Abstract

Magnetic thermotherapy (MT) for treating solid tumors employs magnetic hyperthermia (MH) protocols. Superparamagnetic iron oxide nanoparticles -nowadays considered nanomedicines- are delivered to the tumor to induce temperature elevations upon the application of an external alternating magnetic field in the radiofrequency range. This treatment activates cell death pathways and sensitizes cells to other cancer therapies. Much information is available on magnetic materials preparation and functionalization, physical relaxation mechanisms behind magnetic heating, and designed devices for field generation and magnetic material tracking, but further insight is still needed about the nature and characteristics of the cellular response to MH. In this chapter, the physical foundations of MT are thoroughly discussed and complemented by a general overview of the cellular responses induced to counteract the various sources of stress associated with this technology.

Keywords

Magnetic Hyperthermia, Superparamagnetic Iron Oxide Nanoparticles, Nanomedicines, Cell-Nanoparticle Interactions, Cell Response To Heat Shock, Heat Shock Proteins

Contents

1.   Introduction

1.1   Nanomedicines, nano-associated toxicity and cellular response

Nanomedicine refers to the use of nano-size objects in the medical field. According to ISO/TS 80004-2;2015, *nanotechnology-vocabulary-part2: nanoobjects*, a nanoobject is a discrete piece of material with one, two or three external dimensions in the nanoscale, i.e. in the length range approximately from 1 nm to 100 nm. However, objects falling in an extended size range including hundreds of nm have been described as nanomedical agents. In practice, the most frequently used range is from 5–250 nm [1]. Common types of nanomedicines are the various drug delivery nano-systems (e.g. liposomes) and the magnetic nanomaterials employed as contrast agents in magnetic resonance imaging, as vehicles for (image-guided) drug/gene delivery and, more recently, for the first successful glioblastoma tumor hyperthermia treatment in humans [2–6]. Many magnetic nanomaterials were proposed over the last years as potential nanomedicines. Among them, the superparamagnetic iron oxide nanoparticles (SPIONs) [4,7], in the phases magnetite Fe_3O_4 and maghemite γ- Fe_2O_3, are particularly relevant given their outstanding properties such as: large magnetic moment, null coercive field, which is a biomedical requirement, and amphoteric surface, which allows for a variety of surface functionalization approaches. The magnetic moment magnitude enables their contactless manipulation with external magnetic fields, resulting in a good capability to convert energy field into heat and in their extensive use in the magnetic targeting of drugs and biological entities [7,8]. SPIONs magnetic properties and their responsiveness to alternating magnetic fields in the radiofrequency range will be extensively discussed in section 2. Iron oxide nanoparticles are most frequently produced by aqueous coprecipitation[9], coprecipitation in polyols [10] and thermal decomposition protocols [11,12]. The variation in synthesis conditions allows to purposely achieve a variety of isometric and anisometric single core and multicore nanoparticle shapes, to control size and size distribution, as well as the degree of crystallinity. A recent review that discusses synthesis problems and related nanostructures can be found elsewhere [11]. Besides, various metal substitutions in magnetite and maghemite structures have been produced and analyzed in search for better MH agents, such as Zn [13], Co and Mn [14] and other NPs morphologies have been generated, as core-shell structures like Fe_3O_4-Au for magneto-photothermal therapy [15] or for a multimodal platform including imaging [16] and combined with boride compounds for combination with boron neutron capture therapy [17,18].

The variety of nanomaterials already employed as nanomedicines and also proposed for a potential application in this field is indicative of the idea that nano-sized materials could act as medicines in better ways than the traditional ones, for instance by delivering therapeutic drugs more precisely to the target site, thereby reducing side-effects such as the toxicity to non-target organs, or be unspecifically retained in tumor tissues and thus serve as imaging agents to diagnose and monitor this condition. Although nanomaterials have clear advantages in this respect compared with larger objects since they possess a higher surface-to-volume ratio, they also possess a higher surface reactivity and a lower thermodynamic stability, which altogether may prompt redox reactions, dissolution processes, and the generation of Reactive Oxygen Species (ROS) [19], and may therefore induce cellular responses that differ from those observed for larger objects. In particular, nanoparticles of certain sizes can access the intracellular environment and be potentially toxic [20–22]. These properties, as well as others, such as the size and shape of the nanostructure or the type of surface ligand, could result in unwanted cellular and organ toxicity [23–26]. Once in the body, the control of NP dissolution and its clearance to avoid long-term retention and toxicity remains a challenge. In that respect, non-equilibrium Au-Fe nanoparticles that behave as 4D shaping-morphing nanocrystals have been proposed as self-degradable multifunctional nanomedicines with good prospects for combined MRI and CT imaging [27]. These nanosystems change attributes while they are transformed in the body, a fact that is meant to shape their biodistribution, clearance and eventual toxicity.

In contact with tissues, NPs will encounter immune defense cells and cellular mechanisms responsible for clearing unwanted materials from the body and combating them. In the case of external administration, e.g., intravenously or intratumorally, the immune cells present in muscle, blood, the liver, spleen, and kidney are the ones that will mostly interact with the nanomaterials [28–30]. The immune response involves the occurrence of concatenated events set up to rapidly eliminate the NPs through various uptake and degradation processes in immune cells. The mononuclear phagocytic system (MPS), or otherwise named reticuloendothelial system (RES), is composed of a family of cells involved in defense reactions against foreign objects and in immunity. Blood monocytes, bone marrow progenitors, and macrophages present in organs, such as the spleen, liver, bone marrow, brain, lung and the lymph nodes, are very prominent cells of this system [31,32]. Macrophages reside in most of the body tissues and their number increases following lesions, malignancy, and inflammation [31]. NPs' clearance from the bloodstream is mostly carried out by macrophages found in the liver and spleen [33], which can eliminate NPs from blood circulation within seconds to minutes of intravenous administration [30,34]. Alternatively, if NPs have hydrodynamic sizes below 5.5 nm [35] or, more generally, below 10 nm [30], elimination occurs via renal filtration and urination.

However, macrophages can only recognize specific proteins, named opsonins, which are attached to the surface of nanoparticles after injection [36,37]. Once the adsorption of opsonin proteins (called opsonization) has occurred, phagocytes bind to the foreigner element and clear it. Opsonins are blood serum components that function as markers to label foreigner objects, thereby rendering them "visible" to phagocytes [37,38]. The nanomaterials that are engulfed by the phagocytes may be eventually degraded by these cells, or, if undigested, they may be removed by the renal or hepatobiliary routes or, alternatively, be stored in one of the MPS organs [39]. Opsonization of NPs is a critical limitation for any nano-object intended to be used as a nanomedicine. Modifying the surface properties of NPs to effectively reduce opsonin's adsorption has demonstrated clear advantages at increasing NPs' half lives in circulation and at minimizing clearance by the MPS [40,41].

If opsonization is reduced and the NPs can evade immune clearance through the MPS, these NPs may still need to go through the endothelium barrier present in the vasculature, where junctions are below 2 nm in size, to exit the bloodstream and to reach target organs. In most cancers, the endothelium becomes leaky and presents abnormal lymphatic drainage, thereby enabling the passing of NPs through it and their accumulation in cancer tissue [1,42]. This process was named enhanced permeability and retention (EPR) effect. Once out of circulation, nanomaterials will be in contact with the extracellular matrix, where they may migrate through water channels or be spontaneously internalized by cells via endocytosis [43,44]. If endocytosed, NPs will be finally located in the lysosome, where they will be subjected to high concentrations of hydrolytic enzymes able to cleave nucleic acids, proteins, and polysaccharides. Depending on the physico-chemical characteristics of the nanomaterial, the lysosome may provide the right conditions for a complete degradation of the nanomaterial. Noteworthy, some nanomaterials access the cytoplasm without being endocytosed. This is the case of positively-charged polymers and polyelectrolytes, which may elicit a toxic response as they interact with membrane phospholipids, disrupting the cell membrane structure [45].

In the following sections of this chapter, the physics behind MH thermotherapy and the main processes associated with the cellular response to MH protocols will be described. The specifics of the interactions between cells and nanomaterials and their dependency with the physico-chemical properties of the latter will be discussed in section 3 with emphasis on the inherent characteristics of iron oxide NPs and their influence on normal cell functioning. In particular, the relationship between oxidative stress and iron oxide NPs, the appearance of ferroptosis as a form of cell death, and the triggering of an immune response will be presented. Genotoxicity, inflammation, and cell membrane damage as a result of cell-nanoparticle interactions will also be covered. In section 4, the effects of magnetic hyperthermia (MH) treatment onto cell function and cellular response will be introduced and discussed based on published results considered relevant and illustrative of the concepts here presented. Previously, in section 2, the physical principles that govern magnetic hyperthermia and that are used to characterize its attributes will be thoroughly analyzed and discussed, highlighting potential areas of improvement as well as the need for standardization.

1.2   Magnetic hyperthermia and its effects on cells

Magnetic Hyperthermia (MH) is a minimally invasive oncologic therapy [46] under clinical trials to treat deep-seeded solid tumors like glioblastoma multiforme [47], prostate cancer [48] and more recently moving forwards to include pancreatic cancers [49]. This thermotherapy benefits from the heat released by the magnetic nanoparticles (MNPs) when an alternating magnetic field (AMF) is applied. The MNPs absorb energy from the field to deliver it in the tumor environment via Néel and Brown relaxation mechanisms. By raising tumor temperature above normal physiological values, to the range from 42°C to 45°C, heating modifies the function of many enzymatic and structural proteins within cells, which varies cell growth and differentiation rates and triggers cell death by hyperthermia, potentially achieving tumor reduction [50]. Ablation is achieved at temperatures T >45°C. Among the various hyperthermia modalities that include local, regional, and whole-body temperature elevations, MH has the great advantage of achieving locoregional heating. This spatial selectivity is accomplished when nanoparticles are only located in the tumor, thereby helping to prevent unwanted heating and damage to normal tissue. In a typical MH therapeutic procedure, the MNPs are incorporated into the target tissue and then exposed to an external AMF of frequency of 100 kHz and amplitude in the range from 2 to 18 kA m^{-1} for tens of minutes with different field application protocols. Currently, magnetic

Materials Research Forum LLC
https://doi.org/10.21741/9781644902332-4

thermotherapy (MT) in clinical settings uses intratumoral administration of MNPs. The applied procedure relies on safety considerations, namely that the healthy tissue has to be unaffected, both, under the application of the NPs alone and of the AMF alone. Then, when the NPs are inside the tumor and exposed to the AMF application, they act as energy transducers becoming cytotoxic for the malignant cells. Cell toxicity due to the NPs depends on their physico-chemical characteristics (chemical composition, size, surface ligands, etc., see section 3) and dosage [51]. Regarding their chemical composition, iron oxides, magnetite and maghemite phases, are considered the best choice system for biomedical applications because they are easily synthesized in the nanoscale range, are naturally more biocompatible than other substituted ferrite phases and because iron ions have their own recycling metabolic pathways [51]. Despite these advantages, iron oxide NPs are also known for their capacity to generate ROS that, if left unbalanced by antioxidant enzymes in the cytosol, may result in cell death. Fig. 1 summarizes the events that will be discussed in this chapter.

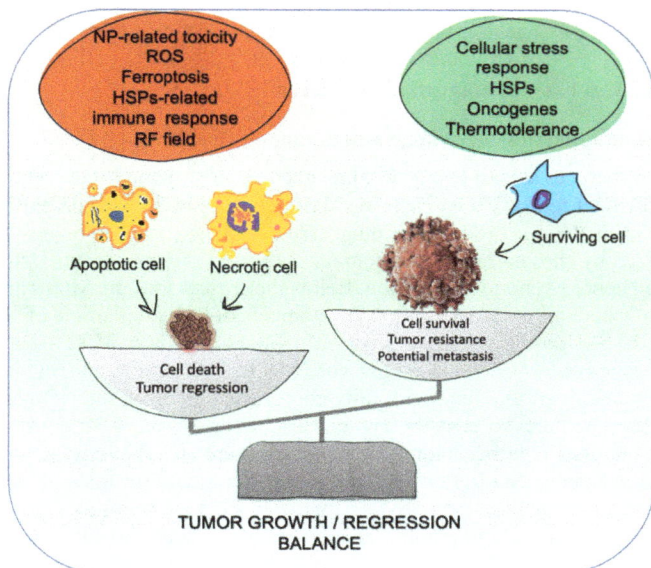

Fig. 1. The phenomena affecting the balance between tumor growth and tumor regression in magnetic hyperthermia applications. RF: radiofrequency field, HSPs: heat shock proteins, ROS: reactive oxygen species.

For MH applications, heat is the cellular stressor usually identified and described, though the nanoparticle itself and the magnetic field also contribute to the type and extent of the observed cellular responses [52–58]. Temperature variations at cellular level, either towards lower or higher temperatures, induce the so-called heat shock response, mainly characterized by the expression of heat shock proteins (HSPs). Nevertheless, the heat shock response is a far more widespread response mechanism that is also activated in the presence of pH variations, toxic ions, reduction in available energy, etc. [59] The heat shock response, and the consequent activation of HSPs, has been vastly employed to denote the success of a MH treatment by referring to HSP expression as a consequence of local temperature elevations. However, as it will be later discussed in section 4, growing evidence in this field recently shed light into the involvement of other cell-dependent mechanisms that also contribute to the success of MH treatments in vivo by activating distinct cellular pathways that result in cell death (e.g., ferroptosis) or in the generation of an immune response [60,61]. Noteworthy, the induction of HSPs has also been linked to thermotolerance and to tumor cell survival, though [62]. The final success of a MH treatment (and of any hyperthermia treatment) then depends on the delicate equilibrium between opposing mechanisms, those pro-tumor and those anti-tumor.

2.   The physical basis of magnetic hyperthermia

2.1   Useful nanoparticle size range and magnetic structures

The nanoscale includes at least two types of magnetic nanostructures that release heat under an AMF by distinct relaxation mechanisms. For magnetic multi-domain particles, of size larger than a critical value (>80 nm for magnetite), the magnetization reversal mechanism is characterized by the motion of the domain walls and pinning occurs, which leads to hysteresis and hence to energy dissipation. Below the critical size, the MNPs are of a single domain (SD), where all the atomic magnetic moments of the nanoparticle of volume V are parallel and add to produce a super moment of magnitude $\mu = M_s V$, where M_s is the saturation magnetization of the magnetic core. As the NP size decreases, its saturation magnetization decreases too due to not fully coordinated atoms at the NP surface, named dead layer. Nanoparticles of sizes below a size limit d_l (of about 30 nm for magnetite) display superparamagnetic behavior at room temperature and above while those of sizes larger than d_l are stable SD [63]. There is also a lower size limit for a SD particle since ferromagnetism is a cooperative behavior that only exists in multiatomic clusters. The minimum size for a particle to retain magnetic order has been theoretically estimated as about 1 nm [64]. Below this value, all the atoms are mostly at the surface of the nanoparticle i.e. forming the magnetically dead layer. Due to this effect, saturation magnetization below 2-3 nm is indeed very low and the efficiency of MT agents of such sizes rather limited, as it will be shown in the next section.

The first pioneer work that reported magnetic hyperthermia on lymph nodes in dogs was carried out using micrometric (multi-magnetic domain) Fe_3O_4 particles where appreciable differential heating was obtained in nodes containing 5 mg of particles per g of tissue [65].

Three decades later, A. Jordan resumed that work and showed that SPIONs (single magnetic domain in the superparamagnetic state) are more effective heaters [66]. This group used aminosilane-coated superparamagnetic iron-oxide NPs specially designed for MH therapy, which later received approval by the U.S. Food and Drug Administration (FDA) [67]. In the superparamagnetic state, the NP moment switches its orientations between opposite directions and the magnetization averages to zero. Then, besides their outstanding capability to convert field energy into heat, SPIONs display null coercive field, which means that magnetization goes to zero when the field is withdrawn. This property is advantageous for medical uses.

2.2   The dynamics of relaxation mechanisms

Since SD magnetic NPs are widely accepted as more effective, we discussed here in detail the relaxation mechanisms of an ensemble of these NPs of volume V and uniaxial anisotropy K homogenously distributed in a medium of viscosity η. The specific heat released by this system at a field $H(t) = H_0 \cos(\omega t)$ of frequency f ($\omega = 2\pi f$) and amplitude H_0 depends on the magnetic relaxation mechanism achieved by the NPs. In a characteristic time τ, the NP moment μ orientation changes from pointing to one direction to the opposite. The already described mechanisms are Néel and Brown processes, which depend on the nanoparticles' properties and the medium condition. The magnetization M, lags the field as

$$M(t) = H_0 \chi_0 \cos(\omega t - \tan^{-1}(\omega \tau))$$

being χ_0 the equilibrium susceptibility, via Néel mechanisms when the nanoparticle magnetic moment switches from one direction to the opposite or via Brown when the particle physically rotates in the medium leading to viscous heating. The Néel relaxation time, at a temperature T, depends on the height of the energy barrier, U, of a double-well potential and on the attempt time $\tau_0 \sim 10^{-13}$-10^{-10} s [68] as $\tau_N = \tau_0 \exp(U/k_B T)$ being $U = KV$ at low field amplitude. The precise value of τ_0 depends on the nanoparticle properties and on its environment. For iron oxide NPs, the use of $\tau_0 = 10^{-10}$ s has been recommended [69] based on the detailed analysis carried out by Dormann et al. who had shown that the dependence of τ with blocking temperature deviates from the Arrhenius law and asymptotically goes to 10^{-10} s for interacting nanoparticles.

An approximation that includes field amplitude dependence is given by the equation $\tau_{N,h} = \tau_0 \exp(U(1-h)^2/k_B T)$ with $h = H_0/H_k$, $H_k = \frac{2K}{\mu_0 M_s}$ the anisotropy field and μ_0 the permeability of the free space ($4\pi\ 10^{-7}$ N A^{-2}) [70]. The Brown relaxation time depends on the interaction between the NPs and the medium mediated by its viscosity η as $\tau_B = \frac{3\eta V_H}{k_B T}$, being V_H the hydrodynamic volume. The dynamics of the particle's dipole moment is driven by the effective relaxation time, τ, given by $\tau^{-1} = \tau_B^{-1} + \tau_N^{-1}$. When relaxation modes are independent [71], the faster process takes place. The Néel mechanism is not achieved if $\tau_N \gg \tau_B$, i.e., when V and/or K are large. The Brown mechanism is not

Materials Research Forum LLC
https://doi.org/10.21741/9781644902332-4

achieved if $\tau_B \gg \tau_N$, which happens when the particles display large hydrodynamic volume, when they cannot freely move (as when NPs get fixed to cell membranes either due to non-specific interactions or to specific targeting), or when the viscosity of the medium is intentionally increased by increasing, for instance, the thixotropic properties of the ferrofluid [72]. As thixotropic substances progressively reduce their viscosity with time when a constant shear stress is applied and gradually recover it when the stress is removed, this property facilitates the injection of the NP suspension in the tumor tissue and their later fixation assuring Néel relaxation for all the particles. Otherwise, when the relaxation modes are not independent, the magnetic moment may switch with a more complex dynamics, starting with one mode and ending the movement with the other.

2.3 Specific absorption rate

The energy dissipation caused by these switching processes is linked to the complex magnetic susceptibility of the NPs given by $\chi = \chi' + i\chi''$. The imaginary part $\chi''(f)$ is proportional to the component of the magnetization induced out of phase with the excitation wave. This component is directly proportional to the dissipated power P [73]. Then, the parameter used to report nanoparticle's magnetic heating efficiency is the specific absorption rate, SAR (also named specific power loss, SLP), being $SAR = P/m$ in W g^{-1}, with m the NP mass. It is customary to report it as W per grams of Fe in the sample. Within the linear response theory ($M \sim H$), valid for $\xi < 1$ with $\xi = \frac{\mu\mu_0 H}{k_B T}$ and the Stoner-Wohlfarth (SW) theory [74], the SAR is given by:

$$SAR = \mu_0 \pi f H_0^2 \chi''(f,\tau) \qquad (1)$$

$$\text{with } \chi''(f,\tau) = \chi_0 \frac{2\pi f\tau}{1+(2\pi f\tau)^2} \qquad (2)$$

and χ_0 can be approximated by the d.c. initial mass susceptibility at low field amplitude. The equilibrium mass susceptibility of the ensemble of magnetic NPs with their easy axes aligned along the magnetic field and mass density δ can be modeled as

$$\chi_{0l} = \frac{\mu_0 M_s^2 V}{3\delta k_B T} \qquad (3)$$

according to a Langevin behavior of the magnetization $M = M_s L(\xi)$, with $L(\xi) = \coth(\xi) - \frac{1}{\xi}$. However, it has been theoretically shown that χ_0 strongly depends on anisotropy, evolving from χ_{0l} to values up to $3\chi_{0l}$ for very anisotropic systems [75]. Note in Eq. 2 that $\chi''(\tau)$ peaks when $2\pi f\tau = 1$ indicating that the optimum NP size at low field amplitudes depends on field frequency and at high field amplitude on both f and H_0.

A real ensemble of NPs displays a NP size distribution and, at useful concentrations, aggregation and magnetic dipolar interactions are significant and affect the relaxation times leading to a distribution of relaxation times, $g(\tau)$, which is a function that peaks at some value, and to the SAR now expressed as:

$$SAR = \mu_0 \pi f H_0^2 \int \chi''(f,\tau) g(\tau) d\tau \qquad (4)$$

Eq. 4 indicates that at a given f, dissipation will be significant only in the τ range where $g(\tau)$ superimposed to a non-null value of $\chi''(\tau)$. Then, in colloids displaying size distribution, aggregation, and attractive magnetic dipolar interactions, only a fraction of the NPs will heat efficiently [69].

Then, within the framework of this theory, the heating efficiency of an ensemble of magnetic NPs at a given temperature T and field parameters f and H_0 depends on nanoparticles' physical and magnetic properties, such as size, size distribution, magnetic anisotropy K and saturation magnetization M_s, on interparticle interactions and on the environment's viscous and rheological properties. Interparticle interactions, which will be discussed in section 2.3.1, are related to concentration because an increase in concentration leads to a decrease in the mean interparticle separation and gives rise to a large increase in the strength of the dipolar interparticle interactions.

From equations 1-3, the SAR increases quadratically with increasing saturation magnetization as M_s^2 and with increasing V, within magnetic single domain size, as $V \chi''(\tau(V))$. The best attainable mass saturation magnetization values are close to the bulk values of 98 and 82 A m^2 kg^{-1} for Fe$_3$O$_4$ and γ- Fe$_2$O$_3$, respectively [76], and SD size range is limited as explained in section 2.1. Besides, depending on all these NP and medium properties, the SAR strongly depends on field parameters as shown in Eq. 1 and 2. The SAR increases quadratically with increasing field amplitude as H_0^2 and varies more complexly with f through the $f \chi''(f)$ term. For that reason, SAR values must always be reported jointly with field settings and the comparison of the SAR values of different NPs is strictly valid at the same field settings.

Noteworthy, careful attention must be paid to the selection of the field parameters because, as tissues are conductive, unwanted Eddy currents are also induced producing unspecific heating. Atkinson et al. [77] estimated the power dissipated per unit volume of tissue in a cylindrical body of radius r and tissue conductivity σ as $P_{eddy} = \sigma (\pi \mu_0)^2 (f H_0)^2 r^2$, showing that P_{eddy} rises with $(f H_0)^2$. The authors also established a safe field range [78] for application in humans narrowing the therapeutic field setting's range to $f H_0 \le$ 4.85×10^8 A m^{-1} s^{-1} [79]. A less conservative limit of $f H_0 \le 4.85 \times 10^9$ A m^{-1} s^{-1} has been proposed and is often used for NPs heating efficiency characterization [80]. The lower the $f H_0$ product, the larger the amount of magnetic material that must be injected to reach the planned temperature increase. One of the main problems of MH therapy is the high dosage of NPs needed. Treatable tumor volumes should be larger than 2 mm in diameter for NPs loading $\ge$ 10 mgFe per g of tumor to accomplish therapeutic temperatures $\ge$ 43°C at a field of 100 kHz and 20 kA m^{-1} [81], i.e. at a field condition of $f H_0$ =20 x10^8 A m^{-1} s^{-1} that is five times larger than the value established as safe. The upper limit of the $f H_0$ product was experimentally established more than 30 years ago by surrounding the thorax of many healthy people with a single-turn induction coil. Field intensities up to 35.8 A m^{-1} at a frequency of 13.56 MHz were thermally tolerated for extended time periods. It is quite surprising that this kind of experiment has not been repeated. It appears to be important to establish this limit with more precision to optimize NPs properties to achieve high SAR values in the fH_0 range useful for medical practice.

2.3.1 Anisotropy sources and dipolar interactions

In section 2.2, Neél relaxation time depends exponentially on the height of the energy barrier U, which for a monodispersed ensemble of uniaxial non-interacting particles of volume V is expressed as KV. However, besides magnetocrystalline anisotropy, which is defined by the crystal structure easy magnetization direction, other contributions, like shape and surface anisotropies, are usually present and give rise to a large increase in the anisotropy constant usually described with K_{eff}, the effective anisotropy constant [82]. Then, U is expressed as $K_{eff} < V >$, with K_{eff} accounting for the various sources of anisotropy that make it highly dependent on nanoparticle size [83], and on the NPs mean volume $< V >$, which considers size distribution. Moreover, interparticle dipolar interaction increases the relaxation time and has been modeled by adding a term to the energy barrier as $U = K_{eff} < V > +E_{int}$, where E_{int} adds dipolar interparticle interaction energies, given by Eq. 5, from all pairs of NPs in the ensemble. The increase in the relaxation time can be considered by an increase in U with a mean field approximation of E_{int} or, alternatively, by a decrease in the thermal energy. The latter is known as the Vogel-Fulcher correction, a well-known approximation to account for weakly interacting nanoparticles [84]. Both approaches are equivalent for uniformly distributed NPs and $\xi <$ 1; otherwise, a mean field interacting superparamagnetic model can be used [85].

The linear relationship between U and V is not valid for dipolar interacting nanoparticles just because the dipolar interaction energy (E_{ij}) between two particles of magnetic moment vectors $\boldsymbol{\mu}_i = \mu\boldsymbol{u}_i$ and $\boldsymbol{\mu}_j = \mu\boldsymbol{u}_j$, oriented according to unitary vectors $\boldsymbol{u}_i$ and $\boldsymbol{u}_j$, and distanced $\boldsymbol{r}_{ij} = r\boldsymbol{u}_{ij}$ is proportional to μ^2 according to

$$E_{ij} = -\frac{\mu^2}{r_{ij}^3}\left[3\big(\boldsymbol{u}_i \cdot \boldsymbol{u}_{ij}\big)\big(\boldsymbol{u}_j \cdot \boldsymbol{u}_{ij}\big) - \boldsymbol{u}_i \cdot \boldsymbol{u}_j\right] \tag{5}$$

Then, $E_{ij}{\sim}\mu^2/L^3 = D^6 M_S^2/L^3$ increases with NP size D and with the closeness among them given by the mean interparticle distance L. The nose-tail configuration minimizes E_{ij} favoring the formation of NP chains [86].

Whether dipolar interaction increases or decreases the SAR has been a source of debate. The effect of dipolar interaction in the SAR magnitudes was theoretically examined by Landi [87] using a random dipolar-field approximation, which is strictly valid for monodisperse NPs for which the mean first order fluctuating dipolar field contribution averages to zero [88]. The results showed that variations in the interparticle distances among NPs have large effects on the magnetization dynamics, being these effects able to increase or decrease the amount of heat released, namely this model theoretically predicted a peak in SAR as a function of interaction strength, ending the controversy derived from the opposite behaviors experimentally observed for SAR against NP concentration in a suspension. These observations were endorsed by kinetic Monte Carlo simulations [89] and by experiments in colloidal suspensions [90,91]. For instance, Coral et al. showed that the SAR at 145 kHz and 35.6 kA m^{-1} of a NP suspension of nearly monodispersed 18 nm size (3054 nm^3) magnetite in water versus NP concentration [x] displays a peak. The experimental behavior was reproduced using Landi's model by writing τ_N as a function of [x] and of interparticle distance L as:

$$\tau_N = \tau_o \, exp\left[\frac{K_{eff}V}{k_BT} + 0.9\,\gamma\left(\frac{K_{eff}V}{k_BT}\right)^2\right]$$ (6)

With $\gamma = \frac{N}{10}\left(\frac{\langle\mu^2\rangle\mu_0}{4\pi KV}\right)^2\langle\frac{1}{L^6}\rangle$, N the total number of NPs in the system, $L = (\rho V/[x])^{\frac{1}{3}}$

and ρ the mass density of magnetite. The SAR peaked at [x]=2 mg$_{Fe}$ mL^{-1} to a value of 110 $W\,g_{Fe}^{-1}$. That work also showed aggregation of randomly oriented NPs to be detrimental for SAR.

Summarizing, the relaxation time depends on NPs mean volume $<V>$, on the effective anisotropy K_{eff} that may have crystallographic, shape and surface contributions, on field amplitude H_0, on NPs concentration through magnetic dipolar interactions among NPs and on the rheological properties of the supporting medium. Then, the SAR takes its maximum value when

$$\tau\left(<V>, K_{eff}, H_0, E_{int}([x]), \eta\right) = 1/2\pi f$$ (7)

The condition given by Eq. 7, clearly indicates that there isn't a simple answer to the question of which is the best NP for magnetic hyperthermia therapy, i.e., which is the one displaying the highest SAR. Once NP chemical composition, f, H_0 and [x] are defined, we can analyze which are the best NP shape and mean size. As an example, kinetic Monte Carlo simulations were recently employed to identify the magnetite NPs that deliver high SAR values for anisotropy values in the range between 11 and 40 kJ m^{-3} at two field settings [92]. The simulations indicated that magnetite nanoparticles of prolate ellipsoidal shape, volume of 3922 ± 35 nm^3 and aspect ratio of 1.56 (a=b=8.4 nm and c=13.2 nm) which yields an effective anisotropy of 20 kJ m^{-3}, constituted the optimum design at H_0 = 18 kA m^{-1} and f = 100 kHz, with a SAR value of 342.0 ± 2.7 W g^{-1}, while NPs of 4147 ± 36 nm^3, aspect ratio of 1.29, and effective anisotropy 20 kJ/m^3 were the most advantageous for a low-field amplitude of H_0 = 4 kA m^{-1} and f = 100 kHz and displayed SAR values of 50.2 ± 0.5 W g^{-1}. The average concentrations of 3.86 ± 0.10 and 0.57 ± 0.01 mg cm^{-3} used at 4 and 18 kA m^{-1}, respectively, were enough to achieve therapeutic temperatures of 42–44°C. Clearly, in this example prolate ellipsoidal NPs are proposed to improve the SAR by increasing the NP anisotropy constant to values larger than that of bulk magnetite 9 kJ m^{-3} [93]. Bulk anisotropy of maghemite is 2 kJ m^{-3} [94].

Materials Research Forum LLC
https://doi.org/10.21741/9781644902332-4

As noted, magnetic colloids are very complex systems, which at higher or lower extents and depending on the used synthesis protocols, display polydispersity and aggregation, and for which dipolar interactions are significant in the useful concentration ranges. The properties that allow to predict the heating efficiency of a given colloidal suspension at a field setting are $<V>$, M_s and the mean activation energy $< U >$ that averages contributions from anisotropy and dipolar interactions [95].

The SAR dependencies with the already mentioned NP physical and magnetic properties have been experimentally tested over the last decades. Unfortunately, data acquisition has been carried out over a wide range of frequencies and field amplitudes, under various experimental conditions and using distinct data analysis methods, which altogether drives to a lack of normalization. To facilitate such comparison, the intrinsic loss parameter has been defined as $ILP = \frac{SAR}{fH_0^2}$ in nH m^2 kg^{-1} units, which is only valid for special cases when $\chi''(f)$ is constant[96]. For further MH experiments, it is strongly advised to analyze challenges and recommendations for MH measurements derived from an interlaboratory study that involved the measurement of identical samples of two stable nanoparticle systems at 21 European sites. Collected SAR data in different experimental facilities was analyzed to assess reproducibility and to identify possible sources of errors [97].

A comprehensive overview has been recently published about already analyzed NPs and their assemblies (produced via different synthetic routes), focusing on the NP properties that have allowed obtaining the highest heating efficiency and highlighting nanoplatforms that prevent magnetic heat loss in the intracellular environment [98]. This review includes, for iron oxide compositions (magnetite and maghemite), the single core NPs, the multicore NPs and assemblies mediated by polymers. In our perspective, among single core NPs, the cubic shape is a good option due to surface anisotropy and because the cubic shape eases the formation of NPs chains [99]. Among the multicore, those aligned following a single crystallographic direction, often named nanoflowers (NFs), have good prospects[100] as well as polymer-mediated assemblies like, for instance, those of cubic NPs [101]. The three most efficient systems mentioned above share therefore the property that the nanoparticles arrange in configurations with anisotropy axes aligned, thereby enabling the formation of nanoclusters that overcome single domain critical volume but still preserve single domain properties.

So far, we have analyzed in depth the dynamics behind the magnetic heat losses of magnetic NPs, pointed out the various contributions to relaxation time and provided a condition for maximum heating at a given frequency (see Eq. 7). However, in spite of nanomaterial optimization, in vitro experiments have shown drastic falls in the heating efficiency of the NPs in cellular environments relative to colloidal suspensions [102,103]. The comparison of SAR measurements of iron oxide NPs between "colloidally dispersed" and "inside live cells" conditions pointed out that the increase in intracellular NPs' clustering, favoring magnetic dipolar interactions, is the major contribution to the intracellular variation of the studied NPs' magnetic field response rather than the nanoparticle immobilization [103].

Altogether, we consider that the actual challenges of nanoparticle design for MH treatments include surface modification and bioconjugation to avoid aggregation and opsonization and to improve NP distribution inside the tumor, as well as the improvement of the nanosystem properties to achieve higher heat efficiency at biomedical field settings with lower dosages compared to those currently used that are extremely high.

2.4   Specific absorption rate measurements

The heating efficiency of a given type of NP, typified by its SAR values, is usually characterized in suspensions, in physiologically relevant liquids (as a ferrofluids), in jelly substances such as hydrogels, gelatin, glycerol [104–106], and agarose [107] which mimic tissues (ferrogels) and less frequently in ex vivo tumor tissues [108]. As the NPs act as energy transducers, two main kinds of measurements are possible. The electromagnetic measurements record the energy the NPs take from the field while the calorimetric measurements report the energy that heats the medium. The first consists in the inductive measurement of the sample magnetic loop when the field is swept as $H(t) = H_0 \cos(\omega t)$ and allows to calculate $SAR = \delta^{-1} f\, A_{loop}$, being the loop area $A_{loop} = \oint \mu_0\, M(H)\, dH$, where M is the magnetization and the integration is done over a period $T = 2\pi/f$. To this end, AC magnetometers based on Faraday's law of induction have been built to measure dynamic hysteresis loops of nanoparticle dispersions. For instance, Garaio et al. built a magnetometer that works at 24 kA m^{-1} in a frequency range between 75 kHz and 1 MHz [109], V. Connord et al. one that works at 80 mT (64 kA m^{-1}) in a frequency range 6 – 56 kHz[110], Bruvera et al. one that operates at various frequencies, 40 – 268 kHz, and at various fields, 5 – 56 kA m^{-1} [111].

In the calorimetric measurement, the sample temperature is recorded as it rises to T_s when a stationary state is reached. Then, the SAR, which is defined as the dissipated power divided by m (the mass of the NPs exposed to the field), is calculated from the heating curve as:

$$SAR = \frac{C\, m_s}{m}\frac{\partial T}{\partial t} \tag{8}$$

where C and m_s are the specific heat capacity (J K^{-1} kg^{-1}) and the mass of the supporting substance (liquid, jelly or tissue), respectively. Mostly in the literature, NPs have been characterized in the ferrofluid state in water. In that case, Eq. 8 can be written as $SAR = \frac{C'}{[x]}\frac{\partial T}{\partial t}$, using C' as the volumetric heat capacity of water (4184 kJ K^{-1}m^{-3}) and $[x]$ the NPs' concentration in the suspension. The temperature rise can be described by the phenomenological Box–Lucas equation: $T = T_s\left(1 - e^{-\beta t}\right)$. The fitting of the whole curve with this function allows to estimate $\frac{\partial T}{\partial t} \cong \beta T_s$ for SAR calculation with Eq. 8. Instead, some authors use the initial slope while others employ the maximum slope, complicating the comparisons among data obtained in different laboratories.

Alternatively, measurements of the complex magnetic susceptibility over the frequency range 500 kHz – 2 MHz have also been used to evaluate the time dependency of temperature increase [112].

Different RF field applicators have been described: a resonant network with a coil made of a hollow-copper-pipe refrigerated with water [113–115], a coil made with Litz wire with a ferromagnetic core achieving a uniform field in a reduced volume [116,117], and one that has been designed to work as a portable device for in vivo application in small animals, replacing water refrigeration by forced air refrigeration [118]. There is only one applicator described with capacity and characteristics for heat treatment in deep regions of the human body, named MFH®300F [119]. This medical device, which works at 100 kHz with field amplitude up to 18 kA m^{-1} in a cylindrical treatment area of 20 cm diameter and aperture height up to 300 mm, is in use for clinical trials.

3. The interactions of iron oxide nanoparticles with cells

Cell-nanomaterial interactions depend on the intrinsic properties of the nano-objects as well as on their altered characteristics resulting upon their interactions with biological species in the body. As noted, NPs are more surface reactive, less thermodynamically stable, and have higher biological reactivity, altogether increasing the likelihood of unwanted pro-oxidant and pro-inflammatory, but also antioxidant reactions [120]. The mechanisms usually described in the literature to account for nanoparticle-related cytotoxicity are oxidative stress (reactive oxygen and nitrogen species), cell membrane damage, inflammation, genotoxicity, autophagy dysfunction, activation of the immune system response, lactate dehydrogenase release, morphological changes in cell or cell organelles, inhibition of cell growth and cell death, among others [121,122]. A brief description of the most relevant mechanisms follows. For a more comprehensive view, the reader is conveyed to other references [123,124]. Besides those nanoparticle-related cytotoxicity mechanisms, cell-nanomaterial interactions also include i) the internalization of those nano-objects by cells through multiple endocytic pathways, ii) the unspecific adsorption of NPs to the cell membrane that can cause membrane structural changes, limiting nutrients and ions intake or can block membrane ducts [125–127] and iii) the cascade of events that is set in place once bioconjugated NPs, functionalized with antibodies or other biomolecules, interact with cellular receptors [128]. These cell-nanomaterial interactions may induce levels of cellular stress that can be counterbalanced by cells or not. If not, cytotoxicity appears.

3.1 The most common nanoparticle-related cytotoxicity mechanisms

3.1.1 Reactive oxygen species

Reactive oxygen species (ROS) are free radicals and reactive molecules derived from the sequential reduction of molecular oxygen (like super oxide radical O_2^-, hydrogen peroxide H_2O_2, or the hydroxyl free radical OH^*) that are produced following cellular respiration and mitochondrial function or by oxidoreductase enzymes. They can also be produced by metal-catalyzed oxidation. These molecules are continuously produced during normal cell metabolism and are normally counter-balanced by antioxidant enzymes and other redox (reduction-oxidation) reactions. An imbalance towards the pro-oxidative state is often named "oxidative stress". Glutathione is one relevant intracellular molecule of small size and non-enzymatic nature that mitigates the deleterious effects of reactive oxygen species in the cell. The generation of ROS and the consequent oxidative stress are usual causes of NP toxicity. Most of the metal-based (Fe, Si, Cu, Cr, Va) NPs induce oxidative stress following Fenton-type reactions [129–131]. Levels of oxidative stress that are not tolerated by the cells result in membrane damage and mitochondrial dysfunction, finally leading to cell death. SPIONs and iron oxide NPs have been linked to four main sources of oxidative stress: ROS generation directly from the NP surface, ROS production due to leaching of iron ions from the NP surface due to enzymatic action, alteration of mitochondrial and other organelles' functions, and activation of inflammatory cells [28,132,133].

3.1.2 Inflammation

Inflammation is the process involved in the repair of damaged tissue and in the defense of the organism against pathogens and external foreigner objects and stressors. When NPs are in the tissues, their interaction with tissue-resident macrophages polarizes the first towards distinct phenotypes, either anti- or pro-inflammatory depending on the NP type [134–136]. This polarization may result in a disruption of the normal levels of pro- and anti-inflammatory cytokines produced by macrophages, a fact that has been associated with various inflammatory disorders. Indeed, the exposure to nanomaterials has been frequently associated to inflammation, being certain NP attributes, like the size, very relevant in the process [137–139].

3.1.3 Cell membrane damage

Nanoparticles participate in cell and organelle membrane damage in various ways, typically depending on nanoparticle properties, such as size, surface charge or degree of hydrophilicity [140–142]. For example, it is well known that cationic NPs can increase membrane permeability through nanoscale holes opening [142]. Also, since the smallest NPs can be rapidly cleared through the kidneys, and the larger ones are engulfed by cells ending into the reticuloendothelial system (RES) organs, to be used as nanomedicines for cancer therapy and to achieve good results, NPs should be in an appropriate size range, typically from 5 to 100 nm [133]. Those NPs that can pass through the cell membrane, either by endocytosis or by direct penetration [141], may produce an irreversible cell damage that strongly depends on NP concentration and properties [143]. Whenever cell damage occurs, the intracellular enzyme lactate dehydrogenase (LDH) leaks from the cells and towards the medium, where its presence can be detected by means of an enzymatic reaction [144].

3.1.4 Genotoxicity

Oxidation of critical cell biomolecules, such as nuclear DNA and membrane lipids, occurs upon alterations of the normal cell redox equilibrium. If left uncorrected, DNA damage can lead to chromosomal aberrations, apoptosis, carcinogenesis, gene mutations or cellular senescence [145]. DNA damage was described upon exposure to various types of NPs, including iron oxides [146–150].

3.2 Nanomaterials properties affecting cellular response

Several nanomaterial properties are critical to determine the potential of that nanomaterial to induce unwanted cellular responses, cellular stress, and even cytotoxicity. It is generally accepted that the interactions of NPs with biological elements (e.g., proteins, opsonins), the cell-nanomaterial interactions and cellular uptake of the nano-object, as well as the in vivo fate and toxicity of nanoparticles are strongly correlated with their physico-chemical characteristics. Nevertheless, once opsonization has occurred, the nanomaterial is transformed into a new entity, whose interactions with cells, tissues and biomolecules will be distinct from that of the bare nano-object. Some relevant NP properties include morphology, composition, degree of crystallinity, shape, size, polydispersity, aspect ratio, surface area, charge and functionalization, state of aggregation, aging in biological media, and the potential to generate ROS [21,28,151]. Particularly, Feng et al. studied the influence of SPIONs size and coating on cell damage. They used commercially available SPIONs of 10 nm and 30 nm coated with polyethylene glycol (PEG) or polyethyleneimine (PEI). The authors found that PEI-coated SPIONs had increased cellular uptake, higher clearance, larger cytotoxicity, and a poorer distribution in the tumor when compared with PEG coated ones [133]. Also, regarding size influence, 10 nm PEG-coated NPs showed proportionally higher cellular uptake and tumor accumulation than PEG-coated NPs of 30 nm [133]. Furthermore, Malvindi et al. published an article comparing the in vitro cytotoxicity of silica shell-coated SPIONs (Fe_3O_4/SiO_2 NPs) versus uncoated SPIONs having a size of ~26 nm on A549 and HeLa cells [152]. Their results showed that surface passivation of SPIONs can diminish oxidative stress and the alteration of iron homeostasis due to the lower in situ degradation of these systems.

Many reviews comprehensively discuss these NP properties and their relationship with toxicological responses [22,28,30,153,154]. For that reason, this is not going to be covered here. Nevertheless, it can be stated that the critical consequences of designing and surface-modifying SPIONs for biological applications have to be judiciously considered so as to increase their biocompatibility, prevent their unwanted and fast clearance, and improve their targeted delivery to the disease site.

In the next section, we will further discuss some of the most relevant cellular responses elicited upon cell interactions with the MNPs and the magnetic field present in hyperthermia treatments. The heat shock response and their major instruments, the heat shock proteins, will be discussed. ROS generation by the MNPs will be associated to a form of cell death named ferroptosis and to immune system activation, though other hyperthermia-induced mechanisms will also be linked to the immune response. Finally, the resistance to further heating, or thermotolerance, will be introduced and briefly referred to.

4. The cellular response to magnetic hyperthermia

As a result of cell-iron oxide nanoparticle interactions, but also as a consequence of MH treatment application, a cascade of cellular events is launched that may result in cell death. This is what one could call the "wanted toxicity", or the toxicity associated to the concomitant use of magnetic NPs and the AMF in a MH treatment that results in localized cell death without major unwanted side-effects. In the following subsections, a brief overview of the most relevant cellular responses observed upon MH application will be provided together with the description of several mechanisms that could be ascribed as the "wanted toxicity" and of others that could incline the balance in the opposite direction, as schematically depicted in Fig. 1.

4.1 Magnetic hyperthermia and the heat shock response

During and following exposure to elevated temperatures in MH applications, cells experience numerous changes in their normal physiology [155] that can go from inhibition of protein and nucleic acid synthesis to drastic alterations in the cytoskeleton and in cell morphology if cytotoxic temperatures are achieved [156]. Normally, the estimated temperature for irreversible cell damage is above 43°C, and cells can die by apoptosis or necrosis depending on the applied protocol and on the cell type [157].

With regards to the question of which is the best cell death pathway in MH applications, there exists a conflict between two postures: some authors claim that only necrosis, associated with ablation, can exterminate cancer cells [158], while others highlight that extensive necrosis can be dangerous for healthy tissues and reduce tissue permeability to drugs and thus apoptosis must be the sought for alternative [159]. On the other hand, unbalanced apoptosis plays a crucial role in many diseases, including degenerative disorders and many cancer types. Indeed, it has been the most targeted mechanism by anti-cancer drugs [160]. However, by augmenting the expression of anti-apoptotic proteins, cancer cells can develop resistance to apoptosis leading to tumor survival and propagation [161].

Therefore, the success of a MH treatment depends on the level of damage imposed to key cellular biomolecules, like proteins or DNA, and key cellular processes, like glycolysis. During heat shock, protein denaturation is the major process accounting for cell death [146]. However, its effects can be, up to a certain extent, counteracted by a family of proteins called heat shock proteins (HSPs), a group of chaperone proteins that bind and refold aberrant peptides, thereby diminishing stress damage and preventing cell death [164]. These proteins are overexpressed under several forms of stress, such as thermal, energy depletion, pH changes, etc. Apart from their chaperone function, it was established that HSPs also participate in many cellular mechanisms, such as apoptosis or cell signaling, thereby constituting a very complex protein family that can be associated with different processes. Furthermore, in cancer cells, the expression of HSPs differs from the one found in normal cells. HSPs are normally overexpressed in tumors and that's a reason why they are thought to act as tumor helpers.

HSPs are therefore a family of key proteins involved in the fight against the consequences of temperature elevations in the cell. As it will be more extensively discussed in the following subsections, HSPs are also involved in the establishment of thermal resistance to further temperature elevations, a phenomenon named thermotolerance. In practical terms, it could be stated that temperature elevations in MH treatments should be sufficient to exceed the response of HSPs while avoiding triggering uncontrolled necrosis, which would affect not only the target organs but their surroundings.

4.2 Reactive Oxygen Species as an ally

As already introduced in section 3.1.1, upon iron oxide NP internalization by cells, intracellular oxidative species are formed that may then lead to lipid and protein peroxidation and, if uncorrected, to cell death. Perillo et al. [163] presented ROS as able to have favorable and unfavorable results in cancer cells since cancer cells can survive and adapt to moderate ROS-inducing oncogenes and, at the same time, are able to inhibit tumor suppressor genes like bcl-2. However, this survival drastically depends on the availability of antioxidant molecules, such as glutathione (GSH). Hence, a promising strategy against cancer could imply the blocking of its mechanism of action or, alternatively, the induction of its depletion [164]. Besides, ROS stimulation could lead to apoptosis via the caspase pathway. Therefore, it can be envisioned that the combined action of MH and GSH depletion would deserve further exploration [165].

4.3 Ferroptosis: an iron dependent type of cell death

Ferroptosis is a cell death mechanism produced by peroxidation and associated to the presence of endogenous iron and of intracellular lipid reactive oxygen species (L-ROS) [166]. Unlike apoptosis, ferroptosis is a mechanism independent from the caspase route but dependent on ROS buildup mediated by iron ions and regulated by glutathione peroxidase 4 (GPX4), an enzyme that protects from lipid oxidation [167]. Since all these molecules are required for the normal functioning of cells, ferroptosis is the consequence of the disproportioned production of L-ROS and of insufficient GPX4 activity [168].

When MH treatments based on iron oxide NPs are applied, ferric or ferrous ions are liberated after NP degradation due to the acidic pH of the tumor environment or in lysosomes. There, they can activate the Fenton reaction in which abundant oxygen peroxide reacts with iron ions and produces hydroxyl radicals. The mechanism of the Fenton reaction is shown in Equations 9 and 10.

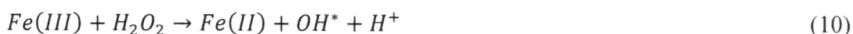

$$Fe(II) + H_2O_2 \rightarrow Fe(III) + OH^- + OH^* \tag{9}$$

$$Fe(III) + H_2O_2 \rightarrow Fe(II) + OH^* + H^+ \tag{10}$$

Therefore, many strategies to induce ferroptosis with iron oxide NPs have been studied. Some authors studied the anticancer effect of ferumoxytol [169], an iron oxide-based drug approved by the FDA to treat anemia, which can induce ferroptosis on cells. Noteworthy, recent studies have compared several iron nanoparticles and found they induced ferroptosis in in vivo experiments, either with the NPs alone or combined with drugs [170–173]. Also, magnetic NPs have been combined to several drugs, such as the anticancer drug Sorafenib; the anti-inflammatory compound Sulfasalazine; and anti-malarial drugs such as Artemisinin and its derivatives, including Artesunate and Dihydroartemisinin [174]. All these drugs induce ROS by promoting ferritin degradation that leads to an increase in free cellular iron.

In spite of the reported anticancer effects associated with the stimulation of ferroptosis, it has been found to also play a role in many diseases, including tumors, kidney injury, heart disease, etc. In this context, it appears important to understand the molecular mechanisms set in place by the cells during ferroptosis to find strategies that will enable the limitation of ferroptosis-related diseases and the fostering of ferroptosis-associated tumor cell death in MH applications [175,176].

4.4   The extra factor: the immune response

At the beginning of this section, we defined the heat shock response as a mechanism by which MH may induce a pro-tumor action through the transcription upregulation of highly conserved HSPs genes, which, at the end, protect cells from heat-induced apoptosis [177,178]. Many decades ago, though, HSPs were associated with specific tumor immunogenicity in mice by Srivastava and coworkers [179], and later, HSP-protein complexes derived from the own primary tumor of patients were used as vaccines in patients for immunotherapy [180]. In 2006, Calderwood et al. investigated the origin of those extracellular peptide–HSP complexes and proposed a mechanism of action for their immunogenic activity (Figure 2) [181]. The HSPs can be located on the cell surface or be released from tumor cells through exocytosis or upon cell lysis. Those HSPs are usually bound to tumor-borne proteins [182,183]. This peptide-HSP complex serves as a tumor antigen (Ag). Dendritic cells (DCs) and antigen presenting cells (APCs) present in the tumor milieu are able to recognize and process those tumor Ags and to finally activate T-lymphocytes against the tumor cells by an adaptive immune response.

Additionally, it has been proposed that heat-damaged cells can expose damage-associated molecular patterns (DAMPs) that will be recognized by specific receptors in natural killer cells (NK) and macrophages [184,185]. Activated NK cells can thereafter recognize tumor cells and trigger an innate immune response.

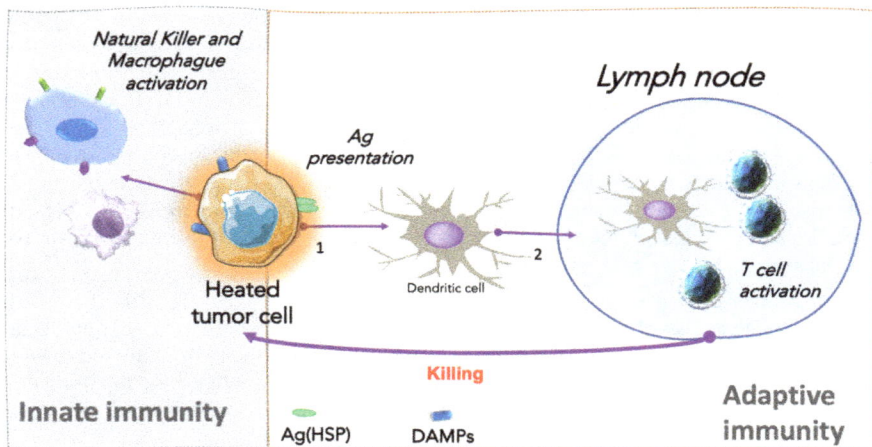

Fig. 2. Schematic representation of MH-driven immunostimulation through HSP-tumor protein complexes. Both adaptive and innate immune responses can be activated when tumor cells express HSP complexes on their surface or release them by tumor exocytosis or upon cell lysis after a heat shock. Here, the cell surface-bound HSPs are represented but the concept is equally valid for released HSP complexes. Ag: Antigen. DAMPs: Damage Associated Molecular Patterns.

Besides these events, others resulting from MH can favor immunity, as the permeability increase of the tumor vasculature and the oxygenation, which can enhance the translation of T cells and DCs from the lymph nodes into tumor locations [186]. In fact, it has been proposed that MH enhances the recruitment of immune cells by other effects, like overexpressed vascular adhesion molecules, incremented expression of major histocompatibility complex (MHC) class I and II antigens, and the up-regulation of co-stimulatory molecules (e.g., CD80, CD86, CD40) on APCs with the following activation of T cells [187,188]. Notice that for immune response activation, MH treatment temperatures must be lower than 60°C, otherwise higher temperatures will cause extensive protein denaturation and coagulation, leading rapidly to coagulative necrosis, which destroys tumor antigens and endothelial cells, cutting off the desired immune response [188].

Thus, MH constitutes a more complex phenomena than just heating up the cells, since many processes occur simultaneously after the application of the NPs and the magnetic field. A deeper understanding of these phenomena may help tune the magnetic NPs and the MH application conditions to achieve personalized treatments according to the tumor characteristics and the requirements to eliminate it, with the possibility to add some extra enhancers through specific immune response activation.

4.5   The dark side of hyperthermia

As it is known, everything is not rosy when fighting against the complex environment of a tumor. When MH is applied and temperatures are kept below 43 °C, the heat stress can result in a protective cellular response, well known as heat stress response, which restrains the effects of hyperthermia and transitorily makes cells less sensitive or even insensitive (i.e., thermotolerant) to the next thermal treatment [189]. According to Glory et al., this thermotolerance can be stimulated by short exposures to HIGH temperatures (42–45 °C for 30 min) or by continuous heating at temperatures between 39.5–41.5 °C for 3–24 h. The main protagonists in this resistance are the HSPs due to their chaperone function. Also, different tissues possess different thermotolerance characteristics [190]. Thus, it is hard to predict if thermotolerance will develop in the treated tumors and how it will affect clinical outcomes. Moreover, if a targeted tumor is non-homogeneously heated, as it occurs with MH due to the irregular distribution of the NPs inside the tumor, thermotolerance develops with different kinetics, depending on the specific areas of the tumor. These variances will then result in diverse sensitivities of tumor cells to a subsequent hyperthermia treatment [191].

To counteract this effect, the most studied strategy is the inhibition of the heat stress response, either by genetic approaches, as HSP genes' disruption [192,193], or by the pharmacological inhibition of various HSPs by different compounds as quercetin[194], Ganetespib [195], or combinations [196]. In pre-clinical studies, these drugs achieved thermotolerance disruption, lower thermal dosages, and enhanced treatment efficacy [197,198].

4.6   The strategies to win

For a long time, scientists from all over the world have devoted efforts to study different synergistic combinations of MH treatment with other therapies, such as chemotherapy radiotherapy, immunotherapy, plasmonic photothermal therapy (PPT), photodynamic therapy (PDT), and gene/cell therapy in order to improve MH treatment outcomes. It is commonly accepted that hyperthermia can selectively kill tumor cells under hypoxic and acidic conditions, since these cells are much more sensitive to heat than those in well-oxygenated environment [157,198,199]. Furthermore, at temperatures between 41 and 43 °C membrane permeability is increased by hyperthermia (thermal sensitization) and DNA damage repairing is reduced, enhancing the efficacy of anti-tumoral drugs. When temperatures between 39 and 43 °C are maintained for 30–60 min, perfusion is increased for the following hours in both, well and poorly-vascularized regions of solid tumors, leading to regularization of oxygen, nutrient, and pH levels [200]. Thus, augmented tumor perfusion and oxygenation are thought to be the principal mechanisms by which mild MH treatment boost chemo and radiotherapeutic treatments [200]. One of the key advantages of combining MH treatment with, for instance, anti-cancer drugs lies in the significant reduction of the amount of drug to be administered and therefore a lowering of its negative secondary effects. Chang et al. summarize in a review many of possible combinations between strategies to eliminate tumor and prevent metastases [201].

In a recent review, Liu *et al.* [202] presented a descriptive analysis of MH treatment principles and the latest information about developments in engineering magnetic NPs' size, composition, shape, and surface to improve their SAR, the study of local induction heat effects on selectively disrupting cells/intracellular structures, and some strategies designed to improve the therapeutics by the combination of MH therapy with chemotherapy, radiotherapy, immunotherapy, PTT, PDT, and gene therapy (Fig. 3).

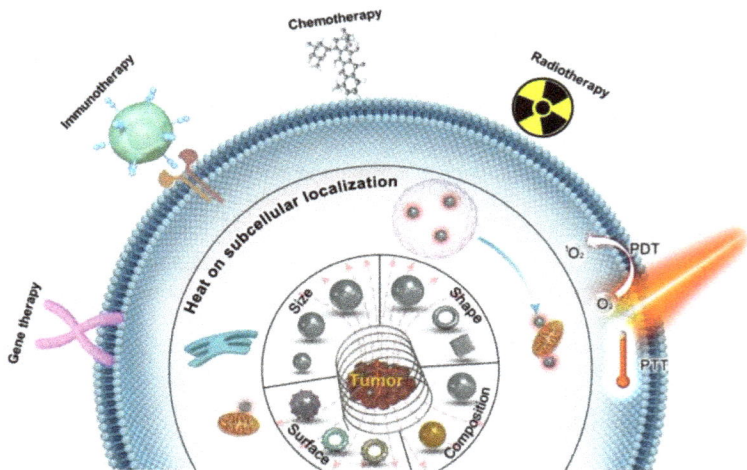

Fig. 3. Representation of the strategies for improving antitumor therapeutic efficacy of MH treatments (center) by optimization of the nanoparticles' properties such as morphology, composition, size, and surface modifications to achieve an effective heating, (inner ring) by a better understanding of the effects of l localized heating in the disruption of cellular/subcellular structures, which increases the effectiveness of MH as antitumor therapy, and (outer ring) by synergistic combination of magnetic NPs and MH with other treatments, such as radiotherapy, immunotherapy, chemotherapy, photothermal therapy (PTT)/photodynamic therapy (PDT), and gene therapy, allowing further improvements of the antitumor efficiency. Reproduced from Liu et al, Theranostics 2020.

All these reports contribute to generate a vast source of significant information regarding different treatment combinations and the results obtained. However, it is important to note that among the many anticancer agents proposed and under study, just a few are ongoing clinical trials.

Conclusions

This chapter begins with a thorough overview of the physical mechanisms associated to Magnetic Hyperthermia, an oncologic therapy under clinical translation that uses iron oxide nanoparticles (NPs) and radiofrequency fields to challenge solid tumors with selective heating. It continues with a general description of the cellular responses observed upon interaction of cells with magnetic NPs and with the magnetic field itself and ends with a summary of the most relevant mechanisms set in place by cells to counteract the stress associated with this technique.

With regards to the physical aspects, the relaxation mechanisms of the NP magnetic moment and their dependence with the nanosystem properties are first outlined, highlighting the role of dipolar interactions. In this respect, we provide a close expression for the relaxation time dependences (Eq. 7) whose solution would point to the best iron oxide nanostructure for MH. The relevance of improving certain NP properties, of controlling the interactions among nanoparticles and their extent and type of aggregation, as well as their irregular distribution in the tumor environment to lower the dosage required to achieve therapeutic temperatures are duly emphasized. In fact, the extremely large amount of magnetic material that must be injected to reach the desired temperatures at the target site is one of the main limitations of the procedure. A final subsection is also dedicated to the instrumentation and to the methodologies used for the characterization of the NP's efficiency as energy transducers, pointing out to the need for methodological standardization.

In the following subsections, the most acknowledged cell-nanoparticle interactions are briefly described focusing on the specificities of iron oxide NPs. The influence of the magnetic field and the cell response mechanisms activated under MH protocols are outlined, highlighting the fact that the magnetic hyperthermia procedure affects cells and organs beyond heating, triggering other molecular events that can either damage the tumor, e.g. heat-induced apoptosis, or incline the balance towards the other side. Iron oxide NPs can dissolve into labile Fe ions that catalyze the Fenton reaction leading to ROS production and to a potential redox imbalance at cell level. This can result in a form of cell death named ferroptosis. Furthermore, heat is behind the induction of the heat shock response, characterized by the expression of HSPs, a family of proteins that can counteract cell damage due to heat, promoting cell survival and tumor resistance, also through the thermotolerance process. Conversely, HSPs have also been associated with specific tumor immunogenicity and have been used as vaccines to trigger an immune response against malignant cells.

Altogether, a better understanding of NP interactions with the applied magnetic field and among themselves, of the effects of NP distribution and their heating capacity within the tumor environment, and of the cellular events associated to this therapeutic methodology may help fostering its translation to the clinics and, in particular, to personalized treatments linked to the specific tumor characteristics and the requirements to eliminate it. We envision that identifying ways to modify the NPs to control their distribution in the tumor and to twist the balance at cellular level, diminishing or eliminating pro-tumor events while favoring those that act synergistically with the heating will be a major area of research in the immediate future.

As future recommendation in the field, the NP properties could be enhanced to reduce the effective dose required for thermal action under the rather restrictive conditions imposed by the low field amplitude, low frequency magnetic field characteristics employed in the clinics. Furthermore, achieving a more uniform NP distribution inside the tumor tissue would be much desirable as well as exploring and exploiting potential synergies between MH and other anti-tumor therapies. Last but not least, unified, standardized protocols both, for lab and clinical translation trials would be very advantageous

Acknowledgements

The authors wish to thank CONICET and Universidad Nacional de La Plata (UNLP) for funding (PIP 0897 and UNLP x807, respectively). All authors are members of CONICET, Argentina".

References

[1] M. C. Garnett and P. Kallinteri, Occupational Medicine 56, 307 (2006). https://doi.org/10.1093/occmed/kql052

[2] Q. A. Pankhurst, J. Connolly, S. K. Jones, and J. Dobson, J. Phys. D: Appl. Phys 36, 167 (2003). https://doi.org/10.1088/0022-3727/36/13/201

[3] R. Banerjee, Y. Katsenovich, L. Lagos, M. Mciintosh, X. Zhang, and C.-Z. Li, Current Medicinal Chemistry 17, 3120 (2010). https://doi.org/10.2174/092986710791959765

[4] K. Wu, D. Su, J. Liu, R. Saha, and J. P. Wang, Nanotechnology 30, 502003 (2019). https://doi.org/10.1088/1361-6528/ab4241

[5] H. Kang, S. Hu, M. H. Cho, S. H. Hong, Y. Choi, and H. S. Choi, Nano Today 23, 59 (2018). https://doi.org/10.1016/j.nantod.2018.11.001

[6] I. Rubia-Rodríguez, A. Santana-Otero, S. Spassov, E. Tombácz, C. Johansson, P. de La Presa, F. J. Teran, M. del P. Morales, S. Veintemillas-Verdaguer, N. T. K. Thanh, M. O. Besenhard, C. Wilhelm, F. Gazeau, Q. Harmer, E. Mayes, B. B. Manshian, S. J. Soenen, Y. Gu, Á. Millán, E. K. Efthimiadou, J. Gaudet, P. Goodwill, J. Mansfield, U. Steinhoff, J. Wells, F. Wiekhorst, and D. Ortega, Materials 14, 706 (2021). https://doi.org/10.3390/ma14040706

[7] Z. Zhou, Z. Shen, and X. Chen, ACS Nano 14, 7 (2020). https://doi.org/10.1021/acsnano.9b06842

[8] P. A. Soto, M. Vence, G. M. Piñero, D. F. Coral, V. Usach, D. Muraca, A. Cueto, A. Roig, M. B. F. van Raap, and C. P. Setton-Avruj, Acta Biomaterialia 130, 234 (2021). https://doi.org/10.1016/j.actbio.2021.05.050

[9] M. E. de Sousa, M. B. Fernández van Raap, P. C. Rivas, P. Mendoza Zélis, P. Girardin, G. A. Pasquevich, J. L. Alessandrini, D. Muraca, and F. H. Sánchez, The Journal of Physical Chemistry C 117, 5436 (2013). https://doi.org/10.1021/jp311556b

[10] W. Cai and J. Wan, Journal of Colloid and Interface Science 305, 366 (2007). https://doi.org/10.1016/j.jcis.2006.10.023

[11] J. G. Ovejero, F. Spizzo, M. P. Morales, and L. del Bianco, Materials 14, 6416 (2021). https://doi.org/10.3390/ma14216416

[12] G. Cotin, C. Kiefer, F. Perton, D. Ihiawakrim, C. Blanco-Andujar, S. Moldovan, C. Lefevre, O. Ersen, B. Pichon, D. Mertz, and S. Bégin-Colin, Nanomaterials 8, 881 (2018). https://doi.org/10.3390/nano8110881

[13] P. M. Zélis, G. A. Pasquevich, S. J. Stewart, M. B. F. van Raap, J. Aphesteguy, I. J. Bruvera, C. Laborde, B. Pianciola, S. Jacobo, and F. H. Sánchez, Journal of Physics D: Applied Physics 46, 125006 (2013). https://doi.org/10.1088/0022-3727/46/12/125006

[14] A. Makridis, K. Topouridou, M. Tziomaki, D. Sakellari, K. Simeonidis, M. Angelakeris, M. P. Yavropoulou, J. G. Yovos, and O. Kalogirou, J. Mater. Chem. B 2, 8390 (2014). https://doi.org/10.1039/C4TB01017E

[15] Q. Lu, X. Dai, P. Zhang, X. Tan, Y. Zhong, C. Yao, M. Song, G. Song, Z. Zhang, G. Peng, Z. Guo, Y. Ge, K. Zhang, and Y. Li, International Journal of Nanomedicine Volume 13, 2491 (2018). https://doi.org/10.2147/IJN.S157935

[16] C. Caro, F. Gámez, P. Quaresma, J. M. Páez-Muñoz, A. Domínguez, J. R. Pearson, M. Pernía Leal, A. M. Beltrán, Y. Fernandez-Afonso, J. M. de la Fuente, R. Franco, E. Pereira, and M. L. García-Martín, Pharmaceutics 13, 416 (2021). https://doi.org/10.3390/pharmaceutics13030416

[17] K. Dukenbayev, I. Korolkov, D. Tishkevich, A. Kozlovskiy, S. Trukhanov, Y. Gorin, E. Shumskaya, E. Kaniukov, D. Vinnik, M. Zdorovets, M. Anisovich, A. Trukhanov, D. Tosi, and C. Molardi, Nanomaterials 9, 494 (2019). https://doi.org/10.3390/nano9040494

[18] V. Torresan, A. Guadagnini, D. Badocco, P. Pastore, G. A. Muñoz Medina, M. B. Fernàndez van Raap, I. Postuma, S. Bortolussi, M. Bekić, M. Čolić, M. Gerosa, A. Busato, P. Marzola, and V. Amendola, Advanced Healthcare Materials 10, 2001632 (2021). https://doi.org/10.1002/adhm.202001632

[19] M. Auffan, J. Rose, J.-Y. Bottero, G. v. Lowry, J.-P. Jolivet, and M. R. Wiesner, Nature Nanotechnology 4, 634 (2009). https://doi.org/10.1038/nnano.2009.242

[20] A. Pietroiusti, L. Campagnolo, and B. Fadeel, Small 9, 1557 (2013). https://doi.org/10.1002/smll.201201463

[21] N. Oh and J. H. Park, International Journal of Nanomedicine 9, 51 (2014). https://doi.org/10.2147/IJN.S26592

[22] A. A. Shvedova, V. E. Kagan, and B. Fadeel, Annual Review of Pharmacology and Toxicology 50, 63 (2010). https://doi.org/10.1146/annurev.pharmtox.010909.105819

[23] N. Oh and J. H. Park, International Journal of Nanomedicine 9, 51 (2014). https://doi.org/10.2147/IJN.S26592

[24] S. C. Sahu and A. W. Hayes, Toxicology Research and Application 1, 239784731772635 (2017). https://doi.org/10.1177/2397847317726352

[25] T. Xia, N. Li, and A. E. Nel, Annual Review of Public Health 30, 137 (2009). https://doi.org/10.1146/annurev.publhealth.031308.100155

[26] G. Oberdörster, Z. Sharp, V. Atudorei, A. Elder, R. Gelein, W. Kreyling, and C. Cox, Inhalation Toxicology 16, 437 (2004). https://doi.org/10.1080/08958370490439597

[27] V. Torresan, D. Forrer, A. Guadagnini, D. Badocco, P. Pastore, M. Casarin, A. Selloni, D. Coral, M. Ceolin, M. B. Fernández van Raap, A. Busato, P. Marzola, A. E. Spinelli, and V. Amendola, ACS Nano 14, 12840 (2020). https://doi.org/10.1021/acsnano.0c03614

[28] C. Buzea, I. I. Pacheco, and K. Robbie, Biointerphases 2, MR17 (2007). https://doi.org/10.1116/1.2815690

[29] V. de Matteis, Toxics 5, 29 (2017). https://doi.org/10.3390/toxics5040029

[30] H. Arami, A. Khandhar, D. Liggitt, and K. M. Krishnan, Chemical Society Reviews 44, 8576 (2015). https://doi.org/10.1039/C5CS00541H

[31] D. A. Hume, K. M. Irvine, and C. Pridans, Trends in Immunology 40, 98 (2019). https://doi.org/10.1016/j.it.2018.11.007

[32] A. Chow, B. D. Brown, and M. Merad, Nature Reviews Immunology 11, 788 (2011). https://doi.org/10.1038/nri3087

[33] W. Poon, Y.-N. Zhang, B. Ouyang, B. R. Kingston, J. L. Y. Wu, S. Wilhelm, and W. C. W. Chan, ACS Nano 13, 5785 (2019). https://doi.org/10.1021/acsnano.9b01383

[34] R. Gref, Y. Minamitake, M. T. Peracchia, V. Trubetskoy, V. Torchilin, and R. Langer, Science 263, 1600 (1994). https://doi.org/10.1126/science.8128245

[35] H. Soo Choi, W. Liu, P. Misra, E. Tanaka, J. P. Zimmer, B. Itty Ipe, M. G. Bawendi, and J. V. Frangioni, Nature Biotechnology 25, 1165 (2007). https://doi.org/10.1038/nbt1340

[36] H. H. Gustafson, D. Holt-Casper, D. W. Grainger, and H. Ghandehari, Nano Today 10, 487 (2015). https://doi.org/10.1016/j.nantod.2015.06.006

[37] D. E. Owens III and N. A. Peppas, International Journal of Pharmaceutics 307, 93 (2006). https://doi.org/10.1016/j.ijpharm.2005.10.010

[38] T. U. Wani, S. N. Raza, and N. A. Khan, Polymer Bulletin 77, 3865 (2020). https://doi.org/10.1007/s00289-019-02924-7

[39] P. Aggarwal, J. B. Hall, C. B. McLeland, M. A. Dobrovolskaia, and S. E. McNeil, Advanced Drug Delivery Reviews 61, 428 (2009). https://doi.org/10.1016/j.addr.2009.03.009

[40] S.-D. Li and L. Huang, Biochimica et Biophysica Acta (BBA) - Biomembranes 1788, 2259 (2009). https://doi.org/10.1016/j.bbamem.2009.06.022

[41] S. Guo and L. Huang, Journal of Nanomaterials 2011, 1 (2011). https://doi.org/10.1155/2011/742895

[42] K. Greish, In: S. Grobmyer, B. Moudgil (eds). Cancer Nanotechnology. Methods in Molecular Biology, vol. 624. Humana Press (2010). https://doi.org/10.1007/978-1-60761-609-2_3

[43] P. Foroozandeh and A. A. Aziz, Nanoscale Research Letters 13, 339 (2018). https://doi.org/10.1186/s11671-018-2728-6

[44] A. B. Engin, D. Nikitovic, M. Neagu, P. Henrich-Noack, A. O. Docea, M. I. Shtilman, K. Golokhvast, and A. M. Tsatsakis, Particle and Fibre Toxicology 14, 22 (2017). https://doi.org/10.1186/s12989-017-0199-z

[45] S. E. A. Gratton, P. A. Ropp, P. D. Pohlhaus, J. C. Luft, V. J. Madden, M. E. Napier, and J. M. DeSimone, Proceedings of the National Academy of Sciences 105, 11613 (2008). https://doi.org/10.1073/pnas.0801763105

[46] K. Maier-Hauff, F. Ulrich, D. Nestler, H. Niehoff, P. Wust, B. Thiesen, H. Orawa, V. Budach, and A. Jordan, J Neurooncol 103, 317 (2011). https://doi.org/10.1007/s11060-010-0389-0

[47] K. Mahmoudi, A. Bouras, D. Bozec, R. Ivkov, and C. Hadjipanayis, International Journal of Hyperthermia 34, 1316 (2018). https://doi.org/10.1080/02656736.2018.1430867

[48] M. Johannsen, B. Thiesen, P. Wust, A. Jordan, International J of Hyperthermia 26, 790 (2010). https://doi.org/10.3109/02656731003745740

[49] F. Brero, M. Albino, A. Antoccia, P. Arosio, A. Avolio, F. Berardinelli, D. Bettega, P. Calzolari, M. Ciocca, M. Corti, A. Facoetti, S. Gallo, F. Groppi, A. Guerrini, C. Innocenti, C. Lenardi, S. Locarno, S. Manenti, R. Marchesini, M. Mariani, F. Orsini, E. Pignoli, C. Sangregorio, I. Veronese, and A. Lascialfari, Nanomaterials 10, 1 (2020). https://doi.org/10.3390/nano10101919

[50] I. Hilger, International Journal of Hyperthermia 29, 828 (2013). https://doi.org/10.3109/02656736.2013.832815

[51] M. Tasso, F. Ghilini, M. Cathcarth, and A. S. Picco, In: S. K. Sharma (eds). Toxicity Assessment of Nanoferrites. Spinel Nanoferrites. Topics in Mining, Metallurgy and Materials Engineering. Springer, Cham. (2021). https://doi.org/10.1007/978-3-030-79960-1_9

[52] R. Vakili-Ghartavol, A. A. Momtazi-Borojeni, Z. Vakili-Ghartavol, H. T. Aiyelabegan, M. R. Jaafari, S. M. Rezayat, and S. Arbabi Bidgoli, Artificial Cells,

Nanomedicine, and Biotechnology 48, 443 (2020). https://doi.org/10.1080/21691401.2019.1709855

[53] S. Alarifi, D. Ali, S. Alkahtani, and M. S. Alhader, Biological Trace Element Research 159, 416 (2014). https://doi.org/10.1007/s12011-014-9972-0

[54] A. Hanini, A. Schmitt, K. Kacem, F. Chau, S. Ammar, J. Gavard, International Journal of Nanomedicine 2011:6, 787 (2011). https://doi.org/10.2147/IJN.S17574

[55] V. Valdiglesias, N. Fernández-Bertólez, G. Kiliç, C. Costa, S. Costa, S. Fraga, M. J. Bessa, E. Pásaro, J. P. Teixeira, and B. Laffon, Journal of Trace Elements in Medicine and Biology 38, 53 (2016). https://doi.org/10.1016/j.jtemb.2016.03.017

[56] J. Shaw, S. O. Raja, and A. K. Dasgupta, Cancer Nanotechnology 5, 2 (2014). https://doi.org/10.1186/s12645-014-0002-x

[57] B. Hajipour Verdom, P. Abdolmaleki, and M. Behmanesh, Scientific Reports 8, 990 (2018). https://doi.org/10.1038/s41598-018-19247-8

[58] X. Tian, D. Wang, M. Zha, X. Yang, X. Ji, L. Zhang, and X. Zhang, Electromagnetic Biology and Medicine 37, 114 (2018). https://doi.org/10.1080/15368378.2018.1458627

[59] M. E. Feder and G. E. Hofmann, Annual Review of Physiology 61, 243 (1999). https://doi.org/10.1146/annurev.physiol.61.1.243

[60] D. Przepiorka and P. K. Srivastava, Molecular Medicine Today 4, 478 (1998). https://doi.org/10.1016/S1357-4310(98)01345-8

[61] S. K. Calderwood, J. R. Theriault, and J. Gong, European Journal of Immunology 35, 2518 (2005). https://doi.org/10.1002/eji.200535002

[62] K. C. Kregel, Journal of Applied Physiology 92, 2177 (2002). https://doi.org/10.1152/japplphysiol.01267.2001

[63] V. Reichel, A. Kovács, M. Kumari, É. Bereczk-Tompa, E. Schneck, P. Diehle, M. Pósfai, A. M. Hirt, M. Duchamp, R. E. Dunin-Borkowski, and D. Faivre, Scientific Reports 7, 45484 (2017). https://doi.org/10.1038/srep45484

[64] Yu. L. Raikher and V. I. Stepanov, Journal of Experimental and Theoretical Physics 112, 173 (2011). https://doi.org/10.1134/S1063776110061160

[65] R. K. Gilchrist, R. Medal, W. D. Shorey, R. C. Hanselman, J. C. Parrott, and C. B. Taylor, Annals of Surgery 146, 596 (1957). https://doi.org/10.1097/00000658-195710000-00007

[66] A. Jordan, P. Wust, R. Scholz, B. Tesche, H. Fähling, T. Mitrovics, T. Vogl, J. Cervós-navarro, and R. Felix, International Journal of Hyperthermia 12, 705 (1996). https://doi.org/10.3109/02656739609027678

[67] D. Bobo, K. J. Robinson, J. Islam, K. J. Thurecht, and S. R. Corrie, Pharmaceutical Research 33, 2373 (2016). https://doi.org/10.1007/s11095-016-1958-5

[68] J. L. Dormann, L. Bessais, and D. Fiorani, Journal of Physics C: Solid State Physics 21, 2015 (1988). https://doi.org/10.1088/0022-3719/21/10/019

[69] D. F. Coral, P. Mendoza Zélis, M. E. de Sousa, D. Muraca, V. Lassalle, P. Nicolás, M. L. Ferreira, and M. B. Fernández van Raap, Journal of Applied Physics 115, 043907 (2014). https://doi.org/10.1063/1.4862647

[70] J. Dieckhoff, D. Eberbeck, M. Schilling, and F. Ludwig, Journal of Applied Physics 119, 043903 (2016). https://doi.org/10.1063/1.4940724

[71] M. I. Shliomis and V. I. Stepanov, In: W. Coffey (ed.). Advances in Chemical Physics. Theory of the Dynamic Susceptibility of Magnetic Fluids (1994). https://doi.org/10.1002/9780470141465.ch1

[72] Z. Li, D. Li, Y. Chen, and H. Cui, Soft Matter 14, 3858 (2018). https://doi.org/10.1039/C8SM00478A

[73] R. E. Rosensweig, Journal of Magnetism and Magnetic Materials 252, 370 (2002). https://doi.org/10.1016/S0304-8853(02)00706-0

[74] C. Tannous and J. Gieraltowski, European Journal of Physics 29, 475 (2008). https://doi.org/10.1088/0143-0807/29/3/008

[75] J. Carrey, B. Mehdaoui, and M. Respaud, Journal of Applied Physics 109, 083921 (2011). https://doi.org/10.1063/1.3551582

[76] M. Colombo, S. Carregal-Romero, M. F. Casula, L. Gutiérrez, M. P. Morales, I. B. Böhm, J. T. Heverhagen, D. Prosperi, and W. J. Parak, Chemical Society Reviews 41, 4306 (2012). https://doi.org/10.1039/c2cs15337h

[77] W. J. Atkinson, I. A. Brezovich, and D. P. Chakraborty, IEEE Transactions on Biomedical Engineering BME-31, 70 (1984). https://doi.org/10.1109/TBME.1984.325372

[78] R. Hergt and S. Dutz, Journal of Magnetism and Magnetic Materials 311, 187 (2007). https://doi.org/10.1016/j.jmmm.2006.10.1156

[79] I. A. Brezovich, In: B. Paliwal, F. W. Hetzel, and M. W. Dewhirst (eds.). Medical Physics Monograph No. 6: Biological, Physical, and Clinical Aspects of Hyperthermia. Medical Physics Publishing (1988).

[80] R. Hergt, S. Dutz, R. Müller, and M. Zeisberger, Journal of Physics: Condensed Matter 18, S2919 (2006). https://doi.org/10.1088/0953-8984/18/38/S26

[81] M. L. Etheridge and J. C. Bischof, Annals of Biomedical Engineering 41, 78 (2013). https://doi.org/10.1007/s10439-012-0633-1

[82] R. Yanes, O. Chubykalo-Fesenko, H. Kachkachi, D. A. Garanin, R. Evans, and R. W. Chantrell, Physical Review B 76, 064416 (2007). https://doi.org/10.1103/PhysRevB.76.064416

[83] F. Luis, J. M. Torres, L. M. García, J. Bartolomé, J. Stankiewicz, F. Petroff, F. Fettar, J.-L. Maurice, and A. Vaurè, and A. Vaurè, Phys. Rev. B 65, 094409 (2002). https://doi.org/10.1103/PhysRevB.65.094409

[84] J. L. Dormann, D. Fiorani, and E. Tronc, In: I. Prigogine and S.A. Rice (eds.). Advances in Chemical Physics, Magnetic Relaxation in Fine-Particle Systems (1997). https://doi.org/10.1002/9780470141571.ch4

[85] F. H. Sánchez, P. Mendoza Zélis, M. L. Arciniegas, G. A. Pasquevich, and M. B. Fernández van Raap, Physical Review B 95, 134421 (2017). https://doi.org/10.1103/PhysRevB.95.134421

[86] M. B. Fernández van Raap, P. Mendoza Zélis, D. F. Coral, T. E. Torres, C. Marquina, G. F. Goya, and F. H. Sánchez, Journal of Nanoparticle Research 14, 1072 (2012). https://doi.org/10.1007/s11051-012-1072-5

[87] G. T. Landi, Physical Review B 89, 014403 (2014). https://doi.org/10.1103/PhysRevB.89.014403

[88] G. T. Landi, Journal of Applied Physics 113, 163908 (2013). https://doi.org/10.1063/1.4802583

[89] R. P. Tan, J. Carrey, and M. Respaud, Physical Review B 90, 214421 (2014). https://doi.org/10.1103/PhysRevB.90.214421

[90] D. F. Coral, P. Mendoza Zélis, M. Marciello, M. del P. Morales, A. Craievich, F. H. Sánchez, and M. B. Fernández van Raap, Langmuir 32, 1201 (2016). https://doi.org/10.1021/acs.langmuir.5b03559

[91] I. Conde-Leboran, D. Baldomir, C. Martinez-Boubeta, O. Chubykalo-Fesenko, M. del Puerto Morales, G. Salas, D. Cabrera, J. Camarero, F. J. Teran, and D. Serantes, The Journal of Physical Chemistry C 119, 15698 (2015). https://doi.org/10.1021/acs.jpcc.5b02555

[92] C. Papadopoulos, A. Kolokithas-Ntoukas, R. Moreno, D. Fuentes, G. Loudos, V. C. Loukopoulos, and G. C. Kagadis, Medical Physics 49, 547 (2022). https://doi.org/10.1002/mp.15317

[93] K. Gilmore, Y. U. Idzerda, M. T. Klem, M. Allen, T. Douglas, and M. Young, Journal of Applied Physics 97, 10B301 (2005). https://doi.org/10.1063/1.1845973

[94] A. I. Figueroa, J. Bartolomé, L. M. García, F. Bartolomé, A. Arauzo, A. Millán, and F. Palacio, Physics Procedia 75, 1050 (2015). https://doi.org/10.1016/j.phpro.2015.12.174

[95] M. B. Fernández van Raap, D. F. Coral, S. Yu, G. A. Muñoz, F. H. Sánchez, and A. Roig, Physical Chemistry Chemical Physics 19, 7176 (2017). https://doi.org/10.1039/C6CP08059F

[96] M. Kallumadil, M. Tada, T. Nakagawa, M. Abe, P. Southern, and Q. A. Pankhurst, Journal of Magnetism and Magnetic Materials 321, 1509 (2009). https://doi.org/10.1016/j.jmmm.2009.02.075

[97] J. Wells, D. Ortega, U. Steinhoff, S. Dutz, E. Garaio, O. Sandre, E. Natividad, M. M. Cruz, F. Brero, P. Southern, Q. A. Pankhurst, and S. Spassov, International Journal of Hyperthermia 38, 447 (2021). https://doi.org/10.1080/02656736.2021.1892837

[98] H. Gavilán, S. K. Avugadda, T. Fernández-Cabada, N. Soni, M. Cassani, B. T. Mai, R. Chantrell, and T. Pellegrino, Chemical Society Reviews 50, 11614 (2021). https://doi.org/10.1039/D1CS00427A

[99] J. M. Orozco-Henao, D. Muraca, F. H. Sánchez, and P. Mendoza Zélis, Journal of Physics D: Applied Physics 53, 385001 (2020). https://doi.org/10.1088/1361-6463/ab9264

[100] A. Gallo-Cordova, J. G. Ovejero, A. M. Pablo-Sainz-Ezquerra, J. Cuya, B. Jeyadevan, S. Veintemillas-Verdaguer, P. Tartaj, and M. del P. Morales, Journal of Colloid and Interface Science 608, 1585 (2022). https://doi.org/10.1016/j.jcis.2021.10.111

[101] D. Niculaes, A. Lak, G. C. Anyfantis, S. Marras, O. Laslett, S. K. Avugadda, M. Cassani, D. Serantes, O. Hovorka, R. Chantrell, and T. Pellegrino, ACS Nano 11, 12121 (2017). https://doi.org/10.1021/acsnano.7b05182

[102] B. Sanz, M. P. Calatayud, E. de Biasi, E. Lima, M. V. Mansilla, R. D. Zysler, M. R. Ibarra, and G. F. Goya, Scientific Reports 6, 38733 (2016). https://doi.org/10.1038/srep38733

[103] R. di Corato, A. Espinosa, L. Lartigue, M. Tharaud, S. Chat, T. Pellegrino, C. Ménager, F. Gazeau, and C. Wilhelm, Biomaterials 35, 6400 (2014). https://doi.org/10.1016/j.biomaterials.2014.04.036

[104] U. M. Engelmann, J. Seifert, B. Mues, S. Roitsch, C. Ménager, A. M. Schmidt, and I. Slabu, Journal of Magnetism and Magnetic Materials 471, 486 (2019). https://doi.org/10.1016/j.jmmm.2018.09.113

[105] S. Dutz, M. Kettering, I. Hilger, R. Müller, and M. Zeisberger, Nanotechnology 22, 265102 (2011). https://doi.org/10.1088/0957-4484/22/26/265102

[106] D. Cabrera, A. Lak, T. Yoshida, M. E. Materia, D. Ortega, F. Ludwig, P. Guardia, A. Sathya, T. Pellegrino, and F. J. Teran, Nanoscale 9, 5094 (2017). https://doi.org/10.1039/C7NR00810D

[107] M. Avolio, A. Guerrini, F. Brero, C. Innocenti, C. Sangregorio, M. Cobianchi, M. Mariani, F. Orsini, P. Arosio, and A. Lascialfari, Journal of Magnetism and Magnetic Materials 471, 504 (2019). https://doi.org/10.1016/j.jmmm.2018.09.111

[108] D. F. Coral, P. A. Soto, V. Blank, A. Veiga, E. Spinelli, S. Gonzalez, G. P. Saracco, M. A. Bab, D. Muraca, P. C. Setton-Avruj, A. Roig, L. Roguin, and M. B. Fernández van Raap, Nanoscale 10, 21262 (2018). https://doi.org/10.1039/C8NR07453D

[109] E. Garaio, J. M. Collantes, J. A. Garcia, F. Plazaola, S. Mornet, F. Couillaud, and O. Sandre, Journal of Magnetism and Magnetic Materials 368, 432 (2014). https://doi.org/10.1016/j.jmmm.2013.11.021

[110] V. Connord, B. Mehdaoui, R. P. Tan, J. Carrey, and M. Respaud, Review of Scientific Instruments 85, 093904 (2014). https://doi.org/10.1063/1.4895656

[111] I. J. Bruvera, D. G. Actis, M. P. Calatayud, and P. Mendoza Zélis, Journal of Magnetism and Magnetic Materials 491, 165563 (2019). https://doi.org/10.1016/j.jmmm.2019.165563

[112] I. Malaescu, P. C. Fannin, C. N. Marin, and D. Lazic, Medical Hypotheses 110, 76 (2018). https://doi.org/10.1016/j.mehy.2017.11.004

[113] M. Beković, M. Trlep, M. Jesenik, and A. Hamler, Journal of Magnetism and Magnetic Materials 355, 12 (2014). https://doi.org/10.1016/j.jmmm.2013.11.045

[114] M. E. Cano, A. Barrera, J. C. Estrada, A. Hernandez, and T. Cordova, Review of Scientific Instruments 82, 114904 (2011). https://doi.org/10.1063/1.3658818

[115] S. Dürr, W. Schmidt, C. Janko, H. P. Kraemer, P. Tripal, F. Eiermann, R. Tietze, S. Lyer, and C. Alexiou, Biomedical Engineering / Biomedizinische Technik (2013). https://doi.org/10.1515/bmt-2013-4129

[116] V. Connord, B. Mehdaoui, R. P. Tan, J. Carrey, and M. Respaud, Review of Scientific Instruments 85, 093904 (2014). https://doi.org/10.1063/1.4895656

[117] L.-M. Lacroix, J. Carrey, and M. Respaud, Review of Scientific Instruments 79, 093909 (2008). https://doi.org/10.1063/1.2972172

[118] S. A. Gonzalez, E. M. Spinelli, A. L. Veiga, D. F. Coral, M. B. F. van Raap, P. M. Zelis, G. A. Pasquevich, and F. H. Sanchez, in 2017 IEEE 8th Latin American Symposium on Circuits & Systems (LASCAS) (IEEE, 2017), pp. 1-4. https://doi.org/10.1109/LASCAS.2017.7948091

[119] U. Gneveckow, A. Jordan, R. Scholz, V. Brüß, N. Waldöfner, J. Ricke, A. Feussner, B. Hildebrandt, B. Rau, and P. Wust, Medical Physics 31, 1444 (2004). https://doi.org/10.1118/1.1748629

[120] G. Oberdörster, E. Oberdörster, and J. Oberdörster, Environmental Health Perspectives 113, 823 (2005). https://doi.org/10.1289/ehp.7339

[121] P. Khanna, C. Ong, B. Bay, and G. Baeg, Nanomaterials 5, 1163 (2015). https://doi.org/10.3390/nano5031163

[122] R. Vakili-Ghartavol, A. A. Momtazi-Borojeni, Z. Vakili-Ghartavol, H. T. Aiyelabegan, M. R. Jaafari, S. M. Rezayat, and S. Arbabi Bidgoli, Artificial Cells, Nanomedicine, and Biotechnology 48, 443 (2020). https://doi.org/10.1080/21691401.2019.1709855

[123] M. Tasso, M. A. Lago Huvelle, I. Diaz Bessone, and A. S. Picco, In: S. K. Sharma, Y. Javed (eds). Toxicity Assessment of Nanomaterials. Magnetic Nanoheterostructures. Nanomedicine and Nanotoxicology. Springer, Cham. (2020). https://doi.org/10.1007/978-3-030-39923-8_13

[124] S. T. Stern, P. P. Adiseshaiah, and R. M. Crist, Particle and Fibre Toxicology 9, 20 (2012). https://doi.org/10.1186/1743-8977-9-20

[125] E. A. K. Warren and C. K. Payne, RSC Advances 5, 13660 (2015). https://doi.org/10.1039/C4RA15727C

[126] D. C. Pan, J. W. Myerson, J. S. Brenner, P. N. Patel, A. C. Anselmo, S. Mitragotri, and V. Muzykantov, Scientific Reports 8, 1615 (2018). https://doi.org/10.1038/s41598-018-19897-8

[127] A. Kurtz-Chalot, J. P. Klein, J. Pourchez, D. Boudard, V. Bin, G. B. Alcantara, M. Martini, M. Cottier, and V. Forest, Journal of Nanoparticle Research 16, 2738 (2014). https://doi.org/10.1007/s11051-014-2738-y

[128] S. Xu, B. Z. Olenyuk, C. T. Okamoto, and S. F. Hamm-Alvarez, Advanced Drug Delivery Reviews 65, 121 (2013). https://doi.org/10.1016/j.addr.2012.09.041

[129] A. Manke, L. Wang, and Y. Rojanasakul, BioMed Research International 2013, 1 (2013). https://doi.org/10.1155/2013/942916

[130] D. Ling and T. Hyeon, Small 9, 1450 (2013). https://doi.org/10.1002/smll.201202111

[131] N. Fernández-Bertólez, C. Costa, M. J. Bessa, M. Park, M. Carriere, F. Dussert, J. P. Teixeira, E. Pásaro, B. Laffon, and V. Valdiglesias, Mutation Research/Genetic Toxicology and Environmental Mutagenesis 845, 402989 (2019). https://doi.org/10.1016/j.mrgentox.2018.11.013

[132] M. Mahmoudi, H. Hofmann, B. Rothen-Rutishauser, and A. Petri-Fink, Chemical Reviews 112, 2323 (2012). https://doi.org/10.1021/cr2002596

[133] Q. Feng, Y. Liu, J. Huang, K. Chen, J. Huang, and K. Xiao, Scientific Reports 8, 2082 (2018). https://doi.org/10.1038/s41598-018-19628-z

[134] L. C. Davies, S. J. Jenkins, J. E. Allen, and P. R. Taylor, Nature Immunology 14, 986 (2013). https://doi.org/10.1038/ni.2705

[135] R. Stevenson, A. J. Hueber, A. Hutton, I. B. McInnes, and D. Graham, The Scientific World Journal 11, 1300 (2011). https://doi.org/10.1100/tsw.2011.106

[136] D. Reichel, M. Tripathi, and J. M. Perez, Nanotheranostics 3, 66 (2019). https://doi.org/10.7150/ntno.30052

[137] B. B. Manshian, J. Poelmans, S. Saini, S. Pokhrel, J. J. Grez, U. Himmelreich, L. Mädler, and S. J. Soenen, Acta Biomaterialia 68, 99 (2018). https://doi.org/10.1016/j.actbio.2017.12.020

[138] A. Gojova, B. Guo, R. S. Kota, J. C. Rutledge, I. M. Kennedy, and A. I. Barakat, Environmental Health Perspectives 115, 403 (2007). https://doi.org/10.1289/ehp.8497

[139] M.-T. Zhu, B. Wang, Y. Wang, L. Yuan, H.-J. Wang, M. Wang, H. Ouyang, Z.-F. Chai, W.-Y. Feng, and Y.-L. Zhao, Toxicology Letters 203, 162 (2011). https://doi.org/10.1016/j.toxlet.2011.03.021

[140] P. R. Leroueil, S. Hong, A. Mecke, J. R. Baker, B. G. Orr, and M. M. Banaszak Holl, Accounts of Chemical Research 40, 335 (2007). https://doi.org/10.1021/ar600012y

[141] C. Contini, M. Schneemilch, S. Gaisford, and N. Quirke, Journal of Experimental Nanoscience 13, 62 (2018). https://doi.org/10.1080/17458080.2017.1413253

[142] J. Chen, J. A. Hessler, K. Putchakayala, B. K. Panama, D. P. Khan, S. Hong, D. G. Mullen, S. C. DiMaggio, A. Som, G. N. Tew, A. N. Lopatin, J. R. Baker, M. M. B. Holl, and B. G. Orr, The Journal of Physical Chemistry B 113, 11179 (2009). https://doi.org/10.1021/jp9033936

[143] H. Nakamura and S. Watano, KONA Powder and Particle Journal 35, 49 (2018). https://doi.org/10.14356/kona.2018011

[144] T. Decker and M.-L. Lohmann-Matthes, Journal of Immunological Methods 115, 61 (1988). https://doi.org/10.1016/0022-1759(88)90310-9

[145] N. Singh, B. Nelson, L. Scanlan, E. Coskun, P. Jaruga, and S. Doak, International Journal of Molecular Sciences 18, 1515 (2017). https://doi.org/10.3390/ijms18071515

[146] K. Kansara, P. Patel, D. Shah, R. K. Shukla, S. Singh, A. Kumar, and A. Dhawan, Environmental and Molecular Mutagenesis 56, 204 (2015). https://doi.org/10.1002/em.21925

[147] D. Y. Seo, M. Jin, J.-C. Ryu, and Y.-J. Kim, Toxicology and Environmental Health Sciences 9, 23 (2017). https://doi.org/10.1007/s13530-017-0299-z

[148] R. Wan, Y. Mo, L. Feng, S. Chien, D. J. Tollerud, and Q. Zhang, Chemical Research in Toxicology 25, 1402 (2012). https://doi.org/10.1021/tx200513t

[149] M. Ahamed, H. A. Alhadlaq, J. Alam, M. A. Majeed Khan, D. Ali, and S. Alarafi, Current Pharmaceutical Design 19, 6681 (2013). https://doi.org/10.2174/1381612811319370011

[150] S. Alarifi, D. Ali, S. Alkahtani, and M. S. Alhader, Biological Trace Element Research 159, 416 (2014). https://doi.org/10.1007/s12011-014-9972-0

[151] X.-Q. Zhang, X. Xu, N. Bertrand, E. Pridgen, A. Swami, and O. C. Farokhzad, Advanced Drug Delivery Reviews 64, 1363 (2012). https://doi.org/10.1016/j.addr.2012.08.005

[152] M. A. Malvindi, V. de Matteis, A. Galeone, V. Brunetti, G. C. Anyfantis, A. Athanassiou, R. Cingolani, and P. P. Pompa, PLoS ONE 9, e85835 (2014). https://doi.org/10.1371/journal.pone.0085835

[153] G. Crisponi, V. M. Nurchi, J. I. Lachowicz, M. Peana, S. Medici, and M. A. Zoroddu, In: A. M. Grumezescu (eds). Toxicity of Nanoparticles: Etiology and Mechanisms. Antimicrobial Nanoarchitectonics. Elsevier (2017). https://doi.org/10.1016/B978-0-323-52733-0.00018-5

[154] P. Rivera Gil, G. Oberdörster, A. Elder, V. Puntes, and W. J. Parak, ACS Nano 4, 5527 (2010). https://doi.org/10.1021/nn1025687

[155] A. Laszlo, Cell Proliferation 25, 59 (1992). https://doi.org/10.1111/j.1365-2184.1992.tb01482.x

[156] J.R. Lepock, in: Advances in Molecular and Cell Biology, Elsevier, 1997, pp. 223–259.

[157] J. L. Roti, International Journal of Hyperthermia 24, 3 (2008). https://doi.org/10.1080/02656730701769841

[158] C. M. Neophytou, I. P. Trougakos, N. Erin, and P. Papageorgis, Cancers 13, 4363 (2021). doi: 10.3390/cancers13174363

[159] M. Sefidgar, E. Bashooki, and P. Shojaee, Journal of Thermal Biology 94, 102742 (2020). https://doi.org/10.1016/j.jtherbio.2020.102742

[160] N. CM, T. IP, E. N, and P. P, Cancers (Basel) 13, (2021).

[161] F. H. Igney and P. H. Krammer, Nature Reviews Cancer 2002 2:4 2, 277 (2002). https://doi.org/10.1038/nrc776

[162] A. Murshid, J. Gong, and S. K. Calderwood, Frontiers in Immunology 3, (2012). https://doi.org/10.3389/fimmu.2012.00063

[163] B. Perillo, M. Di Donato, A. Pezone, E. Di Zazzo, P. Giovannelli, G. Galasso, G. Castoria, and A. Migliaccio, Experimental and Molecular Medicine 52, 192 (2020). https://doi.org/10.1038/s12276-020-0384-2

[164] Y. Huang, S. Wu, L. Zhang, Q. Deng, J. Ren, and X. Qu, ACS Nano 16, 4228 (2022). https://doi.org/10.1021/acsnano.1c10231

[165] X. Cheng, H.-D. Xu, H.-H. Ran, G. Liang, and F.-G. Wu, ACS Nano 15, 8039 (2021). https://doi.org/10.1021/acsnano.1c00498

[166] S. J. Dixon, K. M. Lemberg, M. R. Lamprecht, R. Skouta, E. M. Zaitsev, C. E. Gleason, D. N. Patel, A. J. Bauer, A. M. Cantley, W. S. Yang, B. Morrison, and B. R. Stockwell, Cell 149, 1060 (2012). https://doi.org/10.1016/j.cell.2012.03.042

[167] M. R. Sepand, S. Ranjbar, I. M. Kempson, M. Akbariani, W. C. A. Muganda, M. Müller, M. H. Ghahremani, and M. Raoufi, Nanomedicine 29, 102243 (2020). https://doi.org/10.1016/j.nano.2020.102243

[168] J. Y. Cao and S. J. Dixon, Cellular and Molecular Life Sciences 73, 2195 (2016). https://doi.org/10.1007/s00018-016-2194-1

[169] S. Zanganeh, G. Hutter, R. Spitler, O. Lenkov, M. Mahmoudi, A. Shaw, J. S. Pajarinen, H. Nejadnik, S. Goodman, M. Moseley, L. M. Coussens, and H. E. Daldrup-Link, Nature Nanotechnology 11, 986 (2016). https://doi.org/10.1038/nnano.2016.168

[170] L. Yue, J. Wang, Z. Dai, Z. Hu, X. Chen, Y. Qi, X. Zheng, and D. Yu, Bioconjugate Chemistry 28, 400 (2017). https://doi.org/10.1021/acs.bioconjchem.6b00562

[171] C. Hu, X. Chen, Y. Huang, and Y. Chen, Scientific Reports 8, 2274 (2018). https://doi.org/10.1038/s41598-018-20715-4

[172] Z. Zhou, J. Song, R. Tian, Z. Yang, G. Yu, L. Lin, G. Zhang, W. Fan, F. Zhang, G. Niu, L. Nie, and X. Chen, Angewandte Chemie - International Edition 56, 6492 (2017). https://doi.org/10.1002/anie.201701181

[173] W. P. Li, C. H. Su, Y. C. Chang, Y. J. Lin, and C. S. Yeh, ACS Nano 10, 2017 (2016). https://doi.org/10.1021/acsnano.5b06175

[174] Z. Shen, J. Song, B. C. Yung, Z. Zhou, A. Wu, and X. Chen, Advanced Materials 30, 1704007 (2018). https://doi.org/10.1002/adma.201704007

[175] J. Li, F. Cao, H.-L. Yin, Z.-J. Huang, Z.-T. Lin, N. Mao, B. Sun, and G. Wang, Cell Death and Disease 11, 88 (2020). https://doi.org/10.1038/s41419-020-2298-2

[176] X. Chen, C. Yu, R. Kang, and D. Tang, Frontiers in Cell and Developmental Biology 8, 1 (2020). https://doi.org/10.3389/fcell.2020.590226

[177] P. K. Srivastava, Advances in Cancer Research 62, 153 (1993). https://doi.org/10.1016/S0065-230X(08)60318-8

[178] R. J. Binder, The Journal of Immunology 193, 5765 (2014). https://doi.org/10.4049/jimmunol.1401417

[179] C. Wood, P. Srivastava, R. Bukowski, L. Lacombe, A. I. Gorelov, S. Gorelov, P. Mulders, H. Zielinski, A. Hoos, F. Teofilovici, L. Isakov, R. Flanigan, R. Figlin, R. Gupta, and B. Escudier, The Lancet 372, 145 (2008). https://doi.org/10.1016/S0140-6736(08)60697-2

[180] F. Gong, N. Yang, X. Wang, Q. Zhao, Q. Chen, Z. Liu, and L. Cheng, Nano Today 32, 100851 (2020). https://doi.org/10.1016/j.nantod.2020.100851

[181] S. K. Calderwood, M. A. Khaleque, D. B. Sawyer, and D. R. Ciocca, Trends in Biochemical Sciences 31, 164 (2006). https://doi.org/10.1016/j.tibs.2006.01.006

[182] B. Frey, E. M. Weiss, Y. Rubner, R. Wunderlich, O. J. Ott, R. Sauer, R. Fietkau, and U. S. Gaipl, International Journal of Hyperthermia 28, 528 (2012). https://doi.org/10.3109/02656736.2012.677933

[183] A. Murshid, T. J. Borges, C. Bonorino, B. J. Lang, and S. K. Calderwood, Front Immunol 10, 3035 (2019). https://doi.org/10.3389/fimmu.2019.03035

[184] J. Gong, B. Zhu, A. Murshid, H. Adachi, B. Song, A. Lee, C. Liu, and S. K. Calderwood, The Journal of Immunology 183, 3092 (2009). https://doi.org/10.4049/jimmunol.0901235

[185] J. R. Thériault, H. Adachi, and S. K. Calderwood, The Journal of Immunology 177, 8604 (2006). https://doi.org/10.4049/jimmunol.177.12.8604

[186] S. Persano, P. Das, and T. Pellegrino, Cancers (Basel) 13, (2021). https://doi.org/10.3390/cancers13112735

[187] S. Toraya-Brown, M. R. Sheen, P. Zhang, L. Chen, J. R. Baird, E. Demidenko, M. J. Turk, P. J. Hoopes, J. R. Conejo-Garcia, and S. Fiering, Nanomedicine: Nanotechnology, Biology, and Medicine 10, 1273 (2014). https://doi.org/10.1016/j.nano.2014.01.011

[188] K.-G. Tranberg, Frontiers in Oncology 11, (2021). https://doi.org/10.3389/fonc.2021.708810

Materials Research Forum LLC
https://doi.org/10.21741/9781644902332-4

[189] E. M. Scutigliani, Y. Liang, H. Crezee, R. Kanaar, and P. M. Krawczyk, Cancers (Basel) 13, 1 (2021). https://doi.org/10.3390/cancers13061243

[190] K. Ohnishi, A. Takahashi, S. Yokota, and T. Ohnishi, International Journal of Radiation Biology 80, 607 (2004). https://doi.org/10.1080/09553000412331283470

[191] K. Ohnishi, In: S. Kokura, T. Yoshikawa, T. Ohnishi (eds). Thermo-Tolerance. Hyperthermic Oncology from Bench to Bedside. Springer, Singapore (2016). https://doi.org/10.1007/978-981-10-0719-4_7

[192] J. H. Wang, M. Z. Yao, J. F. Gu, L. Y. Sun, X. Y. Liu, and Y. F. Shen, Biochemical and Biophysical Research Communications 290, 1454 (2002). https://doi.org/10.1006/bbrc.2002.6373

[193] Y. Nakamura, M. Fujimoto, N. Hayashida, R. Takii, A. Nakai, and M. Muto, J Dermatol Sci 60, 187 (2010). https://doi.org/10.1016/j.jdermsci.2010.09.009

[194] E. Sahin, M. Sahin, A. D. Sanlioğlu, and S. Gümüslü, Int J Hyperthermia 27, 63 (2011). https://doi.org/10.3109/02656736.2010.528139

[195] L. E. M. Vriend, N. van den Tempel, A. L. Oei, M. L'Acosta, F. J. Pieterson, N. A. P. Franken, R. Kanaar, and P. M. Krawczyk, Oncotarget 8, 97490 (2017). https://doi.org/10.18632/oncotarget.22142

[196] T. Miyagawa, H. Saito, Y. Minamiya, K. Mitobe, S. Takashima, N. Takahashi, A. Ito, K. Imai, S. Motoyama, and J. Ogawa, Int J Clin Oncol 19, 722 (2014). https://doi.org/10.1007/s10147-013-0606-x

[197] D. Egea-Benavente, J. G. Ovejero, M. D. P. Morales, and D. F. Barber, Cancers (Basel) 13, (2021). https://doi.org/10.3390/cancers13184583

[198] M. Asgari, T. Miri, M. Soleymani, and A. Barati, Journal of Molecular Liquids 324, 114731 (2021). https://doi.org/10.1016/j.molliq.2020.114731

[199] J. R. Lepock, International Journal of Hyperthermia 19, 252 (2003). https://doi.org/10.1080/0265673031000065042

[200] G. C. van Rhoon, M. Franckena, and T. L. M. ten Hagen, Advanced Drug Delivery Reviews 163-164, 145 (2020). https://doi.org/10.1016/j.addr.2020.03.006

[201] M. Chang, Z. Hou, M. Wang, C. Li, and J. Lin, Advanced Materials 33, 2004788 (2021). https://doi.org/10.1002/adma.202004788

[202] X. Liu, Y. Zhang, Y. Wang, W. Zhu, G. Li, X. Ma, Y. Zhang, S. Chen, S. Tiwari, K. Shi, S. Zhang, H. M. Fan, Y. X. Zhao, and X. J. Liang, Theranostics 10, 3793 (2020). https://doi.org/10.7150/thno.40805

Magnetic Nanoparticles for Biomedical Applications   Materials Research Forum LLC
Materials Research Foundations 143 (2023) 140-169   https://doi.org/10.21741/9781644902332-5

Chapter 5

Ferromagnetic Ni Nanostructures via Chemical Reduction Methods

Ramany Revathy[1], Manoj Raama Varma[1, 2]**, Kuzhichalil Peethambharan Surendran[1, 2]*

[1]Materials Science and Technology Division, CSIR- National Institute for Interdisciplinary Science and Technology (CSIR-NIIST), Research Centre, University of Kerala, Thiruvananthapuram -695019, India

[2]Academy of Scientific and Innovative Research (AcSIR), CSIR-HRDC Campus, Ghaziabad, Uttar Pradesh- 201002, India

*kpsurendran@niist.res.in, **manoj@niist.res.in

Abstract

For the last couple of decades, nickel (Ni) nanostructures are in the focus of active research since they are widely applied in countless walks of modern life. But being tiny magnets, their synthesis in monodisperse form is extremely challenging. In this chapter, we review the recent trends in the synthesis of Ni through chemical reduction routes without the aid of magnetic field, hydrothermal environments, and templates. The particle size and morphology of the synthesized Ni nanostructures can easily be tailored by controlling the reaction conditions. Depending on particle size and morphology, these ferromagnetic nanomaterials have a wide variety of applications in various technological fields.

Keywords

Magnetic Nanostructures, Wet Chemical Reduction, Hierarchical Nickel Nanostructures, pH Value, Hydrazine

Contents

1.   Introduction

The scope of nanotechnology has proliferated to many application domains in recent years, resulting in a resurged interest for novel inorganic multifunctional nanomaterials [1]. Among them, ferromagnetic (FM) nanomaterials find active research interest due to their potential applications in various technological fields such as electronics, optics, magnetic data storage, magnetic resonance imaging, catalysis, biomedical etc. [2-4]. The room temperature ferromagnetic Ni nanostructures have been widely investigated due to the large magnetocrystalline anisotropy, excellent electrical and thermal conductivities, high strength, chemical stability, and higher oxidation resistance compared to other FM metals [5-6]. As a result, Ni nanostructures have immense potential applications in magnetic memories [7], catalysis [8], battery manufacturing [9], printable electronics [10], immobilization of biological molecules [11], optical switches [12], dye-sensitized solar cells [13], and magnetic fluids [14].

There are various synthesis methods used for the preparation of nanosized Ni with various morphologies like spherically symmetric structures such as spherical nanoparticles (NPs) or nanospheres (NSs) [15-16] and anisotropic nanostructures such as nanowires (NWs) [17], nanorods [18], nanoflowers [19], nanoshells [20] or nanochains [21]. Among the latter group, NWs were extensively studied due to the shape anisotropy, which can largely improve the magnetic properties [22]. The commonly adopted synthesis protocols of Ni nanostructures include decomposition of organometallic precursors [23-25], electrochemical deposition [26], wet-chemical reduction of nickel salts [26-29], thermal decomposition of metal-surfactant complexes under inert atmosphere [30], green plant-mediated synthesis [31-33], solvothermal [34], hydrothermal [35], microwave-assisted synthesis [36], template-based synthesis [37], and magnetic field-assisted synthesis [38]. Among them, the wet-chemical reduction technique is the economically efficient and easiest method to synthesize Ni nanostructures at low temperatures [39]. However, a serious challenge in wet-chemical reduction protocol is the necessity to control various parameters like concentration of salts, pH, surfactants, concentration of surfactants, temperature of the reaction, strength and amount of reducing agents [40].

The main constraints associated with the synthesis of isolated Ni nanostructures include (i) reduction of Ni(II) to Ni(0) at room temperature is not possible [41], (ii) surface oxidation of reduced Ni(0) to various oxides and hydroxides of Ni [42], (iii) aggregation of Ni NPs due to its FM nature [43], and (iv) the magnetism of Ni NPs obstruct with magnets of microwave systems in microwave-assisted synthesis [44]. To circumvent these issues, the right selection and optimization of the synthesis methods become vital. Several review articles were published in the literature, some of which address various compounds of nickel while others cover their applications. For example, in 2016, Sequeira et al. reported an overview of commercially available corrosion-resistant Ni alloys [45]. In the same year, Imran Din et al. reviewed the green synthesis strategies of Ni and NiO NPs [46]. In 2017, Thellaputta et al. reported a review on the machinability of alloys made of nickel [47]. Later, Kate et al. presented a review on the supercapacitor application of NiO electrodes [48]. In 2019, Zhang et al. reported an overview of nickel-based materials for supercapacitor applications [49]. Meanwhile, a brief summary of the electrodeposited Ni and Ni-Fe alloys was also published [50]. However, the above-mentioned reviews mostly discussed nickel oxides and Ni-based alloys; did not specifically emphasize the production of Ni NPs. An exception is a recent work by Jaji et al., who published an extensive review of the biosynthesis techniques for Ni nanostructures [51]. Other than this, a comprehensive compilation of the production of Ni nanostructures via chemical reduction methods is still missing. Our review attempt is to bridge this gap, which can serve as a precursor to chemical technologists who would like to mass-produce shape-controlled magnetic nanomaterials. In the present chapter, we briefly review state of the art on synthesis of nickel nanostructures via the facile wet-chemical reduction routes and their various reaction mechanisms.

2. Wet-chemical reduction synthesis of Ni nanostructures

Generally, NPs are synthesized using a top-down approach or bottom-up method. In top-down synthesis, bulk materials are broken down into nanosized particles. Mechanical ball milling, sputtering, laser ablation, nanolithography, and thermal decomposition are examples of top-down approaches [52]. In bottom-up synthesis methods, metal salts are used as precursors for chemical reactions. These metal salts are ultimately reduced to metallic nanoparticles with the aid of suitable solvents and reducing agents. In this approach, the morphology and dimension of synthesized nanomaterials can be controlled by varying the concentration of the metal salts and the reducing agent, pH, temperature, reaction time, and the surfactant [53]. The most widely used bottom-up approaches are chemical vapor deposition, sol-gel synthesis, spray pyrolysis, co-precipitation, green synthesis, atomic condensation, molecular condensation, and aerosol process [54]. Among the various bottom-up approaches, the chemical reduction technique is one of the easiest methods for synthesizing nickel nanoparticles, which is the central theme of the present chapter.

There are numerous articles available about the synthesis of nickel nanoparticles using the chemical reduction technique. As a general rule in this technique, reducing agent and surfactant (capping agent/ stabilizing agent) are essential for controlling the morphology and properties of the nanostructures [55]. The role of reducing agents (reductants) is to guide the electrons from the metal salts solution to metal ions and produce insoluble atoms. The aggregation of these nuclei generates metal clusters, which grow further to form nanomaterials [55]. That is, the three steps involved in the wet-chemical reduction of Ni nanostructures are (i) the reduction of the Ni ions in Ni-salts into insoluble Ni nuclei, (ii) the formation of Ni core via aggregation and, (iii) the growth of Ni core to Ni nanostructures [56].

2.1   Reductants and surfactants used for wet-chemical reduction

Since the 1990s, the chemical reduction method has been used for the controlled synthesis of metallic nanoparticles [55]. The chemical compounds that perform the reduction reaction are known as reducing agents. Usually, reducing agents are used for the synthesis of metallic nanomaterials such as Ni, Co, Ag, Au, etc. With the help of reducing agents, metal salts are reduced to pure elemental metals [56]. In a typical wet-chemical synthesis, surfactants are used to stabilize the generated nanoparticles by preventing agglomeration [53]. Thus, surfactants can prevent the growth rate and thereby control the particle size. The commonly employed reductants and surfactants are tabulated in Table 1. In certain reactions, some chemicals have a dual role, which can act as a reducing agent as well as a surfactant at the same time. $NaBH_4$ and PVP are examples of double agents that can act as reductants as well as surfactants [57]. These reductants and surfactants influence the kinetics of chemical reduction reactions and, thereby controlling the properties of the final product [58]. Chemical reduction using medium to strong reductants such as $NaBH_4$, $N_2H_4.H_2O$, and citrate ions in low boiling solvents usually produce magnetic samples with poor crystallinity only. Therefore, subsequent heat treatment at an inert atmosphere is inevitable for getting well-crystallized metals [55].

The commonly used solvents for the wet-chemical synthesis are ethanol, water, toluene, DMF (N, N-dimethyl formamide), etc. The pH of the reaction is mostly controlled using the bases such as NaOH, KOH, ammonia solution, etc. Here, the choice of capping agent, reducing agent, base, solvent, and the parameters such as time and temperature of the reaction greatly influences the morphology of the nanostructures [76].

Table 1. Most used reductants and surfactants for the chemical reduction synthesis

Sl. No.	Reductants	Ref.	Surfactants	Ref.
1	NaBH$_4$ (Sodium borohydride)	[59]	PVP (Poly(vinyl pyrrolidone))	[60]
2	N$_2$H$_4$ (Hydrazine)	[61]	Ethylene glycol (EG)	[61]
3	Sodium formaldehyde sulfoxylate (SFS)	[62]	CTAB (Cetyltrimethylammonium bromide)	[63]
4	CO (Carbon monoxide)	[64]	linoleic acid	[65]
5	Ascorbic acid	[66]	Oleylamine	[67]
6	Hydroquinone	[68]	PAA (Poly(acrylic acid))	[69]
7	TSC (Trisodium citrate dehydrate)	[70]	SDS (Sodium dodecyl sulfate)	[71]
8	Sodium citrate	[72]	OA (Oleic acid)	[73]
9	C$_{12}$H$_{22}$O$_{11}$ (Sucrose)	[74]	PEI (Polyethylene imine)	[75]

2.2 Optimization of reductants and precursors

When it comes to the synthesis of Ni nanostructures, a variety of precursors like nickel chloride, nickel nitrate, nickel acetate, etc., and reducing agents such as NaBH$_4$, N$_2$H$_4$ and C$_{12}$H$_{22}$O$_{11}$ were commonly used. In order to ascertain the most suitable precursor and reducing agent, Raj et al. estimated the heat of reaction for different synthesis methods [74]. The possible chemical reactions for Ni reduction using various precursors are shown in the following Eq. (1) - (4).

$$Ni(NO_3)_2 + 2H_2 \rightarrow Ni + 2NO_2 + 2H_2O$$
$$[\Delta H = -0.27 \text{ kJ/mol at } 25\ ^\circ C] \tag{1}$$

$$Ni(CH_3COO)_2 + H_2 \rightarrow Ni + 4\ C + 4H_2O$$
$$[\Delta H = 84 \text{ kJ/mol at } 25\ ^\circ C] \tag{2}$$

$$NiCl_2 + H_2 \rightarrow Ni + 2HCl$$
$$[\Delta H = 120.7 \text{ kJ/mol at } 25\ ^\circ C] \tag{3}$$

$$NiSO_4 + 2H_2 \rightarrow Ni + SO_2 + 2H_2O$$
$$[\Delta H = 92.4 \text{ kJ/mol at } 25\ ^\circ C] \tag{4}$$

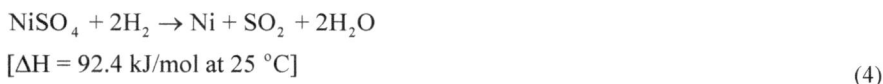

These reactions indicate that the reduction using $Ni(NO_3)_2$ need lower energy than $NiCl_2$, $Ni(CH_3COO)_2$ and $NiSO_4$. Since the calculated heat of reaction for $Ni(NO_3)_2$ is lower, it is preferred for the synthesis of Ni NPs [74]. In order to find the best reducing agent, Raj et al. further elucidated the heat of reaction of Ni reduction using N_2H_4, $NaBH_4$ and $C_{12}H_{22}O_{11}$ (sucrose) and arrived at the following values, as shown in Eq. (5) - (7) [74].

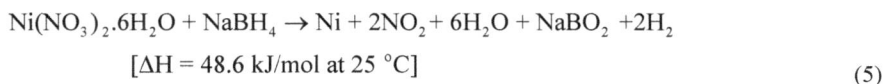

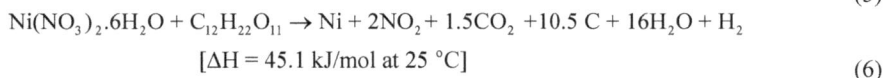

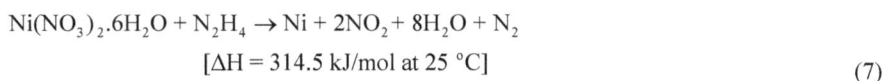

$$Ni(NO_3)_2.6H_2O + NaBH_4 \rightarrow Ni + 2NO_2 + 6H_2O + NaBO_2 + 2H_2$$
$$[\Delta H = 48.6 \text{ kJ/mol at } 25\ ^\circ C] \tag{5}$$
$$Ni(NO_3)_2.6H_2O + C_{12}H_{22}O_{11} \rightarrow Ni + 2NO_2 + 1.5CO_2 + 10.5\ C + 16H_2O + H_2$$
$$[\Delta H = 45.1 \text{ kJ/mol at } 25\ ^\circ C] \tag{6}$$

$$Ni(NO_3)_2.6H_2O + N_2H_4 \rightarrow Ni + 2NO_2 + 8H_2O + N_2$$
$$[\Delta H = 314.5 \text{ kJ/mol at } 25\ ^\circ C] \tag{7}$$

Equations (5) – (7) suggest that $C_{12}H_{22}O_{11}$ is the better reductant for the reduction of Ni since ΔH is low. Even though $NaBH_4$ possesses nearly the same heat of reaction as that of sucrose, complete removal of sodium and boron is too difficult. Hence, sucrose can be considered as the best reductant for the synthesis of nickel [74]. Earlier, Khanna et al. reported a comparative study of the synthesis of Ni NPs using the reductants SFS and $NaBH_4$ and arrived at the conclusion that $NaBH_4$ results in the formation of amorphous Ni NPs. Due to these reasons, the synthesis of Ni using $NaBH_4$ is less recommended [62]. Hence, it can be concluded that nickel nitrate and sucrose are the best chemical reagents for the facile chemical reduction synthesis of nickel.

As shown in Table 1, yet another important player in the reduction synthesis of magnetic nanostructures is hydrazine since the reactions are very simple, and the by-products are typically nitrogen and water. In the case of nickel also, most of the previously reported Ni nanostructures were synthesized using the monohydrate form of hydrazine as the reductant. Therefore, the reduction kinetics of the metallic Ni nanostructures using $N_2H_4.H_2O$ is to be further discussed.

2.3 Reduction mechanism of Ni using $N_2H_4.H_2O$

The formation of Ni via chemical reduction involves various changes in the manifestation of the reaction mixture before it turns into black-colored Ni nanocrystals. During the synthesis of Ni nanostructures, complexation occurs between the Ni^{2+} ions in the nickel salt and the hydrazine. Thus, a series of hydrazine linked complexes such as $[Ni(N_2H_4)_2]Cl_2$, $[Ni(N_2H_4)_3]Cl_2$ and $[Ni(NH_3)_6]Cl_2$ may form [77]. These complexes are very stable in an alkaline medium at room temperature, but they can readily decompose at a critical temperature [78]. The reduction of Ni using $N_2H_4.H_2O$ involve the following chemical reactions [60]:

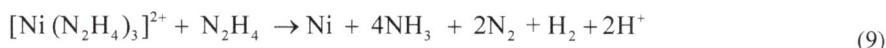

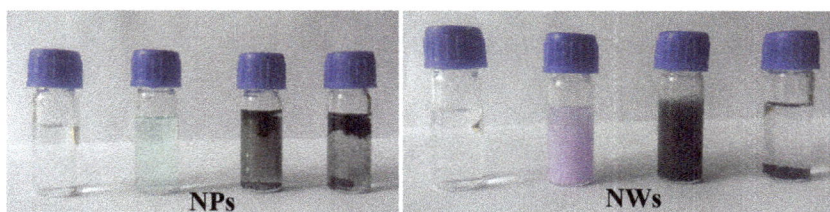

$$Ni^{2+} + N_2H_4 \rightarrow [Ni(N_2H_4)_3]^{2+} \tag{8}$$

$$[Ni(N_2H_4)_3]^{2+} + N_2H_4 \rightarrow Ni + 4NH_3 + 2N_2 + H_2 + 2H^+ \tag{9}$$

Fig. 1. Color change observed during the synthesis of Ni NPs and NWs

In 2019, our research group reported the synthesis of Ni NPs and NWs via the reduction of Ni using $N_2H_4.H_2O$ [61]. We have initially prepared an alkali (NaOH) mixed hydrazine solution, and the Ni salt precursor is finally added to that solution. The photographs of the color change observed during the synthesis of Ni NPs and Ni NWs are shown in Fig.1. In the case of Ni NPs, on the addition of nickel chloride solution to the hydrazine and NaOH mixed solution, the transparent solution changes to green and then turns to grey-black. In contrast, in the case of sample NPW and NW, color changes from transparent to blue and then to grey-black. The formation of Ni $(OH)_2$ is indicated by the green color, while that of nickel-hydrazine complexes by blue color. The color changes from blue to green clearly show the formation of Ni $(OH)_2$. In the case of nickel NPs, there is no visible blue color precipitate formation. i.e., Ni-complex decomposition in the sample NP is very rapid [77]. The situation is slightly different when nanospecies of nickel join to form nanowires. For Ni NWs, there is no green-colored precipitate formation, indicating that the reduction happens so fast that we cannot observe the color changes from blue to green [77]. Also,

Eluri et al. reported the color change observed during the Ni reduction by providing $NiCl_2.6H_2O$ as the starting precursor [16]. They reported that the addition of hydrazine monohydrate to the green-colored $NiCl_2.6H_2O$ solution resulted in a rapid change to blue due to the formation of nickel hydrazine complexes and finally turned to black. The formation of intermediate $Ni(OH)_2$ and nickel hydrazine complexes were largely depending on the molar ratio of N_2H_4 and Ni^{2+}. Roselina et al. studied the dependence of molar ratio $\dfrac{[N_2H_4]}{[Ni^{2+}]}$ with the observed color change during the reduction of Ni NPs, which is shown in Table 2 [77].

Table 2. Observed color change during the reduction of Ni (Reproduced from reference [77] with permission from the Elsevier)

Sl. No.	Molar ratio, $\dfrac{[N_2H_4]}{[Ni^{2+}]}$	Observed color change
1	5	Blue → Green → Black
2	10	Blue → Green → Black
3	20	Blue → Green → Black
4	30	Blue → Black

The sample having the highest molar ratio of 30 shows a sudden color change from blue to black, indicating that the reduction occurs so fast. Thus the color change from blue to green (indication of the formation of $Ni(OH)_2$) cannot be visible. These results suggest that the increased molar ratio of $\dfrac{[N_2H_4]}{[Ni^{2+}]}$ increases the rate of reduction. However, the XRD analysis resulted in the emergence of diffraction peaks of Ni-complexes, and the degree of particle agglomeration also increased [77]. Thus, an optimum concentration of hydrazine is needed for the phase pure formation of Ni metal and, the reduction of Ni using a large amount of hydrazine is less recommended. The schematic representation of the production of Ni metal via the reduction using hydrazine is shown in Fig. 2.

In spite of hydrazine's facile reducing properties, hydrazine-based reactions are discouraged these days since it is recognized as a potential occupational carcinogen. Hence the qualification of alternative protocols using less hazardous reductants became inevitable.

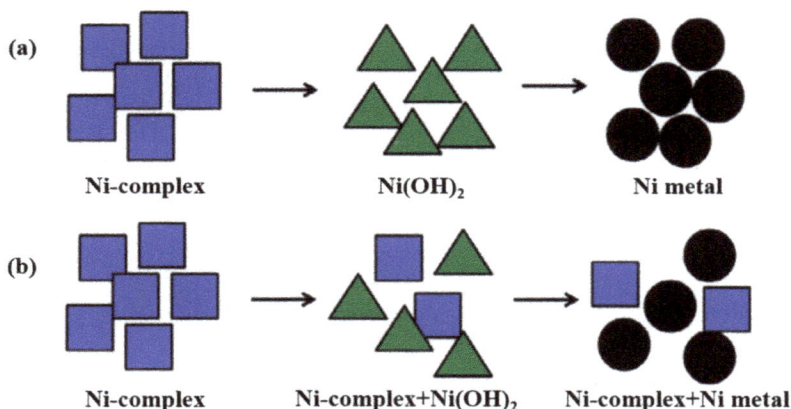

Fig. 2. Schematic representation of reduction of Ni using (a) moderate amount of hydrazine, (b) excess amount of hydrazine (Reproduced from reference [77] with permission from the Elsevier)

2.4   Conventional reduction and dropping reduction using $N_2H_4.H_2O$

Metallic nanostructures, including nickel, can be synthesized via conventional as well as dropping reduction methods. In the conventional reduction reaction, all the precursor solutions are directly mixed and stirred for some time to get a homogeneous solution. The prepared homogeneous solution containing metal salt, solvent, base and reductant is stirred while supplying an optimum temperature to yield the reduced metal nanostructure. On the other hand, the dropping method is a modified conventional reduction, in which reductant is added drop by drop to the alkali mixed solvent under vigorous stirring. Further, the metal precursor solution is added drop-wise into the prepared homogeneous solution containing solvent, base and reductant [79].

In 2016, Zhang et al. reported the synthesis of Ni NWs both by both conventional and dropping reduction routes by providing hydrazine as the reducing agent and nickel chloride as the precursor [79]. The SEM images and schematic illustration of the above-mentioned synthesis of Ni NWs are shown in Fig. 3. The SEM images (see Figs. 3(a) and (b)) show that the diameter of NWs formed by the conventional reduction route (C-NWs) is larger than that of NWs synthesized via dropping reduction (D-NWs). In the illustration of the synthesis process (see Figs. 3(c) and (e)), the yellow-colored dots indicate the formed Ni NPs, and the horizontal dotted lines represent the magnetic field lines. In the conventional synthesis, Ni ions are initially reduced by the reductant and form spherical particles. The applied external magnetic field causes the self-assembly of Ni nanoparticles along the magnetic line of force and forms one dimensional (1D) Ni NWs. To minimize the surface energy, continuously generating Ni nuclei are aggregated on the surface of formed Ni NWs [79]. Thus the diameter of C-NWs increases, and the resultant distribution of C-NWs after the experiment is depicted in Fig. 3(d).

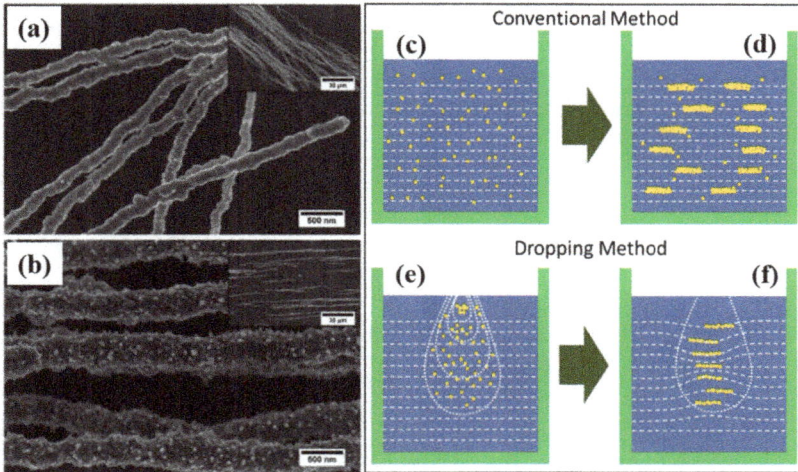

Fig. 3. SEM images of (a) D-NWs, (b) C-NWs, schematic illustration of the synthesis of (c)-(d) C-NWs, (e)-(f)) D-NWs (Reproduced from reference [79] with permission from Springer)

In the dropping method reported by Zhang et al. [79], $NiCl_2.6H_2O$ solution is added drop-wise to the homogeneous solution of NaOH and hydrazine under stirring. Since the density of nickel chloride is larger than the density of the reducing agent, the diffusion of nickel chloride droplets is unlike the conventional reduction. In Fig. 3(e), the area circumscribed by the dotted droplet-shaped regions shows the diffusion zone of Ni precursor in the reductant mixture. Also, it is evident that the diffusion area enlarged with the increase in reaction time. Thus, an efficient metal reduction mechanism happens in the drop-wise addition. Due to the small amount of nickel content in each precursor drop, the possibility of Ni nuclei aggregation can be avoided and the diameter of D-NWs is smaller than C-NWs [79].

2.5   Prickles formation on the surface of Ni nanostructures

In previous studies of Ni nanostructures, hierarchical Ni with numerous prickles (thorns or dendrite) on the surface were mostly reported. These branched growth formation on the surface of nanomaterials were theoretically studied by Witten [80] and Halsey [81]. The boundary-layer model (BLM) is the commonly used model to probe the dendrite growth mechanism. According to the BLM model, the steps involved in the prickle formation are: (i) Ni nuclei generation; (ii) Ni nuclei solidify and result in spherical growth; (iii) dendrite formation originate on the surface and develop prickles [82]. The inverse square relationship of growth rate and the curvature of the tip of prickles was stated by Mullins et al. [83]. Hence, the rapid solidification of nuclei leads to the development of sharp prickles [82]. Due to these prickle formation, nanomaterials with various morphology such as urchin-like NPs, flower-like NPs, and prickly NWs can form [61]. An illustration of the growth of sea urchin-like Ni NPs is depicted in Fig. 4.

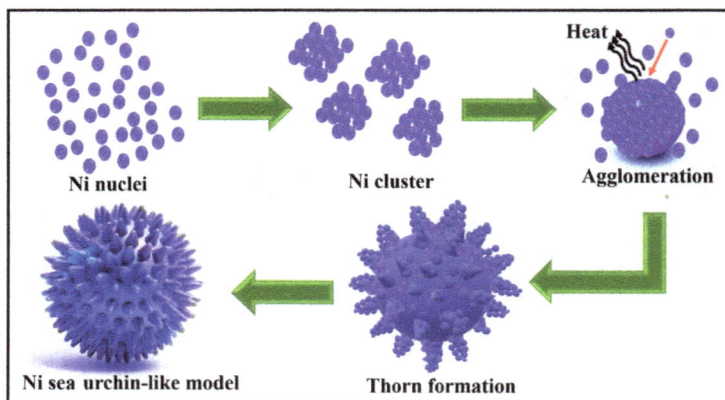

Fig. 4. The schematic representation of the formation of sea-urchin-like Ni particle (Reproduced from reference [61] with permission from the Elsevier)

2.6   Effect of reaction conditions

It was found that the reaction conditions of chemical reduction can greatly influence the crystallinity, particle size and morphology. In the following sections, we discuss the effect of reaction conditions such as addition sequence of chemicals, alkali concentration, pH, Ni^{2+} ion concentration, amount of hydrazine, reduction temperature and reaction time.

2.6.1   Effect of alkali (base) concentration and pH

The pH of the reaction mixture has a significant impact on the redox potential of the solutes, particle size and morphology of reduced Ni nanostructures. pH value can be varied by changing the molar concentration of alkali used for the reaction. In the former studies of nickel reduction, NaOH is mostly used as the base for controlling the pH of the reaction [16, 61, 77, 84-86]. For a solution having acidic pH, hydrazine existed as hydrazinium ion ($N_2H_5^+$ ion). In an acidic medium, H^+ ions dominate, and it follows the equation $N_2 + 5H^+ + 4e^- \rightarrow N_2H_5^+$. But, in an alkaline medium, the probable reaction sequence can be, $N_2H_4 + 4OH^- \rightarrow N_2 + 4H_2O + 4e^-$. Since the reduction potential of the metal ion is greater in the acidic medium than that of the alkaline medium, hydrazine can act as the best reducing agent in the alkaline pH [87]. Even though the pH should not be very high since the abundance of OH$^-$ ions causes the bulk precipitation of $Ni(OH)_2$ [88]. According to previous reports, phase-pure Ni was obtained for samples with pH between 10 and 12, while for pH > 14, the sample contains a secondary phase of $Ni(OH)_2$ [84]. In 2019, our research group also reported the synthesis of Ni nanostructures by changing the pH of the reaction from 9.1 to 10.2 through controlled base addition [61]. The corresponding XRD spectra and SEM images of the Ni samples having pH of 10.2, 9.9 and 9.1 are shown in Fig. 5.

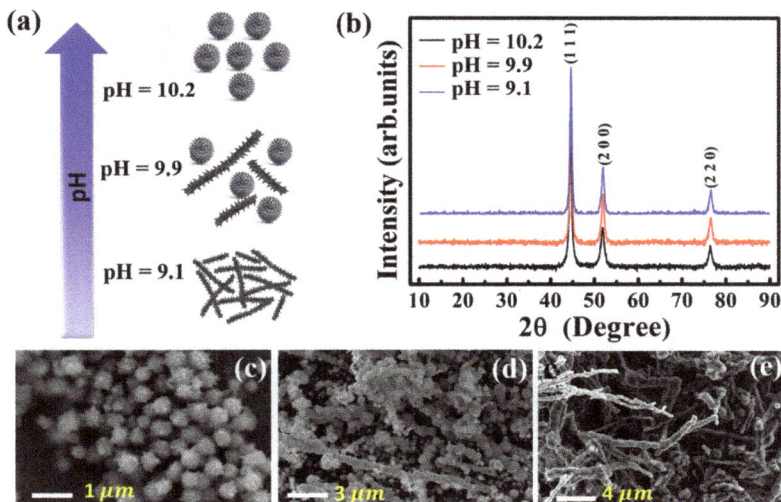

Fig. 5. (a) Schematic representation of morphology change with pH variation, (b) XRD patterns of Ni sample synthesized at different pH, SEM images of samples with a pH value of (c) 10.2, (d) 9.9, (e) 9.1 (Reproduced from reference [61] with permission from the Elsevier)

Figs. 5(c)-(e) shows the variation of morphology from sea urchin-like NPs to prickly NWs by changing the pH of the reaction from 10.2 to 9.1. This suggests that a lower pH promotes nanowire growth, whereas directional growth of the nanospecies is constrained as we increase the pH. Fig. 6 illustrates the self-assembly of sea-urchin-like Ni NPs and the formation of 1D Ni NWs. For large molar concentrations of NaOH, OH⁻ ions were dominating in the reaction medium. For the sample with higher pH, Ni NPs are completely surrounded by the anions and resulting in the repulsion between isolated NPs. Thus the growth mechanism is arrested, and uniform sea-urchin-like NPs are formed [89]. On the other hand, the concentration of OH⁻ ions decreases with the decrease in pH, i.e., a small amount of OH⁻ ions are present on the surface of Ni NPs. Hence, there is a possibility of dipolar interaction between Ni NPs, which causes the growth of ID Ni NWs. These results suggest that any decrease in pH favors the self-assembly of magnetic NPs via dipolar interactions [86, 89]. Hence, the role of NaOH in the synthesis of Ni nanostructures is vital, and the optimum concentration of NaOH can create directional growth of ferromagnetic nanostructures.

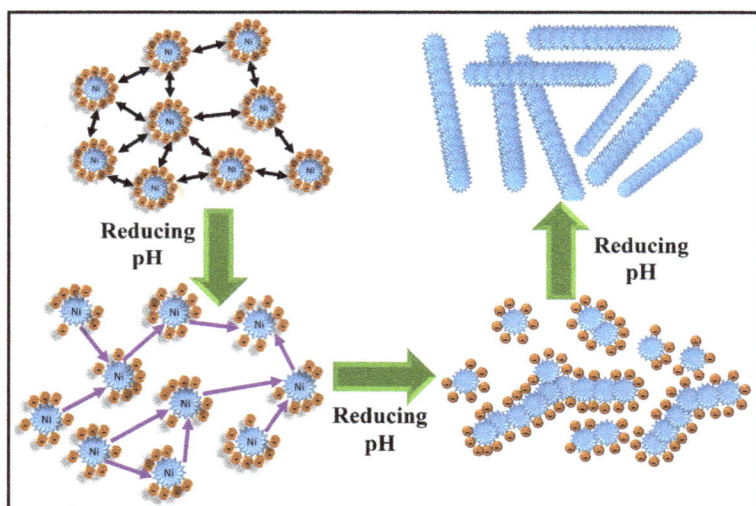

Fig. 6. Schematic illustration of the self-assembly of Ni NPs and the formation of Ni NWs by decreasing the pH of the reaction. (Green-colored arrows show attraction, double-headed black arrows shows repulsion)

In 2007, Zhang et al. investigated the effect of NaOH on the formation of urchin-like Ni and found that the prickle length of the urchin-like Ni increases with the increase in NaOH concentration [84]. Later, Zhao et al. [86] reported that the size of Ni NPs increases with an increase in NaOH concentration. This is due to the alkaline-dependent reducing ability of hydrazine; i.e., the reducing power of hydrazine in a basic medium is higher than that of an acidic medium [86].

For constant Ni^{2+} and N_2H_4 concentration, the Nernst equation can be written as,

$$\Delta E = E_0 + \frac{2.303RT}{4F} \log[OH]^-$$

(10)

Eq. (10) indicates that both the increase in NaOH concentration and temperature increases the ΔE. Eluri et al. [16] studied the influence of NaOH concentration and temperature on the reaction time for the synthesis of nickel, which is shown in Fig. 7. The increase in NaOH from 0.1 M to 0.5 M causes a significant reduction in reaction time from 60 minutes to 15 minutes at a reduction temperature of 60 °C. Also, the variation of NaOH is more significant than the temperature on reaction time [16].

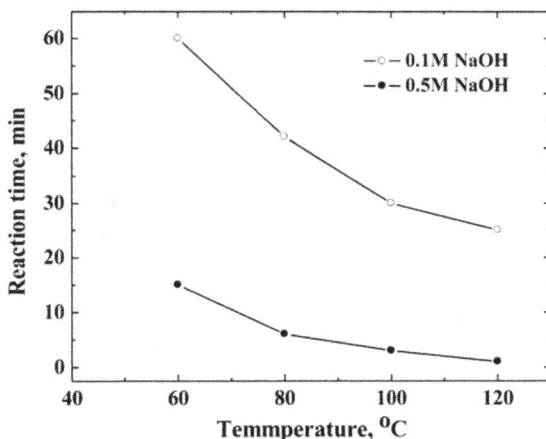

Fig. 7. Effect of [NaOH] and temperature on reaction time (Reproduced from reference [16] with permission from Springer)

2.6.2 Effect of nickel precursor

As hinted before, the nature and concentration of nickel precursors play an important role in the size-controlled reduction and morphology of Ni nanostructures. Tang et al. [90] and Kong et al. [91] reported the effect of initial concentration of Ni^{2+} used for the synthesis of Ni NWs, and both observed that the average diameter of NWs increases linearly with increasing concentration of Ni^{2+}. Also, the grain size of Ni NWs increases with the initial concentration of Ni, as shown in Fig. 8. On increasing the $[Ni^{2+}]$ concentration, excess Ni ions were accumulated on the surface of existing Ni nuclei and caused the formation of NPs or NWs with larger diameters [92]. Tang et al. observed that the surface of Ni NWs was rough for small $[Ni^{2+}]$ concentrations and smoother at higher concentrations [90]. In 2011, Krishnadas et al. [40] analyzed the length and diameter variation of Ni NWs by varying the concentration of Ni ions. They found that both the diameter and length of Ni NWs increased with concentration of Ni^{2+}. However, the increase in length as a function of nickel ion concentration is more prominent than the corresponding increase in diameter [40].

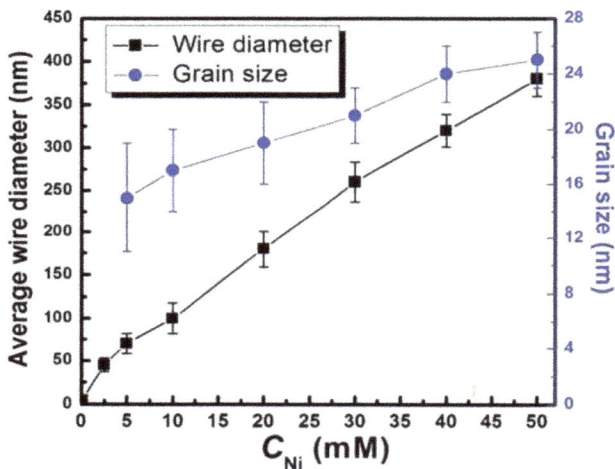

Fig. 8. The average diameter and grain size versus concentration of Ni (Reproduced from reference [90] with permission from the Royal Society of Chemistry)

In a magnetic field-assisted synthesis, Zhang et al. succeeded in producing Ni nanostructures with different Ni ions concentrations and, the corresponding SEM images are shown in Fig. 9 [87]. When Ni ion concentration was 0.1 M, nanowire formation was abruptly broken, and Ni clusters were produced (shown in Fig. 9(c)). Thus, the morphology and particle size of Ni nanostructures are strongly influenced by the concentration of Ni ions.

Fig. 9. SEM images of Ni NWs prepared with different Ni ion concentration (a) 0.01 M, (b) 0.05 M, (c) 0.1 M (Reproduced from reference [87] with permission from Springer)

2.6.3 Effect of hydrazine hydrate

The difference in hydrazine concentration used for the synthesis of Ni can make changes in the surface morphology. Krishnadas et al. synthesized the Ni NWs by using 80 μL, 200 μL, and 500 μL of $N_2H_4.H_2O$ solution [40]. For 80 μL, the surface of NW was smooth, while the surface became rough and nonuniform for higher concentrations of hydrazine. This was evident from the SEM images of samples shown in Fig. 10. However, in a study by Tang et al., the surface roughness of Ni NWs increased up to a critical concentration of hydrazine, and above that, NWs with smooth surfaces were formed [90].

In 2003, Wu et al. studied the effect of hydrazine hydrate on the particle size of Ni NPs [93]. On increasing the ratio of $\dfrac{[N_2H_4.H_2O]}{[NiCl_2]}$, the average diameter was decreased, and the diameter remains constant for a ratio greater than 12 [93]. Later in 2014, Chandra et al. also reported the decreasing trend of the average diameter of Ni NPs with increasing concentration of hydrazine [94]. These results ensure that the particle size and surface morphology can be greatly controlled by the amount of reducing agent.

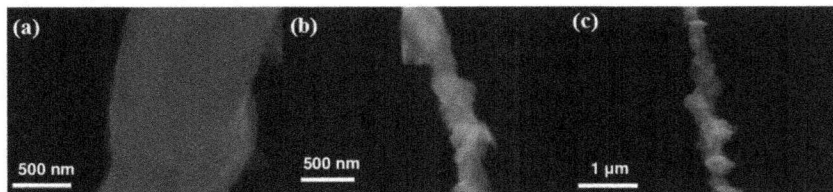

Fig. 10. SEM images of Ni NWs synthesized using hydrazine concentration of (a) 80 μL, (b) 200 μL, (c) 500 μL (Reproduced from reference [40] with permission from American Chemical Society)

2.6.4 Effect of reduction temperature

The role of temperature in the wet-chemical reduction of Ni NPs and NWs was extensively investigated in the last two decades. In 2008 Hu et al. reported the chemical reduction synthesis of Ni NPs and optimized the threshold reduction temperature needed for the phase-pure synthesis of Ni to be at 55 °C [82]. The resultant XRD patterns are shown in Fig. 11(a), in which an impurity phase of Ni(OH)$_2$ is formed below 55 °C. The variation of particle size with reduction temperature is depicted in Fig. 11(b) that indicates the increase in particle size with reduction temperature [82]. At higher reduction temperatures, the chemical potential increased that did generate a large number of nuclei, which is favorable to the production of small particle size. However, the increase in reduction temperature can accelerate the velocity of reaction, which leads to the aggregation of nuclei on the surface of Ni NPs and increases the size of NPs [82]. Similar results were also reported by He et al., who found that the particle size of Ni NPs was increased from 23 nm to 114 nm when the reaction temperature was increased from 240 °C to 285 °C [95].

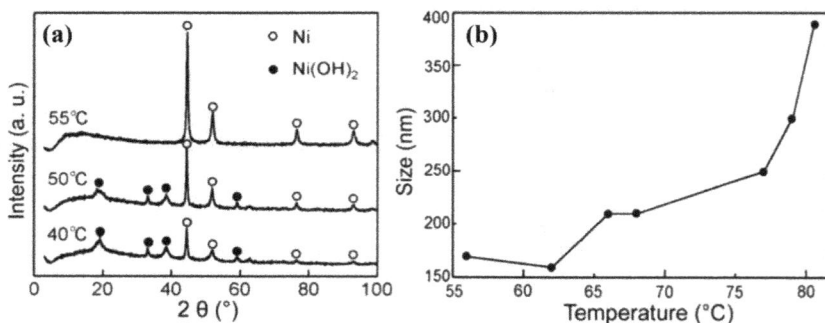

Fig. 11. (a) XRD patterns of Ni NPs synthesized at different reduction temperatures, (b) dependence of particle size on the reduction temperature (Reproduced from reference [82] with permission from Elsevier)

The influence of reduction temperature on the morphology of Ni nanostructures was reported by Jia et al., and the SEM images of samples obtained at 160 °C and 180 °C are shown in Fig. 12 [96]. Ni samples prepared at 160 °C have no defined morphology, even though some parts of the sample exhibit 1D morphology. The increase of reduction temperature to 180 °C yields Ni NWs of uniform morphology with an average diameter of 400 nm. It can be further understood that the particle size decreases with the increase of reduction temperature [96].

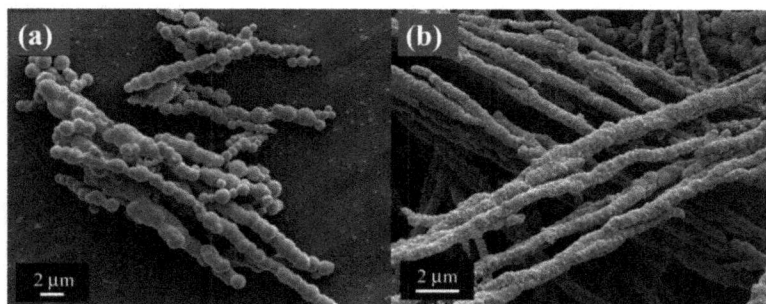

Fig. 12. SEM images of the Ni samples synthesized at (a) 160 °C and (b) 180 °C
(Reproduced from reference [96] with permission from Wiley)

Table 3. Variation of NW dimension with reduction temperature (Reproduced from
reference [40] with permission from American Chemical Society)

Sl. No.	Reduction Temperature [°C]	Average length of Ni NWs [μm]	Average diameter of Ni NWs [nm]
1	90	20 - 25	180 ± 7
2	120	5 - 6	130 ± 5
3	160	0.5 – 1.5	100 ± 4

In order to examine the variation of Ni NW dimension with reduction temperature, Krishnadas et al. [40] synthesized Ni NWs at three different temperatures, where the observed variations in the dimension of Ni NWs were incorporated in Table 3. From the table, it is evident that both the length and diameter of Ni NWs decrease with the increase in reduction temperature. An increase in chemical potential with increasing reduction temperature causes the generation of smaller nuclei, which is subsequently self-assembled via dipolar interaction to form one dimensional Ni NWs [40, 82]. However, the above-mentioned study probing the dimension–temperature correlation of Ni NW was limited to three temperatures only. So, the accuracy of data solicits further meticulous investigation by measuring the length and diameter of NWs at more consecutive temperatures.

In 2016, Xia et al. [97] conducted the Ni NW synthesis via chemical reduction at various temperatures between 60 °C and 150 °C, and the SEM images of NWs synthesized at three different temperatures are shown in Figs. 13 (a)-(c). SEM images clearly indicate a gradual change in the diameter of Ni NWs; i.e., the average diameter of NWs decreases with an increase in reduction temperature. The SEM image of the sample synthesized at 70 °C shows the formation of nanoprickles on the surface of NWs, and the prickle formation disappeared with the increase in reduction temperature. The variation in the dimension of NWs with reduction temperature is depicted in Figs. 13 (d)-(f). While increasing the temperature above 100 °C, length suddenly decreases, and the average width (average diameter) also decreases with reduction temperature. As a result, the length-to-width ratio was maximum at 100 °C, so that the shape anisotropy is larger for NWs prepared at 100 °C [97].

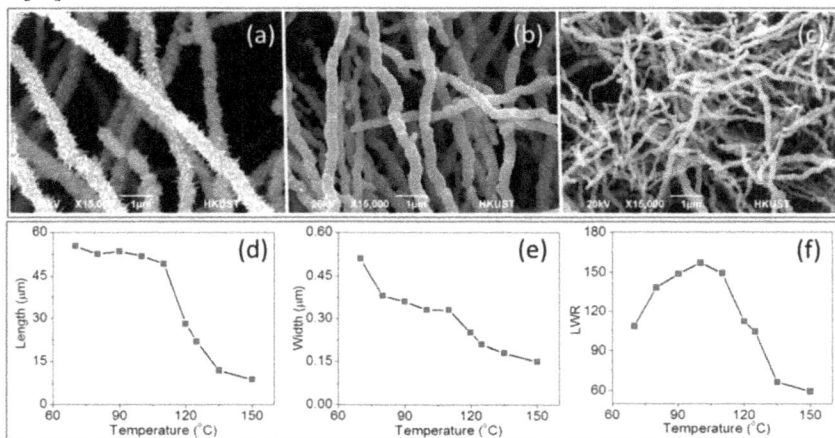

Fig. 13. SEM images of Ni NWs synthesized at (a) 70 °C, (b) 110 °C, and (c) 150 °C, variation of (d) length, (e) width, and (f) length-to-width ratio (LWR) of Ni NWs with reduction temperature (Reproduced from reference [97] with permission from the Multidisciplinary Digital Publishing Institute)

From the above-cited works, it is evident that the reduction temperature can strongly dictate the morphology and particle size of the material. In the case of NPs, reduction temperature and particle size have direct relation, whereas the average diameter of NWs decreases with an increase in reduction temperature [40, 82, 95-97].

2.6.5 Effect of reaction time

In 2007, Zhang et al. [84] made a detailed probe on the effect of reaction time on the Ni NPs synthesis at a temperature of 120 °C. The influence of reaction time on the field-assisted synthesis of Ni NWs was studied by Zhang et al. in 2009 [87]. Later, Zhao et al. investigated the variation in the average diameter of Ni chains at different reaction times [86]. The effect of reaction time on the morphology and particle size of Ni nanostructures were incorporated in Table 4.

Table 4. Effect of reaction time on the synthesis of Ni nanostructures

Sl. No.	Reaction time [Minute]	Observation	Ref.
1	1	Quasi-spherical NPs	[84]
	5	Mixture of studded spheres and urchin-like particles with very short prickles	
	10	Urchin-like particles with long prickles	
	40	Urchin-like particles with a length of prickles up to 300 nm	
2	30	Ni NWs were floated on the surface solution	[87]
	120	Ni NWs joined to form silvery-white flakes	
	180	Grey-coloured Ni NWs disappeared, and bright silvery-white Ni films were formed on the beaker wall	
3	5	Ni chains of the average diameter of 200 – 300 nm, and some of the chains with 500 – 600 nm	[86]
	10	Ni chains of the average diameter of 500 – 700 nm, and some of the chains with 300 – 400 nm	
	60	Ni chains of the average diameter of 500 – 700 nm	

In the same report, Zhang et al. observed a morphology change from quasi-spherical to the urchin-like structure while increasing the reaction time from 1 minute to 5 minutes. Further increase in reaction time causes the increase in prickle length of urchin-like particles, as well as the particle size, was increased [84]. For a longer reaction time, Ni NW powder formation was arrested, and the Ni films were formed on the walls of the beaker [87]. Zhao et al. reported the increase in average diameter of Ni chains with the increase in reaction time [86].

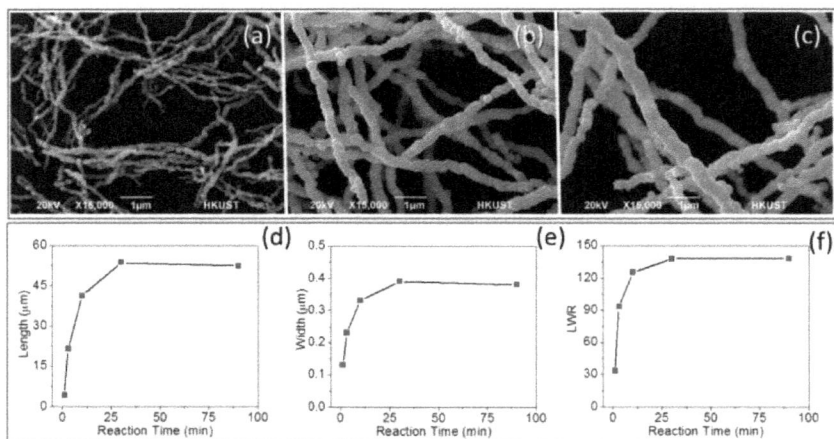

Fig. 14. SEM images of Ni NWs synthesized with a reaction time of (a) 1 minute, (b) 10 minute, and (c) 90 minute, variation of (d) length, (e) width, and (f) LWR of Ni NWs with reaction time (Reproduced from reference [97] with permission from the Multidisciplinary Digital Publishing Institute)

Fig. 14 shows the SEM images of Ni NWs prepared at different reaction times, which indicates that the diameter of NWs increases with reaction time. Both the length and width of NWs increase abruptly while changing the reaction time from 1 minute to 10 minutes, which is due to the sharp increase of shape anisotropy of NWs. After 30 minutes, reaction time had no influence on the dimension of Ni NWs, and there was no further increase of particle size [97]. These observations point to the fact that the particle size increases with the increase in reaction time up to a specific duration, and there is no further growth as a function of reaction time.

Conclusion

Nickel nanostructures, thanks to their myriad applications in the fields of magnetic fluids, solid oxide fuel cells, automotive catalytic converters, and catalysis, are catching both academic and industrial attention in recent years. The synthesis of nanosized nickel with uniform size and desirable morphology is truly a challenging task due to its close dependence on experimental conditions. The wet-chemical reduction route is one of the best synthesis strategies for the development of Ni nanostructures. This chapter provides an overview of the chemical reduction of nickel producing various morphologies. The reduction mechanism of Ni using hydrazine monohydrate and the effect of various reaction conditions on the particle size and morphology of reduced nickel are profoundly discussed.

The decrease in the overall pH value of the reaction induces the self-assembly of Ni NPs and the formation of 1D nickel nanostructures. Thus the concentration of alkali used in the reduction reaction has a unique role in controlling the pH value and the morphology thereafter. The particle size and grain size of Ni nanostructures increase with the increase in the concentration of Ni^{2+} ions. Also, a larger concentration of Ni^{2+} constrains the oriented growth of nanospecies, resulting in the breakage of 1D morphology. The average diameter of reduced Ni nanostructures decreases with the increased concentration of hydrazine. In addition to that, the amount of hydrazine can affect the surface roughness of nanostructures. The influence of reduction temperature is different for NPs and NWs; with the increase of reduction temperature, the particle size of NPs increases while the average diameter of NWs decreases. On increasing the reaction time, particle size increases up to a critical duration, above which the size remains constant. In a nutshell, each reaction condition can strongly affect the particle size and morphology of Ni nanostructures and thereby critically tune the magnetic properties of the system.

References

[1] L.A. Paramo, A.A. Feregrino-Perez, R. Guevara, S. Mendoza, K. Esquivel, Nanoparticles in agroindustry: applications, toxicity, challenges, and trends, Nanomaterials. 10 (2020) 1654. https://doi.org/10.3390/nano10091654

[2] D.E. Laughlin, D.N. Lambeth, Microstructural and crystallographic aspects of thin film recording media, IEEE Transactions on Magnetics. 36 (2000) 48-53. https://doi.org/10.1109/20.824424

[3] B. Issa, I.M. Obaidat, B.A. Albiss, Y. Haik, Magnetic nanoparticles : surface effects and properties related to biomedicine applications, Int. J. Mol. Sci. 14 (2013) 21266-21305. https://doi.org/10.3390/ijms141121266

[4] S. Schrittwieser, D. Reichinger, J. Schotter, Applications, Surface modification and functionalization of nickel nanorods, Materials. 11 (2018) 45. https://doi.org/10.3390/ma11010045

[5] A. Sagasti, V. Palomares, J.M. Porro, I. Orue, M.B. SanchezIlarduya, A.C. Lopes, J. Gutiérrez, Magnetic, magnetoelastic and corrosion resistant properties of (Fe-Ni)-based metallic glasses for structural health monitoring applications, Materials. 13 (2020) 57-70. https://doi.org/10.3390/ma13010057

[6] F. Ma, Q. Li, J. Huang, J. Li, Morphology control and characterizations of nickel sea-urchin-like and chain-like nanostructures, J. Cryst. Growth. 310 (2008) 3522-3527. https://doi.org/10.1016/j.jcrysgro.2008.04.044

[7] E. Verrelli, D. Tsoukalas, D. Giannakopoulos, K Kouvatsos, P. Normand, D.E. Ioannou, Nickel nanoparticle deposition at room temperature for memory applications, Microelectron. Eng. 84 (2007) 1994-1997. https://doi.org/10.1016/j.mee.2007.04.078

[8] I. Bibi, S. Kamal, A. Ahmed, M. Iqbal, S. Nouren, K. Jilani, N. Nazar, M. Amir, A. Abbas, S. Ata, F. Majid, Nickel nanoparticle synthesis using camellia sinensis as reducing and capping agent: growth mechanism and photocatalytic activity evaluation, Int. J. Biol Macromol. 103 (2017) 783-790. https://doi.org/10.1016/j.ijbiomac.2017.05.023

[9] Y. Cheng, M. Guo, M. Zhai, Y. Yu, J. Hu, Nickel nanoparticles anchored onto Ni foam for supercapacitors with high specific capacitance, J. Nanosci . Nanotechnol. 20 (2020) 2402-2407. https://doi.org/10.1166/jnn.2020.17377

[10] A.R. Abdel Fattah, T. Majdi, A.M. Abdalla, S. Ghosh, I.K. Puri, Nickel nanoparticles entangled in carbon nanotubes: novel ink for nanotube printing, ACS Appl. Mater. Interfaces. 8 (2016) 1589-1593. https://doi.org/10.1021/acsami.5b11700

[11] M.M. Barsan, T.A. Enache, N. Preda, G. Stan, N.G. Apostol, E. Matei, A. Kuncser, V. Diculesc C., Direct immobilization of biomolecules through magnetic forces on Ni electrodes via Ni nanoparticles: applications in electrochemical biosensors, ACS Appl. Mater. Interfaces. 11 (2019) 19867-19877. https://doi.org/10.1021/acsami.9b04990

[12] A.P. Reena Mary, C.. S.S. Sandeep, T.. N. Narayanan, R. Philip, P. Moloney, P.M. Ajayan, M.. Anantharaman, Nonlinear and magneto-optical transmission studies on magnetic nanofluids of non-interacting metallic nickel nanoparticles, Materials. 22 (2011) 375702-375708. https://doi.org/10.1088/0957-4484/22/37/375702

[13] R. Krishnapriya, S. Praneetha, A. V Murugan, Microwave-solvothermal synthesis of various TiO_2 nano-morphologies with enhanced efficiency by incorporating Ni nanoparticles in an electrolyte for dye-sensitized solar cells, Inorg. Chem. Front. 4 (2017) 1665-1678. https://doi.org/10.1039/C7QI00329C

[14] K. Raj, B. Moskowitz, R. Casciari, Advances in ferrofluid technology, J. Magn. Magn. Mater. 149 (1995) 174-180. https://doi.org/10.1016/0304-8853(95)00365-7

[15] S.H. Wu, D.H. Chen, Synthesis and characterization of nickel nanoparticles by hydrazine reduction in ethylene glycol, J. Colloid Interface Sci. 259 (2003) 282-286. https://doi.org/10.1016/S0021-9797(02)00135-2

[16] R. Eluri, B. Paul, Synthesis of nickel nanoparticles by hydrazine reduction: mechanistic study and continuous flow synthesis, J. Nanopart. Res. 14:800 (2012) 1-14. https://doi.org/10.1007/s11051-012-0800-1

[17] W. Wernsdorfer, K. Hasselbach, A. Benoit, B. Barbara, B. Doudin, J. Meier, J. Ansermet, D. Mailly, Measurements of magnetization switching in individual nickel nanowires, Phy. Rev. B. 55 (1997) 11552-11559. https://doi.org/10.1103/PhysRevB.55.11552

[18] Z. Huajun, Z. Jinhuan, G. Zhenghai, W. Wei, Preparation and magnetic properties of Ni nanorod arrays, J. Magn. Magn. Mater. 320 (2008) 565-570. https://doi.org/10.1016/j.jmmm.2007.07.018

[19] X. Ni, Q. Zhao, H. Zheng, J. Song, D. Zhang, X. Zhang, A novel chemical reduction route towards the synthesis of crystalline nickel nanoflowers from a mixed source, Eur. J. Inorg. Chem. 23 (2005) 4788-4793. https://doi.org/10.1002/ejic.200500453

[20] M. Sanles-Sobrido, M. Bañobre-López, V. Salgueiriño, M.A. Correa-Duarte, B. Rodríguez-González, J. Rivas, L.M. Liz-Marzán, Tailoring the magnetic properties of nickel nanoshells through controlled chemical growth, J Mater. Chem. 20 (2010) 7360-7365. https://doi.org/10.1039/c0jm01107j

[21] S.H. Xu, G.T. Fei, H.M. Ouyang, Y. Zhang, P.C. Huo, L. De Zhang, Controllable fabrication of nickel nanoparticle chains based on electrochemical corrosion, Journal of Materials Chemistry C. 3 (2015) 2072-2079. https://doi.org/10.1039/C4TC02450H

[22] Nicola A Spaldin, Magnetic Materials: Fundamentals and applications, 2nd ed., Cambridge University Press, New York, 2011.

[23] Y. Koltypin, A. Fernandez, T.. Rojas, J. Campora, P. Palma, R. Prozorov, A. Gedanken, Encapsulation of Nickel Nanoparticles in Carbon Obtained by the Sonochemical Decomposition of Ni(C$_8$H$_{12}$)$_2$, Chem. Mater. 11 (1999) 1331-1335. https://doi.org/10.1021/cm981111o

[24] T.O. Ely, C. Amiens, B. Chaudret, E. Snoeck, M. Verelst, M. Respaud, J.-M. Broto, Synthesis of nickel nanoparticles. Influence of aggregation induced by modification of poly(vinylpyrrolidone) chain length on their magnetic properties, Chem. Mater. 11 (1999) 526-529. https://doi.org/10.1021/cm980675p

[25] N. Cordente, M. Respaud, F. Senocq, M.-J. Casanove, C. Amiens, B. Chaudret, Synthesis and magnetic properties of nickel nanorods, Nano Lett. 1 (2001) 565-568. https://doi.org/10.1021/nl0100522

[26] A.-G. Boudjahem, S. Monteverdi, M. Mercy, D. Ghanbaja, M.M. Bettahar, Nickel nanoparticles supported on silica of low surface area. hydrogen chemisorption and TPD and catalytic properties, Catal. Lett. 84 (2002) 115-122. https://doi.org/10.1023/A:1021093005287

[27] Y.-P. Sun, H.. Rollins, R. Guduru, Preparations of Nickel, Cobalt, and Iron Nanoparticles through the Rapid Expansion of Supercritical Fluid Solutions (RESS) and chemical reduction, Chem. Mater. 11 (1999) 7-9. https://doi.org/10.1021/cm9803253

[28] A. Duteil, G. Schmid, W. Meyer-Zaika, Ligand stabilized nickel colloids, J. Chem. Soc. Chem. Commun. (1995) 31-32. https://doi.org/10.1039/c39950000031

[29] C.. Murray, S. Sun, H. Doyle, T. Betley, Monodisperse 3d transition-metal (Co,Ni,Fe) nanoparticles and their assembly into nanoparticle superlattices, MRS Bull. 26 (2001) 985-991. https://doi.org/10.1557/mrs2001.254

[30] J. Park, E. Kang, S.U. Son, H.M. Park, M.K. Lee, J. Kim, K.W. Kim, H.-J. Noh, J.-H. Park, C.J. Bae, J.-G. Park, T. Hyeon, Monodisperse Nanoparticles of Ni and NiO.

Synthesis, characterization, self-assembled superlattices, and catalytic applications in the suzuki coupling reaction., Adv. Mater. 17 (2005) 429-434. https://doi.org/10.1002/adma.200400611

[31] C.J. Pandian, R. Palanivel, S. Dhanasekaran, Screening antimicrobial activity of nickel nanoparticles synthesized using ocimum sanctum leaf extract, J. Nanopart. 2016 (2016) 1-13. https://doi.org/10.1155/2016/4694367

[32] H. Chen, J. Wang, D. Huang, X. Chen, J. Zhu, D. Sun, J. Huang, Q. Li, Plant mediated synthesis of size-controllable Ni nanoparticles with alfalfa extract, Mater. Lett. 122 (2014) 166-169. https://doi.org/10.1016/j.matlet.2014.02.028

[33] S.A. Mamuru, A.S. Bello, S.B. Hamman, Annona squamosa leaf extract as an efficient bioreducing agent in the synthesis of chromium and nickel nanoparticles, IJASBT. 3 (2015) 167-169. https://doi.org/10.3126/ijasbt.v3i2.11651

[34] H. Guo, B. Pu, H. Chen, J. Yang, Y. Zhou, J. Yang, B. Bismark, H. Li, X. Niu, Surfactant assisted solvothermal synthesis of pure nickel submicron spheres with microwave-absorbing properties, Nanoscale Res Lett. 11 (2016) 352-367. https://doi.org/10.1186/s11671-016-1562-y

[35] Z. Liu, S. Li, Y. Yang, S. Peng, Z. Hu, Y. Qian, Complex- surfactant-assisted hydrothermal route to ferromagnetic nickel nanobelts, Adv. Mater. 15 (2003) 1946-1948. https://doi.org/10.1002/adma.200305663

[36] W. Xu, K.Y. Liew, H. Liu, T. Huang, C. Sun, Y. Zhao, Microwave-assisted synthesis of nickel nanoparticles, Mater. Lett. 62 (2008) 2571-2573. https://doi.org/10.1016/j.matlet.2007.12.057

[37] J. Bao, C. Tie, Z. Xu, Q. Zhou, D. Shen, Q. Ma, Template synthesis of an array of nickel nanotubules and its magnetic behavior, Adv. Mater. 13 (2001) 1631-1633. https://doi.org/10.1002/1521-4095(200111)13:21<1631::AID-ADMA1631>3.0.CO;2-R

[38] J. Wang, L.Y. Zhang, P. Liu, T.M. Lan, J. Zhang, L.M. Wei, E.S.-W. Kong, C.H. Jiang, Y.F. Zhang, Preparation and growth mechanism of nickel nanowires under applied magnetic field, Nano-Micro Lett. 2 (2010) 134-138. https://doi.org/10.1007/BF03353631

[39] A. Pandey, R. Manivannan, A Study on synthesis of nickel nanoparticles using chemical reduction technique, Recent Pat. Nanotechnol . 5 (2015) 33-37. https://doi.org/10.2174/1877912305666150417232717

[40] K.R. Krishnadas, P.R. Sajanlal, T. Pradeep, Pristine and hybrid nickel nanowires: Template, magnetic field, and surfactant-free wet chemical synthesis and raman studies, J. Phys. Chem. C. 115 (2011) 4483-4490. https://doi.org/10.1021/jp110498x

[41] M.D. Hossain, R.A. Mayanovic, S. Dey, R. Sakidja, M. Benamara, Room-temperature ferromagnetism in Ni(ii)-chromia based core-shell nanoparticles:

experiment and first principles calculations, Phys. Chem. Chem. Phys. 20 (2018) 10396-10406. https://doi.org/10.1039/C7CP08597D

[42] D.H. Chen, S.H. Wu, Synthesis of nickel nanoparticles in waterin-oil microemulsions, Chem. Mater. 12 (2000) 1354-1360. https://doi.org/10.1021/cm991167y

[43] L. Karam, J. Reboul, N. El Hassan, J. Nelayah, P. Massiani, Nanostructured nickel aluminate as a key intermediate for the production of highly dispersed and stable nickel nanoparticles supported within mesoporous alumina for dry reforming of methane, Molecules. 24 (2019) 4107-4119. https://doi.org/10.3390/molecules24224107

[44] W. You, R. Che, Excellent NiO-Ni nanoplate microwave absorber via pinning effect of antiferromagnetic-ferromagnetic interface, ACS Appl. Mater. Interfaces. 10 (2018) 15104-15111. https://doi.org/10.1021/acsami.8b03610

[45] C.A. Sequeira, D.S. Cardoso, L. Amaral, B. Šljukić, D.M. Santos, On the performance of commercially available corrosion resistant nickel alloys: a review, Corros. Rev. 34 (2016) 187-200. https://doi.org/10.1515/corrrev-2016-0014

[46] M. Imran Din, A. Rani, Recent advances in the synthesis and stabilization of nickel and nickel oxide nanoparticles: a green adeptness, Int. J. Anal .Chem. 2016 (2016) 1-4. https://doi.org/10.1155/2016/3512145

[47] G.R. Thellaputta, P.S. Chandra, C. Rao, Machinability of nickelbased superalloys: a review, Mater. Today Proc. 4 (2017) 3712-3721. https://doi.org/10.1016/j.matpr.2017.02.266

[48] R.S. Kate, S.A. Khalate, R.J. Deokate, Overview of nanostructured metal oxides and pure nickel oxide (NiO) electrodes for supercapacitors: a review, J. Alloys Compd. 734 (2018) 89-111. https://doi.org/10.1016/j.jallcom.2017.10.262

[49] L. Zhang, D. Shi, T. Liu, M. Jaroniec, J. Yu, Nickel-based materials for supercapacitors, Mater. Today. 25 (2019) 35-65. https://doi.org/10.1016/j.mattod.2018.11.002

[50] H. Ni, J. Zhu, Z. Wang, H. Lv, Y. Su, X. Zhang, A brief overview on grain growth of bulk electrodeposited nanocrystalline nickel and nickel-iron alloys, Rev. Adv. Mater. Sci. 58 (2019) 98-106. https://doi.org/10.1515/rams-2019-0011

[51] N.-D. Jaji, H.L. Lee, M.H. Hussin, H.M. Akil, M.R. Zakaria, M.B.H. Othman, Advanced nickel nanoparticles technology: From synthesis to applications, Nanotechnol. Rev. 9 (2020) 1456-1480. https://doi.org/10.1515/ntrev-2020-0109

[52] A.M. Ealias, M. Saravanakumar, A review on the classification, characterisation, synthesis of nanoparticles and their application, IOP Conf. Ser. Mater. Sci . Eng. 263 (2017) 32019-32032. https://doi.org/10.1088/1757-899X/263/3/032019

[53] J.K. Basu, S. Sengupta, Catalytic reduction of nitrobenzene using silver nanoparticles embedded calcium alginate film, J. Nanosci . Nanotechnol. 19 (2019) 7487-7492. https://doi.org/10.1166/jnn.2019.16669

[54] Y. Hou, H. Kondoh, T. Ohta, S. Gao, Size-controlled synthesis of nickel nanoparticles, Appl. Surf. Sci. 241 (2005) 218-222. https://doi.org/10.1016/j.apsusc.2004.09.045

[55] G. Villaverde-Cantizano, M. Laurenti, J. Rubio-Retama, R. Contreras-Cáceres, Reducing agents in colloidal nanoparticle synthesis - an Introduction, in: Nanoscience & Nanotechnology Series, RSC, 2021: pp. 1-27. https://doi.org/10.1039/9781839163623-00001

[56] M.T. Rahman, E. V. Rebrov, Microreactors for gold nanoparticles synthesis: from faraday to flow, Processes. 2 (2014) 466-493. https://doi.org/10.3390/pr2020466

[57] K.M. Koczkur, S. Mourdikoudis, L. Polavarapu, S.E. Skrabalak, Polyvinylpyrrolidone (PVP) in nanoparticle synthesis, Dalton Trans. 44 (2015) 17883-17905. https://doi.org/10.1039/C5DT02964C

[58] T.S. Rodrigues, M. Zhao, T.H. Yang, K.D. Gilroy, A.G.M. da Silva, P.H.C. Camargo, Y. Xia, Synthesis of Colloidal Metal Nanocrystals: A Comprehensive Review on the Reductants, Chem. - A Eur. J. 24 (2018) 16944 -16963. https://doi.org/10.1002/chem.201802194

[59] D. Gozzi, A. Latini, G. Capannelli, F. Canepa, M. Napoletano, M.R. Cimberle, M. Tropeano, Synthesis and magnetic characterization of Ni nanoparticles and Ni nanoparticles in multiwalled carbon nanotubes, J. Alloys Compd. 419 (2006) 32-39. https://doi.org/10.1016/j.jallcom.2005.10.012

[60] C.-M. Liu, L. Guo, R.-M. Wang, Y. Deng, H.-B. Xu, S. Yang, Magnetic nanochains of metal formed by assembly of small nanoparticles, Chem. Commun. (2004) 2726-2727. https://doi.org/10.1039/b411311j

[61] R. Revathy, M.R. Varma, K.P. Surendran, Effect of morphology and ageing on the magnetic properties of nickel nanowires, Mater. Res. Bull. 120 (2019) 110576. https://doi.org/10.1016/j.materresbull.2019.110576

[62] P.K. Khanna, P. V. More, J.P. Jawalkar, B.G. Bharate, Effect of reducing agent on the synthesis of nickel nanoparticles, Mater. Lett. 63 (2009) 1384-1386. https://doi.org/10.1016/j.matlet.2009.02.013

[63] H. Niu, Q. Chen, M. Ning, Y. Jia, X. Wang, Synthesis and one-dimensional self-assembly of acicular nickel nanocrystallites under magnetic fields, J. Phys. Chem. B. 108 (2004) 3996-3999. https://doi.org/10.1021/jp0361172

[64] O. Antola, L. Holappa, P. Paschen, Nickel ore reduction by hydrogen and carbon monoxide containing gases, Miner. Process. Extr. Metall. Rev. 15 (1995) 169-179. https://doi.org/10.1080/08827509508914195

[65] J.W. Ju, M.Y. Jung, Formation of conjugated linoleic acids in soybean oil during hydrogenation with a nickel catalyst as affected by sulfur addition, J. Agric. Food Chem. 51 (2003) 3144-3149. https://doi.org/10.1021/jf0259213

[66] A.B. Elena, A. Cally, A. Allagui, S. Ntais, R. Wüthrich, Nickel particles with increased catalytic activity towards hydrogen evolution reaction, C. R. Chim. 16 (2013) 28-33. https://doi.org/10.1016/j.crci.2012.02.003

[67] M. Heilmann, H. Kulla, C. Prinz, R. Bienert, U. Reinholz, A.G. Buzanic, F. Emmerling, advances in nickel nanoparticle synthesis via oleylamine route, Nanomaterials. 10 (2020) 1-13. https://doi.org/10.3390/nano10040713

[68] T.G. Stuart, J.F. Stephen, R. Krchnavek, Controlled particle growth of silver sols through the use of hydroquinone as a selective reducing agent, Langmuir. 25 (2009) 2613-2621. https://doi.org/10.1021/la803680h

[69] S. Vivekanandhan, V. Manne, N. Satyanarayana, Effect of ethylene glycol on polyacrylic acid based combustion process for the synthesis of nano-crystalline nickel ferrite (NiFe$_2$O$_4$), Mater. Lett. 58 (2004) 2717-2720. https://doi.org/10.1016/j.matlet.2004.02.030

[70] M. Maize, H.A. El-Boraey, I.A. Mohamed, J.D. Holmes, G. Collins, Controlled morphology and dimensionality evolution of NiPd bimetallic nanostructures, J. Colloid Interface Sci. 585 (2021) 480-489. https://doi.org/10.1016/j.jcis.2020.10.030

[71] A. R, D. H, Sodium-dodecyl-sulphate-assisted synthesis of Ni nanoparticles: electrochemical properties, Bull. Mater. Sci. 40 (2017) 1361-1369. https://doi.org/10.1007/s12034-017-1500-3

[72] Z. Zhang, H. Chen, C. Xing, M. Guo, F. Xu, X. Wang, H.J. Gruber, B. Zhang, J. Tang, Sodium citrate: A universal reducing agent for reduction / decoration of graphene oxide with au nanoparticles, Nano Res. 4 (2011) 599-611. https://doi.org/10.1007/s12274-011-0116-y

[73] M. Ali, N. Remalli, V. Gedela, B. Padya, P.K. Jain, A. Al-Fatesh, U. Ali Rana, V.V.S.S. Srikanth, Ni nanoparticles prepared by simple chemical method for the synthesis of Ni/NiO-multi-layered graphene by chemical vapor deposition, Solid State Sci. 64 (2017) 34-40. https://doi.org/10.1016/j.solidstatesciences.2016.12.007

[74] K.J.A. Raj, B. Viswanathan, Synthesis of nickel nanoparticles with fcc and hcp crystal structures, Indian J. Chem. - Inorg. Phys. Theor. Anal. Chem. 50 (2011) 176-179.

[75] X. Zhou, Z. Chen, D. Yan, H. Lu, Deposition of Fe-Ni nanoparticles on polyethyleneimine-decorated graphene oxide and application in catalytic dehydrogenation of ammonia borane, J. Mater. Chem. 22 (2012) 13506-13516. https://doi.org/10.1039/c2jm31000g

Materials Research Forum LLC
https://doi.org/10.21741/9781644902332-5

[76] Y. Li, Y. Cao, D. Jia, A general strategy for synthesis of metal nanoparticles by a solid-state redox route under ambient conditions, J. Mater. Chem. A. 2 (2014) 3761 - 3765. https://doi.org/10.1039/c3ta14427e

[77] N.R. Nik Roselina, A. Azizan, Ni nanoparticles: Study of particles formation and agglomeration, Procedia Eng. 41 (2012) 1620-1626. https://doi.org/10.1016/j.proeng.2012.07.359

[78] Y.D. Li, C.W. Li, H.R. Wang, L.Q. Li, Y.T. Qian, Preparation of nickel ultrafine powder and crystalline film by chemical control reduction, Mater. Chem. Phys. 59 (1999) 88-90. https://doi.org/10.1016/S0254-0584(99)00015-2

[79] J. Zhang, W. Xiang, Y. Liu, M. Hu, K. Zhao, Synthesis of high-aspect-ratio nickel nanowires by dropping method, Nanoscale Res. Lett. 11 (2016) 1-5. https://doi.org/10.1186/s11671-015-1209-4

[80] W. T. A, Jr., L.M. Sander, Diffusion-limited aggregation, a kinetic critical phenomenon, Phys. Rev. Lett. 47 (1981) 1400. https://doi.org/10.1103/PhysRevLett.47.1400

[81] T.C. Halsey, Diffusion-limited aggregation as branched growth, MRS Online Proceedings Library. 367 (1994) 23-32. https://doi.org/10.1557/PROC-367-23

[82] H. Hu, K. Sugawara, Selective synthesis of metallic nickel particles with control of shape via wet chemical process, Mater. Lett. 62 (2008) 4339-4342. https://doi.org/10.1016/j.matlet.2008.07.033

[83] W.W. Mullins, R.F. Sekerka, Stability of a planar interface during solidification of a dilute binary alloy, J. Appl. Phys. 35 (1964) 444-451. https://doi.org/10.1063/1.1713333

[84] G. Zhang, T. Zhang, X. Lu, W. Wang, J. Qu, X. Li, Controlled synthesis of 3d and 1d nickel nanostructures using an external magnetic field assisted solution-phase approach, J. Phys. Chem. C. 111 (2007) 12663-12668. https://doi.org/10.1021/jp073075z

[85] D.W. Ã, D. Sun, H. Yu, H. Meng, Morphology controllable synthesis of nickel nanopowders by chemical reduction process, J. Cryst. Growth. 310 (2008) 1195-1201. https://doi.org/10.1016/j.jcrysgro.2007.12.052

[86] B. Zhao, B. Fan, G. Shao, B. Wang, X. Pian, W. Li, R. Zhang, Investigation on the electromagnetic wave absorption properties of Ni chains synthesized by a facile solvothermal method, Appl. Surf. Sci. 307 (2014) 293-300. https://doi.org/10.1016/j.apsusc.2014.04.029

[87] L.Y. Zhang, J. Wang, L.M. Wei, P. Liu, H. Wei, Y.F. Zhang, Synthesis of Ni nanowires via a hydrazine reduction route in aqueous ethanol solutions assisted by external magnetic fields, Nano-Micro Lett. 1 (2009) 49-52. https://doi.org/10.1007/BF03353607

[88] S. Sarkar, A.K. Sinha, M. Pradhan, M. Basu, Y. Negishi, T. Pal, Redox transmetalation of prickly nickel nanowires for morphology controlled hierarchical synthesis of nickel / gold nanostructures for enhanced catalytic activity and SERS responsive functional material, J. Phys. Chem. C. 115 (2011) 1659-1673. https://doi.org/10.1021/jp109572c

[89] R. Revathy, A.A. Nair, M. Raama Varma, K.P. Surendran, Magnetism of cobalt during oxidative ageing: A theory supported experimental investigation, Mater. Sci. Eng. B. 273 (2021) 115453. https://doi.org/10.1016/j.mseb.2021.115453

[90] S. Tang, S. Vongehr, H. Ren, X. Meng, Diameter-controlled synthesis of polycrystalline nickel nanowires and their size dependent magnetic properties, Cryst. Eng. Comm. 14 (2012) 7209. https://doi.org/10.1039/c2ce25855b

[91] Y.Y. Kong, S.C. Pang, S.F. Chin, Facile synthesis of nickel nanowires with controllable morphology, Materials Lett. 142 (2015) 1-3. https://doi.org/10.1016/j.matlet.2014.11.140

[92] A. Mathew, N. Munichandraiah, G.M. Rao, Synthesis and magnetic studies of flower-like nickel nanocones, Mater. Sci. Eng. B. 158 (2009) 7-12. https://doi.org/10.1016/j.mseb.2008.12.032

[93] S.H. Wu, D.H. Chen, Synthesis and characterization of nickel nanoparticles by hydrazine reduction in ethylene glycol, J. Colloid Interface Sci. 259 (2003) 282-286. https://doi.org/10.1016/S0021-9797(02)00135-2

[94] S. Chandra, A. Kumar, P.K. Tomar, Synthesis of Ni nanoparticles and their characterizations, J. Saudi Chem. Soc. 18 (2014) 437-442. https://doi.org/10.1016/j.jscs.2011.09.008

[95] X. He, W. Zhong, C.T. Au, Y. Du, Size dependence of the magnetic properties of Ni nanoparticles prepared by thermal decomposition method, Nanoscale Res. Lett. 8 (2013) 1-10. doi:10.1186/1556-276X-8-446. https://doi.org/10.1186/1556-276X-8-446

[96] F. Jia, L. Zhang, X. Shang, Y. Yang, Non-aqueous sol-gel approach towards the controllable synthesis of nickel nanospheres, nanowires, and nanoflowers, Adv. Mater. 20 (2008) 1050-1054. https://doi.org/10.1002/adma.200702159

[97] Z. Xia, W. Wen, Synthesis of nickel nanowires with tunable characteristics, Nanomaterials. 6 (2016) 19. https://doi.org/10.3390/nano6010019

Materials Research Forum LLC
https://doi.org/10.21741/9781644902332-6

Chapter 6

Electromagnetic Properties of Cobalt Ferrite (CoFe$_2$O$_4$) with and without Addition of Niobium Pentoxide

F.E. Carvalho[1*], A.C.C. Migliano[1], J.P.B. Machado[2], R.C. Pullar[3,4**], R.B. Jotania[5]

[1]Instituto de Estudos Avançados – IEAv, Trevo Cel Aviador José Alberto Albano do Amarante 01 – Putim, São José dos Campos, SP

[2]Instituto Nacional de Pesquisas Espaciais – INPE, Av. Dos Astronautas, 1758 - Jardim da Granja, São José dos Campos, SP

[3]Universidade de Aveiro, Campus Universitário de Santiago, 3810 - 193, Aveiro, Portugal

[4]Department of Molecular Sciences and Nanosystems (DSMN), Ca' Foscari University of Venice, Scientific Campus, Via Torino 155, 30172 Venezia Mestre (VE), Italy

[5]Department of Physics, Electronics and Space science, University school of sciences, Gujarat University, Ahmedabad – 380 009, India

*f.carvalho@ieav.cta.br; **robertcarlye.pullar@unive.it

Abstract

This chapter compares the electromagnetic behavior of CoFe$_2$O$_4$ with and without addition of Nb$_2$O$_5$ in order to find biomedical applications of these compositions. Taking into account that their electromagnetic behavior is affected by their morphological structure, a structural analysis of samples prepared with different stoichiometries and sintered at different temperatures was performed. The formation of a new phase and the presence of lamellar veins crystallographically oriented by their magnetic domains were identified. In addition, complex measurements of electrical permittivity and magnetic permeability allowed us to infer that maximum absorption frequency increases with the addition of niobium pentoxide, while hysteresis cycles indicated a decrease in saturation magnetization and an increase in coercive force.

Keywords

Electromagnetic Behavior, Niobium Pentoxide, Cobalt Ferrite, Biomedical Application

Contents

1.   Introduction

The increasing utilization of the electromagnetic spectrum imposed by the technological evolution of the last decades requires that we use shields to minimize the harmful effects of non-ionizing radiation caused by it. Spinel ferrite nanoparticles (FNP) are used for anti-cancer activity, recently $[Mn_{0.5}Zn_{0.5}](Eu_xNd_xFe_{2-2x})O_4$ ferrite nanoparticles (FNP) were investigated on cancerous cells, human adenocarcinoma cells and human colorectal carcinoma cells [1], and Fe_3O_4 nanoparticles are of great interest especially for cancer therapy [2]. This chapter presents the development of research designed to investigate the performance of electromagnetic shielding for biomedical devices coated with layers of cobalt ferrite ($CoFe_2O_4$) with and without addition of niobium pentoxide (Nb_2O_5) when exposed to electromagnetic fields generated by signals in the radio frequency (RF) and microwave range [3]. Taking into account that the electromagnetic behavior of materials is intrinsically related to their morphological microstructure, the improvement of microstructural properties in these materials can raise their level of reliability and allow for more refined applications required by modern technologies [4,5]. To obtain shields that meet this demand, it is essential to prepare advanced materials that allow the design of new multifunctional devices and systems that adapt in real time to the scenario to which they are exposed [6]. Nevertheless, the microelectronics sector has announced a significant increase in new applications of niobium as alternative solutions for obtaining devices [7] such as semiconductors, superconductors, SMD capacitors, etc. Thus, with the purpose of expanding the application potential of cobalt ferrite, its electromagnetic properties were investigated with the addition of niobium pentoxide (Nb_2O_5) in its final composition. The great interest in spinel ferrites, including $CoFe_2O_4$ (cobalt ferrite), is partially due to their potential applications in electromagnetic (EM) shielding for radio frequency (RF), and microwave (GHz) spectra, which motivated the characterization of their properties in high frequency aiming initially its application in cardiac pacemaker. Thus, work on $CoFe_2O_4$

with the addition of a few mol% of another element was carried out to study the effects on the resistivity, high frequency magnetic properties of prepared materials [8]. $CoFe_2O_4$ is a spinel ferrite and it possesses a high saturation magnetization (in the order of 80-90 Am^2kg^{-1}) at room temperature; it is particularly used in applications where minimum volume or weight is essential [9]. As a ceramic, $CoFe_2O_4$ is usually sintered between 1000 °C, and 1200 °C [10], although its melting point is higher ($\sim$ 1600 °C) [11]. In the majority of the works found in the current literature about cobalt ferrites with the addition of other substances, it has been done to adapt or improve their properties as magnetic and magnetostrictive materials [10], usually substituting Mn, Cu or Ni ions. There has been little research on niobium-doped spinel ferrites, apart from those such as Nb_2O_5-doped NiZn spinel ferrite [12], where it was observed that the addition of niobium considerably increased the grain size, as well as increasing the sintering temperature to 1250 °C in NiZn ferrite [12], and the formation of orthorhombic Nb_2FeO_6 was described as a secondary phase [13] with less than 1% niobia, or niobium-doped copper ferrite $Nb_xCu_{1-x}Fe_2O_4$, with $x \geq 0.1$ [14]. Previous research on niobium-doped $CoFe_2O_4$ is scarce, with the exception of studies related to two-phase magnetoelectric ceramic composites with cobalt ferrite and ferroelectric niobates [15–17].

Nevertheless, it was stated that the addition of niobium can increase the permeability of NiZn spinel ferrites [13]. An increase in the magnetic permeability of hexagonal Co_2Z ferrite ($Ba_3Co_2Fe_{24}O_{41}$) doped with 0.8% by weight of Nb_2O_5 and sintered at 1260 °C [18,19] was also reported. To expand the possibilities of choosing the most adequate stoichiometry for shielding devices in the biomedical area, niobium pentoxide (Nb_2O_5) was added to the cobalt spinel ferrite powders ($CoFe_2O_4$) in varying amounts of 5%, 10%, 15% and 20% of its final weight. Samples were prepared using the standard ceramic method with suitable geometry to assess whether the crystalline phases, microstructure and electromagnetic behavior of these compositions were significantly affected. Samples were heated at 1200 °C, 1300 °C, 1400 °C and 1500 °C. Due to limitations in the maximum sintering temperature [9], samples sintered at 1475 °C were designed specifically for the evaluation of losses in the material studied. The microstructural evaluation made it possible to observe a crystalline phase constituted by $CoFe_2O_4$. However, the presence of a non-crystalline phase, rich in niobium, densified predominantly in the intergranular region was confirmed as already evidenced [6,20]. Additionally, a lamellar morphology evidenced motivated an investigation using AFM images to be associated with its electromagnetic evaluation. Next, the graphs of the hysteresis cycles of the various compositions considered are presented, in order to verify the coercive force and magnetic saturation associated with them[21]. At the end, the electrical permittivity (ε^*) and magnetic permeability (μ^*) were characterized in their complex form to observe their dispersion in the frequency domain and from these results it was possible to evaluate the electromagnetic losses associated with the material, where graphically it's possible to measure its reflectivity [22].

1.1   Eletromagnetic behavior of $CoFe_2O_4$

The evaluation of the electromagnetic behavior of $CoFe_2O_4$ samples was carried out through the characterization of their magnetic hysteresis cycles, and measurements in the complex form of their magnetic permeability and electrical permittivity. Taking into account the magnetic properties of the materials of the investigated compositions, the interaction between the moments of induced magnetic dipoles are more significant in cobalt ferrite since its intrinsic magnetic dipole is much greater than in niobium pentoxide. Thus, in the presence of a magnetic induction field, the elementary magnetic dipoles of $CoFe_2O_4$ will react to produce their own induction field that will modify the original field. Therefore, the moment of magnetic dipoles can be understood as microscopic currents that have the effect of magnetic induction sources B, also known as flux density, and that can be represented by the mathematical expression:

$$B = \mu_0.(H + M) \tag{1}$$

Where, M is magnetization associated with the material, μ_0 is the volumetric density at the moment of the magnetic dipole, and H is the magnetic field strength. The magnetic moment of the sample per unit volume has the same dimension as H. In certain magnetic materials, it is empirically observed that the magnetization M is expressed in Gauss and proportional to H [23]:

$$M = \chi H \tag{2}$$

Where χ is the susceptibility, that is ability of a material to magnetize itself under the action of a magnetic stimulation of a magnetizing field to which it is subjected. M is an intrinsic property of material and represents the magnetic moment of the dipoles per unit of mass, as described in the following expression:

$$M = \frac{magnetic\ moment}{mass} \tag{3}$$

Where, magnetization refers to the magnetic moment (emu) and mass is given in grams (g).

Fig. 1 represents graphically the magnetic hysteresis loop, where the magnetization saturation zone (M_s) and the coercive force (H_c) are highlighted for further analysis of the results obtained.

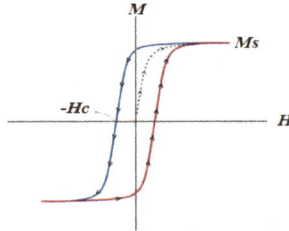

Fig. 1. Graphical representations of the magnetic hysteresis cycle, M_s, the saturation magnetization, and H_c, the coercive force.

1.2 The complex magnetic permeability and complex permittivity in the frequency domain

The scattering parameters (S-parameters) of a dielectric material used as a DUT (Device Under the Test) to be inserted in a coaxial line as shown in Fig. 2 (a) can be determined by using the voltage interactions of a two-port network [24] as shown in (b), where Z_{in} is the characteristic input impedance, Z_{out} is the characteristic output impedance, V_1^+ is the amplitude of the voltage of the incident wave at the input, V_1^- is the amplitude of the voltage of the wave reflected in the input, V_2^+ is the amplitude of the voltage of the incident wave at the output, V_2^- is the amplitude of the wave voltage reflected at the output [24].

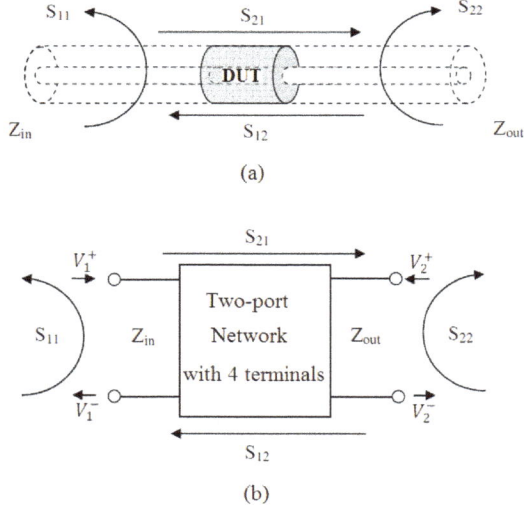

(a)

(b)

Fig. 2(a) Pictorial representations of sample inserted in the coaxial line; (b) two-port network

When we use two ports of a Vector Network Analyzer (VNA), the S-parameters can be defined by the relationships between incident and reflected waves [25], where S_{11} is the input reflection coefficient at port 1, S_{22} is the output reflection coefficient, S_{21} is the transmission insertion loss, and S_{12} is the reverse reflection coefficient associated with transmission isolation. Then, based on the NRW (Nicolson-Ross-Weir) method, the S-parameters can be associated [6,26] to determine the transmission coefficient (T), and reflection coefficient (Γ) as follows:

$$T = \frac{(S_{11}+S_{21})-\Gamma}{1-(S_{11}+S_{21})\Gamma} \tag{4}$$

and

$$\Gamma = k \pm \sqrt{k^2 + 1} \tag{5}$$

Where,

$$k = \frac{(S_{11}^2 - S_{21}^2)+1}{2S_{11}} \tag{6}$$

After that, we can determine the value of the complex relative permeability (μ^*) and complex relative permittivity (ε^*) of the material by the following expressions.

$$\mu^* = \frac{1+\Gamma}{(1-\Gamma)\sqrt{(\frac{1}{\lambda_0^2})-(\frac{1}{\lambda_c^2})}} \tag{7}$$

and,

$$\varepsilon^* = \frac{\lambda_0^2}{\mu_r}\left[\frac{1}{\lambda_c^2} - \left(\frac{1}{2\pi l} \cdot ln\left(\frac{1}{T}\right)\right)^2\right] \tag{8}$$

Where, μ_r is the relative permeability of the material (and is $= (\mu^*/\mu_0)$), μ_0 is the permeability of free space ($4\pi \times 10^{-7}$ H/m), λ_0 is the wavelength in free space, λ_c is the cut-off wavelength of the straight section of the sample holder in the transmission line, and l is the difference between the internal and external diameters in the sample with toroidal geometry (the width of the toroidal ring).

1.3 Determination the reflectivity of absorbing waves

When electromagnetic waves are incident in free space on a material, they can be partially reflected or absorbed by the medium that they cross. The attenuation caused by absorption of the electromagnetic waves depends on the composition and the thickness of the material. According to transmission line theory [27], the loss of energy by reflectivity (RL) of absorbing waves across the material can be determined in (dB) by using equation (9):

$$Rl(dB) = 20 \log \left| \frac{Z_{in} - Z_0}{Z_{in} + Z_0} \right| \tag{9}$$

Where Z_0 is the impedance on free space and Z_{in} is the input impedance on the material surface that could be expressed by:

$$Z_{in} = \sqrt{\frac{\mu_r}{\varepsilon_r}} \tanh \left[j \left(\frac{2\pi f d}{c} \right) \sqrt{\mu_r \varepsilon_r} \right] \tag{10}$$

Where ε_r is the relative permittivity of the material and is $= (\varepsilon^*/\varepsilon_0)$, ε_0 is the permittivity of free space $(8.85 \times 10^{-12}$ F/m), c is the velocity of light, f is the frequency of the electromagnetic wave, and d is the thickness of the material.

2. Materials and methods

Samples in pellet and toroid shape were obtained by the conventional ceramic method prepared by the pressing of the mixture of $CoFe_2O_4$ powders with the addition of Nb_2O_5 in the proportion of 0%, 5%, 10%, 15%, and 20% of the final weight of the mixture. After pressing the powders at 50 MPa with pellet and toroid shape, they were sintered at 1200 °C, 1300 °C, 1400 °C, and 1500 °C. For the structural evaluation, samples in pellet form received adequate surface treatment to reveal the internal phases present by means of the X-ray diffraction (XRD) technique. In addition, to visualize the size of grains and pores distributed along the arrangement of the final morphological structure, images were obtained through SEM (Scanning Electron Microscopy) which was complemented with EDS (Energy Dispersive Spectroscopy) using a- FEG (Field Emission Gun) -SEM, Model MIRA3-TESCAN. The SEM images revealed the presence of lamellar morphology, which motivated a detailed analysis of them by means of Atomic Force Microscopy (5420 AFM, Agilent). However, it was necessary that the samples previously undergo a thermal treatment for 30 minutes at a temperature of 100 °C below the sintering temperature. The electromagnetic evaluation was carried out, initially via the characterization of the room temperature magnetic hysteresis loops of the powders of the studied compositions, which were measured in a commercial SQUID (Superconducting Quantum Interference Device) magnetometer with an applied field of 7 Tesla [28]. Then, using the transmission/reflection technique (T/R method), the electromagnetic evaluation of the toroidal geometry samples was carried out, through the characterization in the complex form of the electrical

Materials Research Forum LLC
https://doi.org/10.21741/9781644902332-6

permittivity (ε^*) and the magnetic permeability (μ^*) [6,27], which are respectively related to the electrical conductivity as $\sigma = \varepsilon''/2\pi f$, and the magnetic induction vector ($\vec{B}$). Fig. 3 shows a coaxial line coupling a pair of sample holders (APC-7 connectors) to the ports of a Vector Network Analyzer (VNA) from Agilent (model N5231 PNA-L), used to measure the s-parameter with the toroid samples being inserted into an APC-7 sample holder. The VNA has been calibrated to take measurements in the frequency range from 300 MHz to 12 GHz.

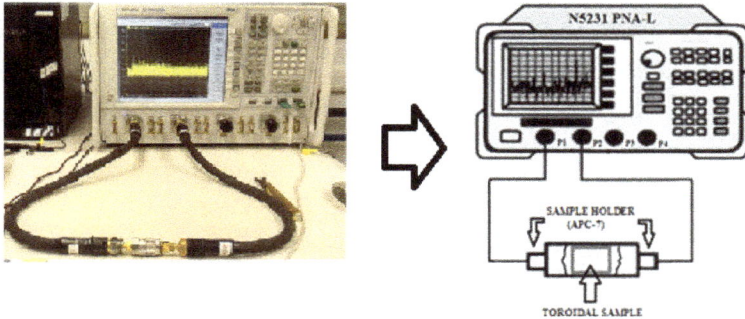

Fig. 3. Network analyzercoupled via coaxial line by a sample holder pair (APC-7 connectors)

Through a GPIB/USB/82357 interface (Agilent), the S-parameters were sent to the Agilent VEE Pro 9.3 program environment, which allowed the complex characterization of magnetic permeability (μ^*) and electrical permittivity (ε^*), as represented by the following expressions:

$$\mu^* = \mu' - j\mu'' \tag{11}$$

and,

$$\varepsilon^* = \varepsilon' - j\varepsilon'' \tag{12}$$

On the same VEE Pro 9.3 platform, reflectivity graphs were obtained that allowed the evaluation of the attenuation of an electromagnetic wave incident orthogonally on a homogeneous layer of the material studied [29].

3. Results and dicussions

The structural evaluation was initiated by the XRD technique, seeking to show the presence of a new phase in the crystal structure of the analyzed samples. Fig. 4 shows an XRD pattern revealing the presence of a new phase, $FeNbO_4$, in a sample with 20% added niobia, and sintered at 1400 °C.

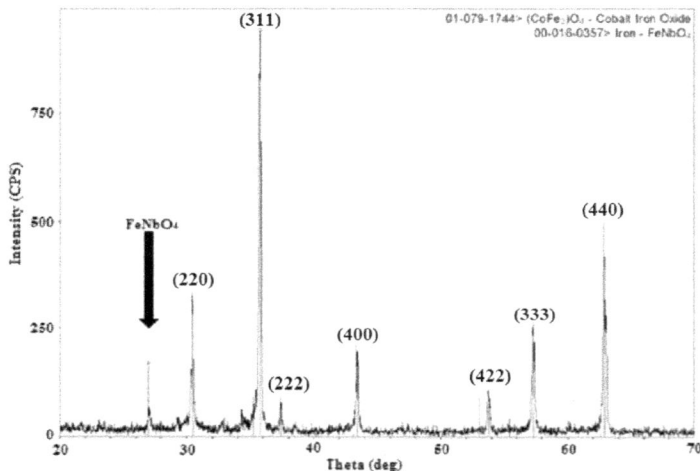

Fig. 4. XRD pattern revealing the presence of FeNbO₄ observed in a sample with the addition of 20% niobia, and sintered at 1400 °C.

However, the presence of this new phase was not observed in all samples, as shown in X-ray diffractograms of Fig.5 (a) 1200 °C; (b) 1300 °C; (c) 1400 °C and (d) 1500 °C. Initially, the diffractograms of the doped samples are similar to that of ferrite without Nb-oxide addition, with peaks similar to the characteristic peaks of $CoFe_2O_4$ ferrite, suggesting that there was no structural interaction with the oxide of niobium or even that the pentoxide has been absorbed by the ferrite structure. However, when checking the individual standard patterns of each phase of the composition, it is observed that niobium oxide peaks that may be superimposed by peaks of the ferrite. That said, it became necessary to prove the presence of the new phase through the SEM, as follows. We then sought to confirm the densification of the new phase through Scanning Electron Microscopy (SEM).

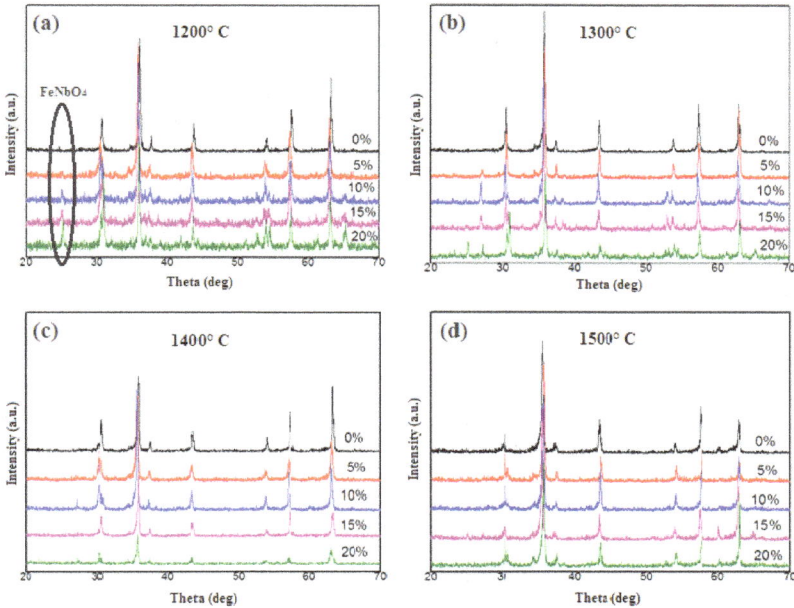

Fig. 5. X-ray diffractograms obtained from samples (pellets) sintered at: (a) 1200 °C; (b) 1300 °C; (c) 1400 °C, and (d) 1500 °C.

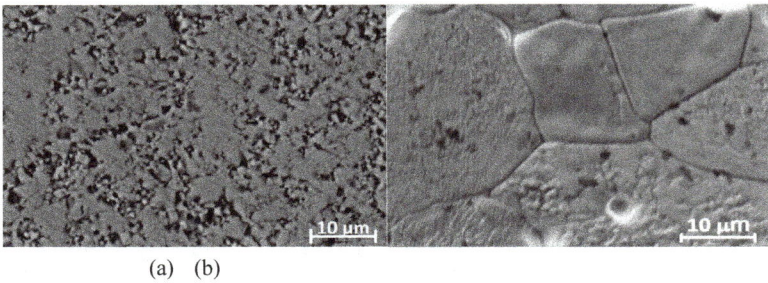

(a)   (b)

Fig. 6. SEM images (magnified 5000 times) of (CoFe$_2$O$_4$ + 0% Nb$_2$O$_5$) samples, sintered at (a) 1200 °C, and (b) 1500 °C

The images obtained by SEM were initially taken on the surface of the samples in SE mode with the purpose of verifying the morphological aspect of the studied compositions. Fig. 6 (a) and 6 (b) below, represent images magnified 5000 times obtained from ferrite samples without the addition of niobium ($CoFe_2O_4$ + 0% Nb_2O_5), sintered at 1200° C, and 1500° C, respectively. The images show the occurrence of grain growth, but the spinel structure remains throughout the temperature transients.

Fig. 7(a) shows an SEM image of pure cobalt ferrite sintered at 1400 °C where pores with different depths and similar characteristics are observed, suggesting the occurrence of a homogeneous morphology in its interior. Fig. 7 (b) shows the qualitative and quantitative evaluation of the elemental chemical composition of the cobalt ferrite sample without Nb addition through EDS analysis.

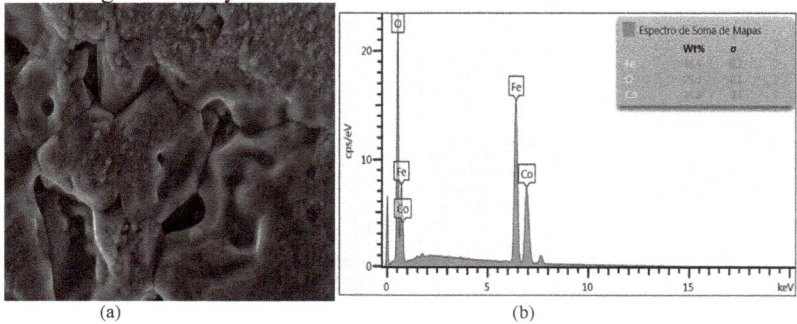

(a)   (b)

Fig.7(a) SEM image; (b)EDS spectrum of cobalt ferrite ($CoFe_2O_4$ + 0% Nb_2O_5) sample, sintered at 1400 °C.

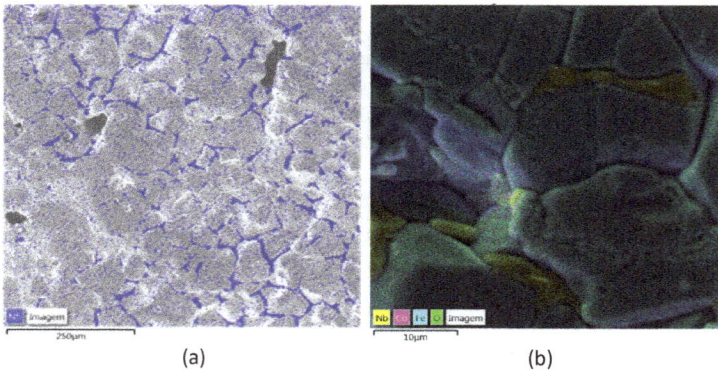

(a)   (b)

Fig. 8(a) EDS layered of a sample with 15% addition of niobium pentoxide ($CoFe_2O_4$ + 15% Nb_2O_5), sintered at 1400 °C; (b) EDS layered image discriminating in different colors the elements of ferrite formation with Nb_2O_5 densification at the grain boundaries.

Fig. 8(a) shows a layered image performed by EDS of a sample with 15% addition of niobium pentoxide ($CoFe_2O_4$ + 15% Nb_2O_5), sintered at 1400 °C, where the presence of niobia (in blue) can be observed on the entire surface of the sample, with greater concentration in the intergranular region, suggesting the occurrence of the diffusion phenomenon with densification in the grain boundaries.Through the mapping of the chemical composition by EDS for a region of the sample, it was possible to identify the elements present, as shown in the Fig. 8 (b); where the EDS layered image is discriminating in different colors the elements of ferrite formation with Nb_2O_5 densification at the grain boundaries.

Fig.9 illustrates the SEM image obtained from a sample with 10% Nb_2O_5($CoFe_2O_4$ + 10% Nb_2O_5), sintered at 1500 °C. Four distinct regions were chosen for analysis with EDS.

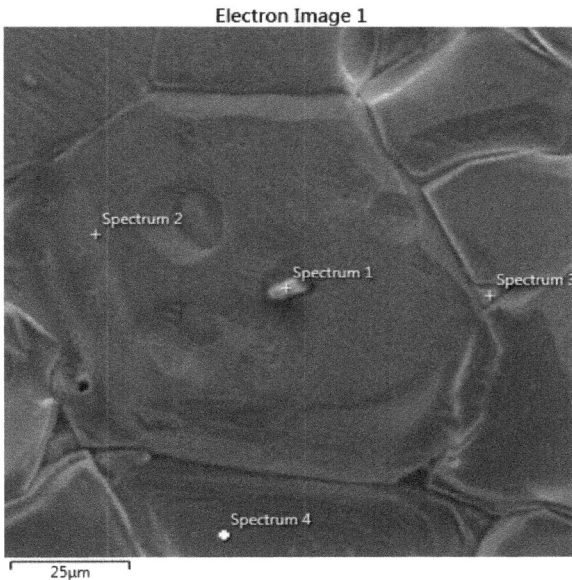

Fig. 9. Four regions to obtain SEM images with EDS from a sample with 10% Nb_2O_5($CoFe_2O_4$ + 10% Nb_2O_5), sintered at 1500 °C

EDS spectra of regions 1, 2, 3, and 4 are represented in Figs. 10, 11, 12, and 13, respectively. Spectrum 1 is located on an irregularity in the intragranular region, where the presence of niobium (38.5%) was observed.

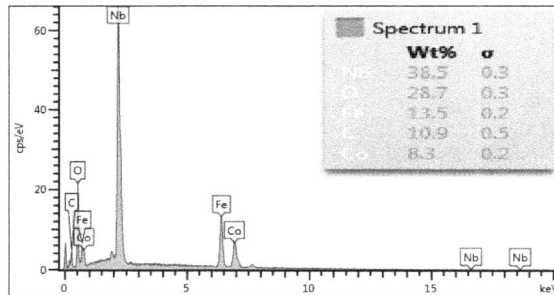

Fig. 10. EDS spectra located on an irregularity in the intragranular region (Spectrum- 1) shows 38.5% niobium

In the Spectrum 2 shown in Fig. 11, located in the intragranular region, the presence of niobium was not seen for the interior of the grain.

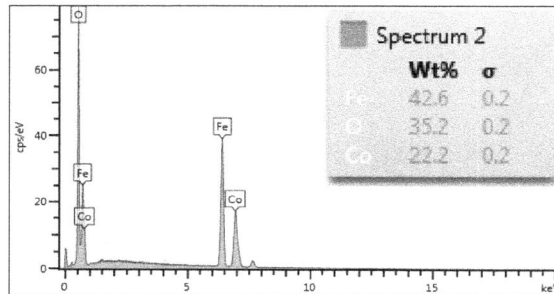

Fig. 11. Spectrum 2 located in the intragranular region without niobium present.

In Fig. 12, the presence of niobium was quantified at 30.8%, according to the table in Spectrum 3, which is located in the intergranular region, at a grain boundary.

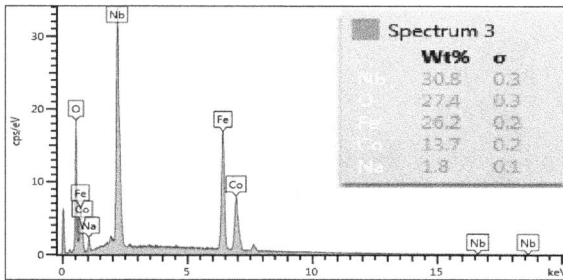

Fig. 12. Spectrum 3, located in the intergranular region containing 30.8% niobium

Finally, Fig/ 13 shows us Spectrum 4 which represents by graph 10 and located in the intragranular region containing only 3.90 % niobium in the grain interior.

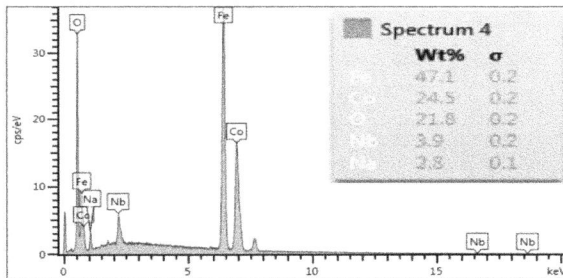

Fig. 13. Spectrum 4, located in the intragranular region with 3.90 % niobium

Another indication of the presence of niobium can be verified by observing Fig.14, extracted from a sample with the addition of 10% of the doping element. In it, surfaces with lighter regions can be highlighted, superimposing the formation of equidistant lines created during sample processing. This formation suggests that a structural strengthening of the composition may have occurred by filling in the grain boundaries and wrinkling.

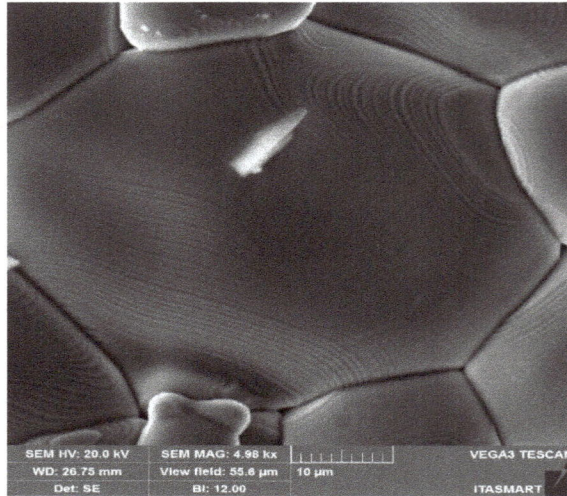

Fig. 14. Sample with addition of 10% dopant niobium oxide with lamellar morphology

In order to associate the randomness of crystallographic and dimensional orientations of crystallites with processing conditions and grain growth, an analysis on a nanometric scale was carried out through images obtained by the Atomic Force Microscope (AFM) that could investigate the arrangement of magnetic domains that could significantly influence the electromagnetic behavior of the studied ceramics. Fig.15 shows us from a sample doped with 10% Nb_2O_5 and sintered at 1400 °C: (a) shows lateral force, (b) is relative to deflection; and (c) the topography.

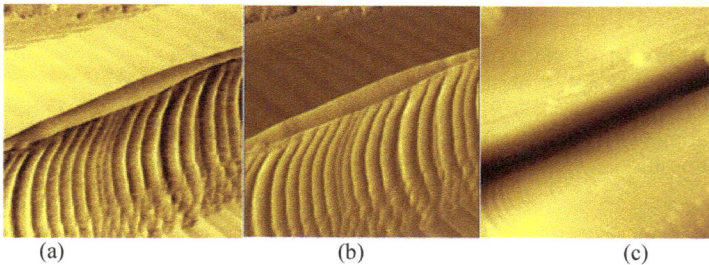

(a)　　　　　　　　　　(b)　　　　　　　　　　(c)

Fig.15. A sample doped with 10% Nb_2O_5 and sintered at 1400 °C: (a) shows lateral force, (b) is relative to deflection; and (c) the topography.

From the topography image it is possible to show the result of the AFM image performed in the intergranular region, where two different types of structure of each grain can be seen, as illustrated in Fig 16. In Fig. 16 (a) an image of 1.5 μm x 1.5 μm is presented, showing different microstructures of the two grains in the sample. The upper grain has a flat surface in the intragranular region, but it changes as it approaches the edge. The grain on the side shows a different microstructure, periodically showing structural growth in the intragranular region, with a change of direction near the edge.

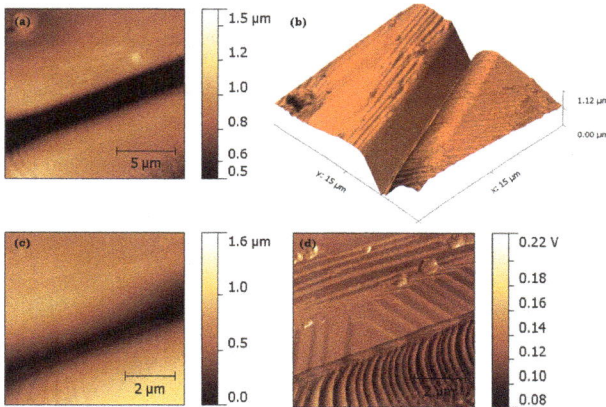

Fig. 16(a) Topographic image of 15 μm x 15 μm, (b) 3D digitized area representation, (c) topography, and (d) area friction (7 μm x 7 μm) of a sample with 10 % Nb$_2$O$_5$

The growth of this grain is stopped when it is positioned perpendicular to the structure of another grain, as shown in Fig.16(b), with a 3D representation of the digitized area (note that the Z axis is not at the same scale as the X and Y axis).Additionally, an AFM image with smaller scanning area (7 μm x 7 μm) was acquired to obtain a detailed image of the intergranular area, where (c) represents the topography, and (d) the friction images.

The addition of niobium pentoxide in the spinel structure of cobalt ferrite makes it possible to adapt its properties to specific applications as long as the phases that appear as a result of the percentage of niobia added and the sintering temperature are known. For the percentages and temperature cycles to be determined, it was necessary to confirm the presence of the new phase corresponding to iron niobate (FeNbO$_4$). After its diffusion between the grains, the new phase that established itself was densified, with the appearance of a glassy phase, in the intragranular region and in the grain boundaries, where it became more evident. This fact has a direct effect on its microstructure and can significantly affect its ferromagnetic properties. The increase in the size of the grains due to the increase in temperature introduces a deformation in them that can cause an increase in the interstitial region and in the size of the pores. Although the internal energy of the grains decreases with its growth, the new phase acquired an amorphous and pore-free appearance after its densification, which may represent a structural reinforcement due to the anchoring of the grains to which it is attached. However, this phase may not be continuous along its length caused by these discontinuities and evaluation by Raman spectroscopy (Raman spectrometer, LabRam HR, Horiba Scientific) can detect these discontinuities. Fig. 17 shows the results obtained at different points in the intragranular, and intergranular regions for a sample sintered at 1500 °C, identifying two distinct sub-phases. Four Raman spectra were obtained in different regions. In regions 1 and 4, the characteristic bands of cobalt ferrite were observed (200 cm^{-1}, 300 cm^{-1}, 500 cm^{-1}, and 700 cm^{-1}). However, in the spectra of regions 2 and 3, iron niobate (FeNbO$_4$) bands were observed in addition to cobalt ferrite, (150 cm^{-1}, 600 cm^{-1}, and 900 cm^{-1}).

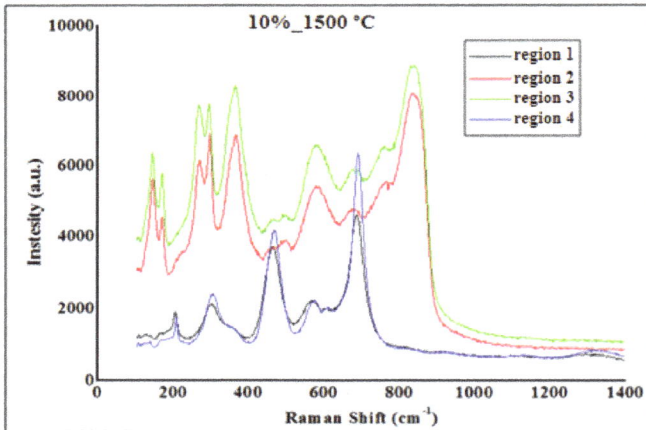

Fig. 17.   *Raman spectra at 4 different points of the sample sintered at 1500 °C with addition of 10% dopant niobium oxide*

Subsequently, the electromagnetic evaluation was carried out, starting with the characterization of the complex measurements of electrical permittivity (ε_r, ε_r'') and magnetic permeability (μ_r', μ_r'') in the frequency domain. Fig.18 illustrates the dispersion of the real part of the complex permittivity in the frequency domain.

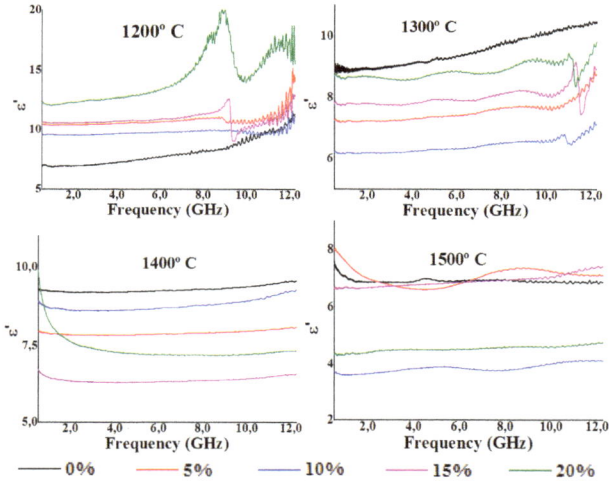

Fig. 18. Dispersion of the real part of the complex electrical permittivity (ε_r') in the frequency domain

Fig. 19 shows the dispersion of the imaginary part (ε_r") that is associated with electrical losses in the material.

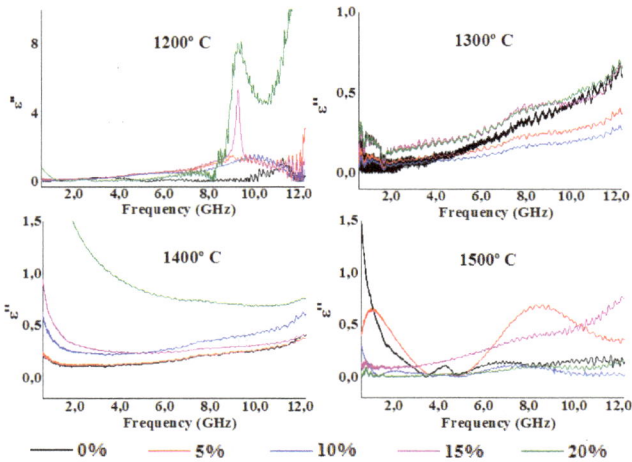

Fig. 19. Dispersion of the imaginary part of the complex electrical permittivity (ε_r")in the frequency domain

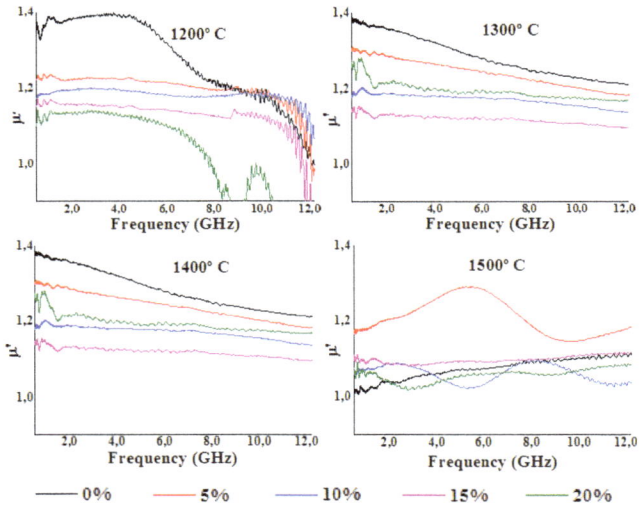

Fig. 20. Dispersion of the real part of the complex magnetic permeability in the frequency domain

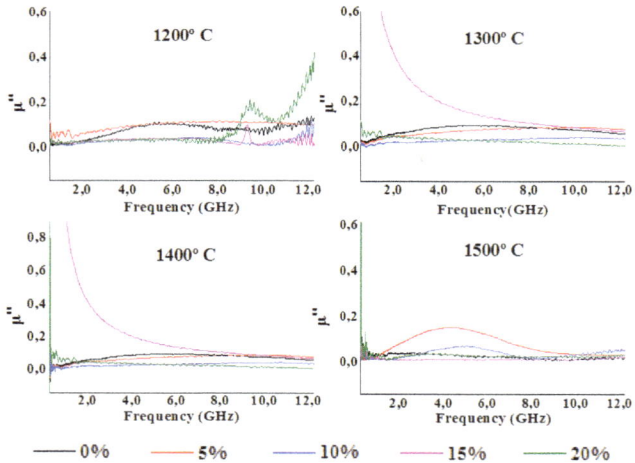

Fig. 21. Dispersion of the imaginary part of thecomplex magnetic permeability in the frequency domain

Evenly, Fig. 20 shows the frequency domain of the dispersion of the real part of the complex magnetic permeability (μ_r'), while Fig. 21 illustrates the dispersion of the imaginary part of the complex magnetic permeability (μ_r'') that is associated with magnetic losses in the composition under study.

From the data associated with these curves, through the Agilent VEE Pro 9.3 platform, losses were simulated as a function of the thickness of the material studied. Fig. 22 presents a reflectivity graph that shows the attenuation peaks (dB) of an electromagnetic wave incident orthogonally to a 1 mm thick layer of a sample sintered at 1475 °C. This behavior is presented in the frequency domain and each curve is associated with different percentage of Nb_2O_5(0, 5, 10, 15%) added.

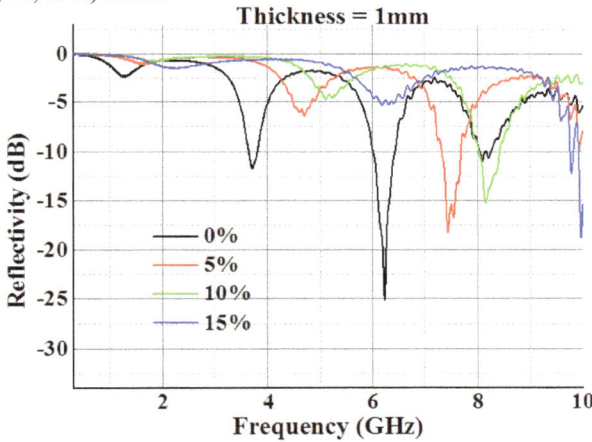

Fig. 22. Reflectivity showing the attenuation peaks for a 1 mm thick samplesintered at 1475° C

Figs. 23 (a), (b), (c) and (d) respectively represent the losses associated with pure cobalt ferrite (0% Nb_2O_5), and ferrites doped with 5% Nb_2O_5, 10% $Nb_2O_5$and 15% Nb_2O_5, sintered at 1500 °C.

Materials Research Forum LLC
https://doi.org/10.21741/9781644902332-6

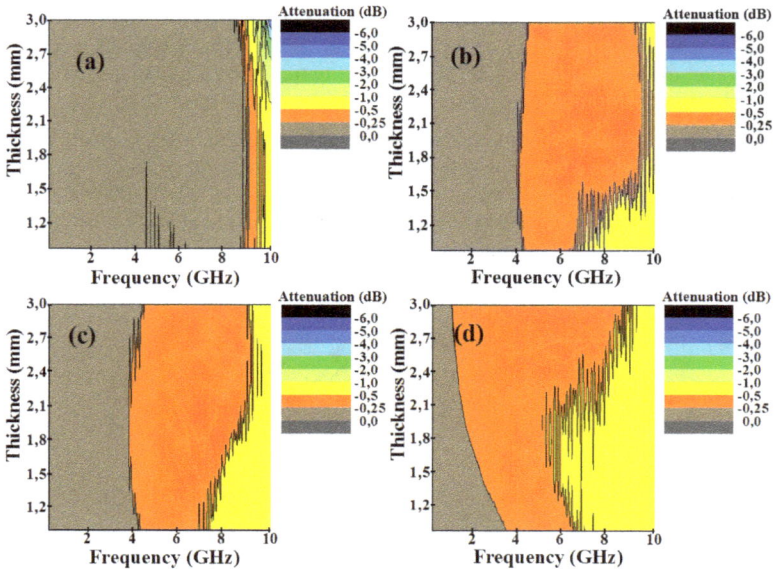

Fig. 23. Reflectivity in the frequency domain representing the losses as a function of the variation of the layer thickness of the investigated material-graphs of the compositions, with different additions of Nb_2O_5: (a) 0%; (b) 5%; (c) 10%, and (d) 15% (sintered at 1500 °C)

Finally, Magnetization measurements were performed as a function of the applied magnetic field. The samples were subjected to a magnetizing field of 7 Tesla and the compiled results relate the total magnetic moment (emu/g) as a function of the magnetizing field (kG, 10 kG = 1 T)). Figs. 24(a), and 24(b) show the hysteresis loops of samples sintered at 1400 °C without addition of Nb_2O_5 and with addition of 15% oxide, respectively.

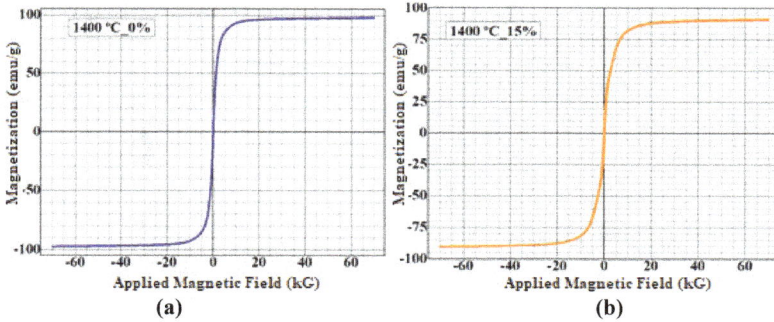

Fig. 24 Hysteresis loops of samples sintered at 1400 °C: (a) with no addition of Nb₂O₅;
(b) with addition of 15% Nb₂O₅

Comparing the values of the saturation magnetization (M_s) of the figures above, there is a small decrease (< 5 emu/g) in the sample with the addition of niobium while the variation of the coercive force (H_c) is very small. In order to verify the variation of the coercive field, the hysteresis curves of the same samples were enlarged, and represented respectively in Fig. 25 (a), and (b) for samples with 0%, and 15% Nb₂O₅.

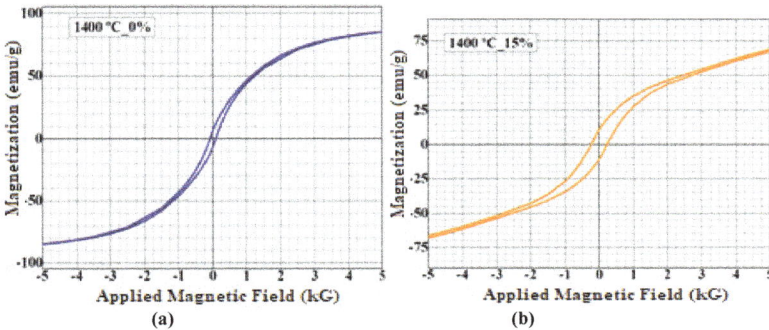

Fig. 25. Variation of coercive force (H_c) in samples sintered at 1400 °C with: (a) 0% of
Nb₂O₅; (b) 15% of Nb₂O₅

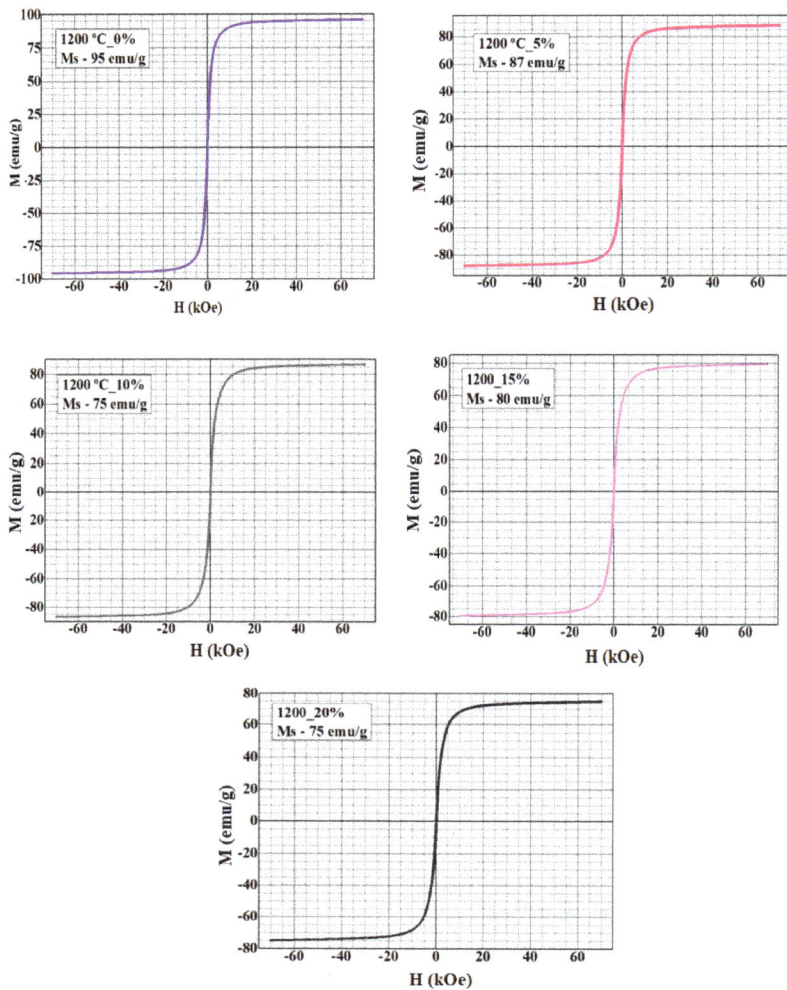

Fig 26 (a). Hysteresis curves integrated from samples prepared with 0%, 5%, 10%, 15% and 20% niobium pentoxide addition; sintered at 1200° C

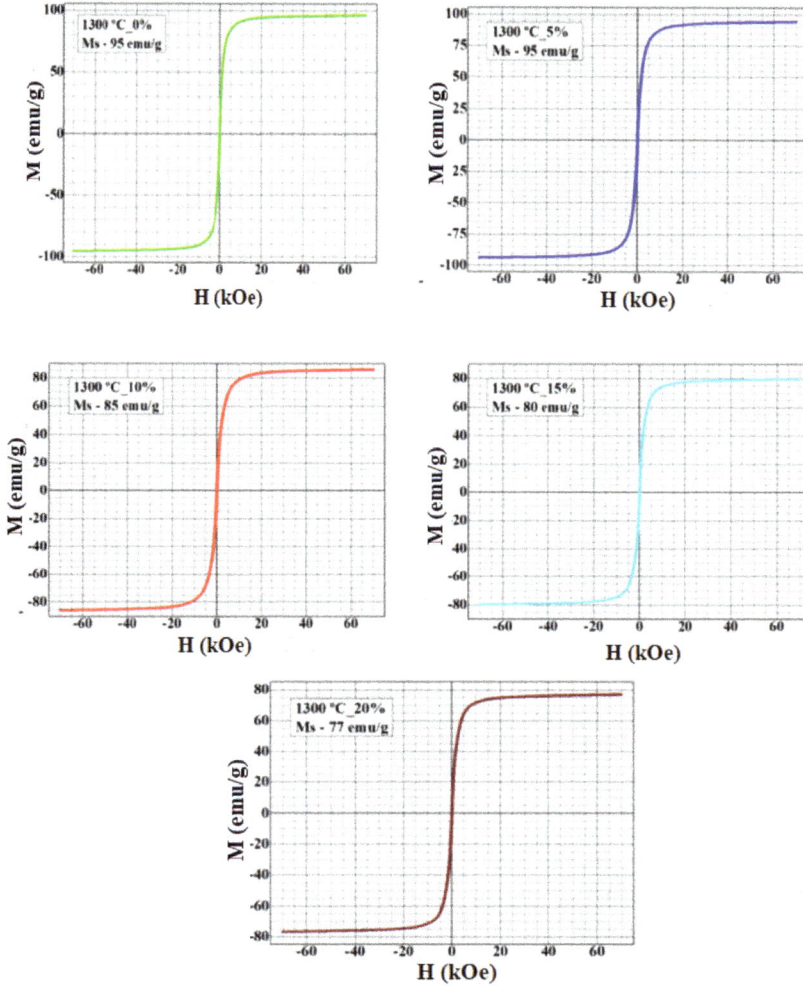

Fig 26 (b). Hysteresis curves integrated from samples prepared with 0%, 5%, 10%, 15% and 20% niobium pentoxide addition; sintered at 1300° C

Fig. 26 (a, b) integrates the hysteresis curves obtained from samples sintered at 1200 °C, and 1300 °C with the different percentages of niobium pentoxide addition (0%, 5%, 10%,

15% and 20%); while Fig. 26 (c) represents hysteresis curves of the samples prepared with 0%, 5%, 10%, 15% of Nb_2O_5 and sintered at 1400 °C, revealed that the saturation magnetization varies from values close to 100 emu/g down to 75 emu/g. The magnetic behavior of this material shows a high saturation magnetization and a very small coercive field, which makes possible the application of this compound in systems that work at higher relaxation frequencies.

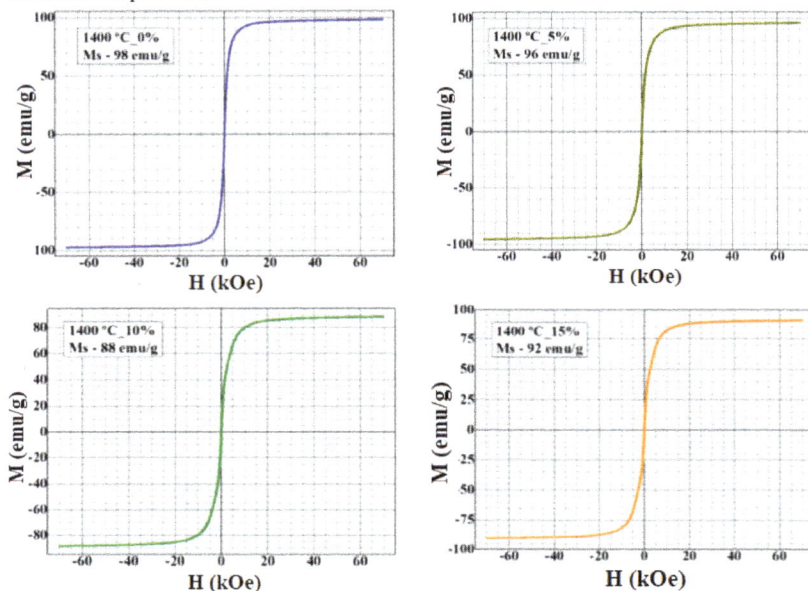

Fig 26 (c). Hysteresis curves integrated from samples prepared with 0%, 5%, 10%, 15% of Nb_2O_5 and sintered at 1400 °C

Conclusions and comments

The addition of niobium pentoxide in the structure of cobalt ferrite made it possible to adapt its properties to specific applications by manipulating the phases that appear as a result of the percentage of niobium added and the sintering temperature.

The dispersion of fine particles of the new phase ($FeNbO_4$) densified in the intragranular region influences the movement of low and high angle grain boundaries (Zenner pinning) through the polycrystalline material. Small particles act to prevent the movement of such boundaries, exerting a clamping pressure that counteracts the driving force pushing the boundaries, and causes a grain boundary pinning effect [30].

Although the dielectric losses are not significant, addition of niobium pentoxide to cobalt ferrite reduces the attenuation peaks of the electromagnetic wave by deallocating them to higher frequencies.

The evaluation results from the procedures described in this work show high magnetization saturation and low coercive field, revealing a characteristic behavior of a paramagnetic material. The total electric dipoles in one crystallite can interact with the total electric dipole in other neighboring crystallites forming domains (Weiss domain) that justify the structural formation of crystallites since the vector sum of these domains can interact with the total electric dipole in neighboring crystallites [31].

Acknowledgements

This study was financed in part by the Coordenação de Aperfeiçoamento de Pessoal de Nível Superior – Brasil (CAPES) – Finance Code 001 – (Pró-Estratégia n. 050/2011), and grant #2012/01448-2, São Paulo Research Foundation (FAPESP) for financial support. The authors also acknowledge the support received from Dr Rodrigo GabasAmaro de Lima from Institute for Advanced Studies – IEAv-Brazil and CompanhiaBrasileira de Metalurgia e Mineração – CBMM.

References

[1] M A. Almessiere, A. V. Trukhanov, F.A. Khan, Y. Slimani, N. Tashkandi, V. A. Turchenko, T. I. Zubar, D.I. Tishkevich, S. V. Trukhanov, L. V. Panina, A. Baykal, Correlation between microstructure parameters and anti-cancer activity of the $[Mn_{0.5}Zn_{0.5}](Eu_xNd_xFe_{2-2x})O_4$nanoferrites produced by modified sol-gel and ultrasonic methods, Ceram Int. 46 (2020) 7346-7354 https://doi.org/10.1016/j.ceramint.2019.11.230

[2] D. I. Tishkevich, I. V. Korolkov, A. L. Kozlovskiy, M. Anisovich, D. A. Vinnik, A. E. Ermekova, A. I. Vorobjova, E. E. Shumskaya, T.I. Zubar, S. V. Trukhanov, M. V. Zdorovets, A. V. Trukhanov, Immobilization of boron-rich compound on Fe_3O_4 nanoparticles: Stability and cytotoxicity, J Alloys Compd. 797 (2019) 573-581 https://doi.org/10.1016/j.jallcom.2019.05.075

[3] J. Smit, H. P. J. Wijn, Ferrites: Physical Properties of Ferrimagnetic Oxides in Relation to Their Technical Application, Philips Technical Library, Eindhoven, The Netherlands, 1959

[4] D. A. Vinnik, V. E. Zhivulin, D. P. Sherstyuk, A. Y. Starikov, P. A. Zezyulina, S. A. Gudkova, D. A. Zherebtsov, K. N. Rozanov, S. V. Trukhanov, K. A. Astapovich, V. A. Turchenko, A.S.B. Sombra, D. Zhou, R.B. Jotania, C. Singh, A. V. Trukhanov, Electromagnetic properties of zinc-nickel ferrites in the frequency range of 0.05-10 GHz, Mater Today Chem.20 (2021) 100460 https://doi.org/10.1016/j.mtchem.2021.100460

[5] F. Chen, X. Wang, Y. Nie, Q. Li, J. Ouyang, Z. Feng, Y. Chen, V.G. Harris, Ferromagnetic resonance induced large microwave magnetodielectric effect in cerium doped $Y_3Fe_5O_{12}$ ferrites, Scientific Reports 2016, 6(2016) 1-8 https://doi.org/10.1038/srep28206

[6] F.E. Carvalho, L. V. Lemos, A. C .C. Migliano, J. P. B. Machado, R. C. Pullar, Structural and complex electromagnetic properties of cobalt ferrite ($CoFe_2O_4$) with an addition of niobium pentoxide, Ceram Int. 44 (2018) 915-921 https://doi.org/10.1016/j.ceramint.2017.10.023

[7] A. S. Sedra, K. C. Smith, Microelectronic Circuits, Oxford series in Electrical and Computer Engineering, 2000

[8] A. R. Hippel, Dielectric Materials and Applications, The MIT Press, Cambridge, USA, 1966.

[9] R M Bozorth, Ferromagnetism, Van Nostrand, New York, USA, 1951.

[10] I.C. Nlebedim, N. Ranvah, P. I. Williams, Y. Melikhov, F. Anayi, J. E. Snyder, A. J. Moses, D. C. Jiles, Influence of vacuum sintering on microstructure and magnetic properties of magnetostrictive cobalt ferrite, J MagnMagn Mater. 321 (2009) 2528-2532 https://doi.org/10.1016/j.jmmm.2009.03.021

[11] D.P. Masse, A. Muan, Phase Equilibria at Liquidus Temperatures in the System Cobalt Oxide-Iron Oxide-Silica in Air, J. Am. Ceram. Soc.48 (1965) 466-469 https://doi.org/10.1111/j.1151-2916.1965.tb14800.x

[12] B. ParvatheeswaraRao, C. Kim, Effect of Nb_2O_5 additions on the power loss of NiZn ferrites, J. Mater. Sci.42 (2007) 8433-8437 https://doi.org/10.1007/s10853-007-1789-1

[13] K. Sun, Z. Lan, Z. Yu, L. Li, J. Huang, Grain growth and magnetic properties of Nb_2O_5-doped NiZn ferrites, Jpn. J. Appl. Phys.47 (2008) 7871-7875 https://doi.org/10.1143/JJAP.47.7871

[14] V. Ribeiro, C. S. P. Mendonca, v. D. De Oliveira, A. C. Baldim. M. R. Da Silva, A. F. Oliveira, Investigation of the microstructure and the magnetic properties of the copper and niobium ferrite, CiênciaTecnol.Mater.27 (2015) 1-80

[15] R. Rakhikrishna, J. Isaac, J. Philip, Magneto-electric coupling in multiferroicnanocomposites of the type $x(Na_{0.5}K_{0.5})_{0.94}Li_{0.06}NbO_{3-(1-x)}CoFe_2O_4$: Role of ferrite phase, Ceram. Int. 43 (2017) 664-671 https://doi.org/10.1016/j.ceramint.2016.09.212

[16] C. E. Ciomaga, C. Galassi, F. Prihor, I. Dumitru, L. Mitoseriu, A.R. Iordan, M. Airimioaei, M.N. Palamaru, Preparation and properties of the $CoFe_2O_4$-Nb-Pb (Zr,Ti)O_3 multiferroic composites prepared in situ by gel-combustion method, J. Alloy. Compd. 485 (2009) 372-378 https://doi.org/10.1016/j.jallcom.2009.05.101

[17] C. S. C. Lekha, A. S. Kumar, S. Vivek, U. P. M. Rasi, K. V. Saravanan, K. Nandakumar, S. S. Nair, High voltage generation from lead-free magnetoelectric

coaxial nanotube arrays and their applications in nano energy harvesters, Nanotechnology 28 (2016) 055402 https://doi.org/10.1088/1361-6528/28/5/055402

[18] R.C. Pullar, Hexagonal ferrites: A review of the synthesis, properties and applications of hexaferrite ceramics, Prog. Mater. Sci.57 (2012) 1191-1334 https://doi.org/10.1016/j.pmatsci.2012.04.001

[19] L. Jia, H. Zhang, Z. Zhong, Y. Liu, Effects of different sintering temperature and Nb_2O_5 content on structural and magnetic properties of Z-type hexaferrites, J. Magn. Magn.Mater.310 (2007) 92-97 https://doi.org/10.1016/j.jmmm.2006.07.034

[20] V. L. O . de Brito, S. A. Cunha, L. V. Lemos, C. Bormio-Nunes, Magnetic Properties of Liquid-Phase Sintered $CoFe_2O_4$ for Application in Magnetoelastic and Magnetoelectric Transducers, Sensors 12 (2012) 10086-10096 https://doi.org/10.3390/s120810086

[21] Z. Ullah, S. Atiq, S. Naseem, Influence of Pb doping on structural, electrical and magnetic properties of Sr-hexaferrites, J. Alloy. Compd. 555 (2013) 263-267 https://doi.org/10.1016/j.jallcom.2012.12.061

[22] A. Cortes, A. Carlos, C. Migliano, V. Lúcia, O. de Brito, A. Côrtes, A.C.C. Migliano, V.L.O. Brito, A.J.F. Orlando, Practical aspects of the characterization of ferrite absorber using one-port device at RF frequencies,Proceedings of Progress In Electromagnetics Research Symposium, Beijing, 2007

[23] D. Jiles, Introduction to Magnetism and Magnetic Materials, 2nd edition, CRC Press, 2015 https://doi.org/10.1201/b18948

[24] L.F. Chen, C. K. Ong, C. P. Varadan, V. V. Varadan, V. K. Varadan, Microwave electronics : measurement and materials characterization, John Wiley and Sons, 552 2004 https://doi.org/10.1002/0470020466

[25] S-Parameter Measurements, Basics of High Speed Digital Engineers, Keysight Technologies Manual. https://www.keysight.com/us/en/assets/7018-06743/application-notes/5952-1087.pdf

[26] R. N. Clarke, A. P. Gregory, D. Cannell, M. Patrick, S. Wylie, I.Youngs, G.Hill, A guide to the characterisation of dielectric materials at RF and microwave frequencies, National Physical Laboratory 2003

[27] E. J. Rothwell, J. L. Frasch, S. M. Ellison, P. Chahal, R. O. Ouedraogo, Analysis of the Nicolson-Ross-Weir method for characterizing the electromagnetic properties of engineered materials, Prog. Electromag. Res. 157 (2016) 31-47. https://doi.org/10.2528/PIER16071706

[28] J. Clarke, A. I. Braginski, The SQUID handbook: Fundamentals and Technology of SQUIDs and SQUID systems, Willey 2 2004 https://doi.org/10.1002/3527603646

[29] D. Zhang, Z. Hao, Y. Qian, Y. Huang, Bizeng, Z. Yang, W. Qibai, Simulation and measurement of optimized microwave reflectivity for carbon nanotube absorber by

controlling electromagnetic factors, Sci. Rep.7 (2017) 1-8
https://doi.org/10.1038/s41598-016-0028-x

[30] P. R. Rios, G. S. Fonseca, Grain Boundary Pinning by Particles, Mater. Sci. For.
638-642 (2010) 3907-3912 https://doi.org/10.4028/www.scientific.net/MSF.638-
642.3907

[31] A. C. C Migliano, Y. C. De Polli, F. R. Daro, A. K. Hirata, F.E.Carvalho,
G.P.Zanella, V.L.O Brito, F.F. Araujo, M.C.Salvadori, Microstructural Analysis of
Ceramics with Applications in Sensors and Biosensors. In: III IEAV Science and
Technology Symposium 2014, São José dos Campos. Proceedings of the III
Symposium on Science and Technology of the Institute for Advanced Studies, São
José dos Campos: Instituo de EstudosAvançados, 2014

Materials Research Forum LLC
https://doi.org/10.21741/9781644902332-7

Chapter 7

Magnetic Nanoparticles: Fabrications and Applications in Cancer Therapy and Diagnosis

T.S. Shrirame[1,2], P. Bhilkar[1], A.R. Chaudhary[3], A.R. Rai[4], R.P. Singh[5], P.R. Dhongle[6], S.R. Thakare[7*], A.A. Abdala[8**] and R.G. Chaudhary[1***]

[1]P. G. Department of Chemistry, S. K. Porwal College of Arts, Science and Commerce, Kamptee-441001, India

[2]Department of Chemistry, Shri Shivaji Science and Arts College, Chikhli-443201, India

[3]Post Graduate Teaching Department of Botany, R.T. M. Nagpur University, Nagpur-440033, India

[4]P.G. Department of Microbiology, S.K. Porwal College of Arts, Science &Commerce, Kamptee-441001, India

[5]Department of Research & Development, Biotechnology,Uttaranchal University, Uttarakhand, India

[6]Department of Mathematics, S. K. Porwal College of Arts, Science and Commerce, Kamptee-441001, India

[7]Department of Chemistry, Govt. Institute of Forensic Science, Nagpur, India

[8]Chemical Engineering Program, Texas A&M University at Qatar, Doha, Qatar

*Corresponding Authors:

* sanjaythakareisc@gmail.com, ** ahmed.abdala@qatar.tamu.edu, *** chaudhary_rati@yahoo.com

Abstract

Cancer is a global epidemic disease. Millions of people are affected by this disease every year. It affects humans including, children to old age. Nonetheless, there are several therapeutic treatments and drugs available to cure cancer. Recently, the use of smart nanoparticles (NPs) has been a promising technique to eradicate cancer. Among these NPs, magnetic NPs are efficient in treating cancer. Targeted drug-delivery based on 'smart' nanoparticles is the next step towards more efficient oncologic therapies, by delivering a minimal dose of drug only to the vicinity of the target. In the present chapter the methods used for fabrication and engineering of magnetic NPs is discussed. We have focused on current state of research on various applications of magnetic nanoparticles as well as recent

Materials Research Forum LLC
https://doi.org/10.21741/9781644902332-7

breakthrough in the development of these NPs in cancer therapy, diagnosis, and targeted delivery.

Keywords

Magnetic Nanoparticles, Metal/Metal Oxide Nanoparticles, Magnetic Resonance Imaging, Oncology, Cancer Therapy, Cancer Diagnosis

Contents

1. Introduction

In the beginning of the 21st century, treatment of cancer is contemplated to be a vital public health concern. Cancer is one of the foremost roots of worldwide death, even though thorough research efforts are performed over the past several decades because it is malignant tumor causes unusual proliferation and differentiation of cell, metastasis, and infiltration. Several methods have been evolved to upgrade effectiveness of cancer diagnosis and treatment, which was showing promising results in the beginning, yet poses limitations in the course of applications [1]. Therapy and diagnosis of cancer is an innovative field. The main objectives of proposed book chapter are to review the application use of MNPs in cancer diagnosis and treatment. Nanoparticles possess large surface area to volume ratio and its functionalization ability makes them competent efficient for high drug binding and loading capacities. Owing to these features, functional NPs can detect and destroy cancer cells prior to tumors formation. Moreover, NPs can limit the damage to healthy organs and reduce the side effects compared to other traditional cancer therapeutic drug [2].

Over the past years, substantial practices have been established to decreasing the progression and seriousness of cancer. Chemotherapy involves use of chemotherapeutic agents such as paclitaxel, carboplatin, docetaxel, methotrexate, cisplatin etc. Although chemotherapy is effective in cancer treatment, the poor selectivity of most therapeutic agents leads to significant side effects. Apart from this chemotherapy, targeted therapy, radiation therapy, hormone therapy, palliative therapy, immunotherapy and cell therapy have been proven effective in cancer treatment. Radiotherapy use radiations to remove lesions. But, similar to chemotherapy, radiotherapy produces prominent side effects [3,4]. Another targeted therapy is more efficient and effective at cellular level and it poses less toxicity and damage to normal cell, but it has a limited application due to its high cost [5]. Other therapies are not that advanced to be promoted globally. The abovementioned therapeutic techniques are successful only to a certain degree and have limitations such as myelosuppression, infertility, alopecia, anemia, mucositis, cardiotoxicity, high cost, incomplete treatment and high toxicity. To overcome the limitations of conventional cancer treatments, constant efforts are building to develop modern nanotechnology-based formulations [6].

Recently, the interest in NPs applications in cancer treatment has been increasing due to the unique, appealing features of NPs in drug delivery, diagnosis, imaging, and therapy by developing vaccine and microscopic medical devices. Currently, the field of nanomedicine is emerging expeditiously. Several years ago, various therapeutic NPs, including albumin, dendrimers, liposomes NPs, and micelles [7], were accepted in cancer therapy. Apart from these organic NPs, variety of inorganic NPs with more diverse and well defined physical properties like magnetic iron oxide NPs were investigated have higher biocompatibility and also capable for cell targeting, diagnosis and treatment [8]. NPs with dimension of 10-100 nm have attained noteworthy curiosity in bioengineering. Magnetic NPs (MNPs) is one of the principal class of nanocarriers that are typically synthesized from pure metals like Co, Fe, Ni and some rare earth metals or a mixture of metals with polymers [9].

Fabrication of nanodimension particles is one of the foremost challenges. Synthesis method determines particles shape and size, surface chemistry and hence magnetic properties. Moreover, the synthesis method impacts the magnetic behavior due to the extent of structural impurities and imperfections of the particles [10,11].

NPs can be synthesize by both physical (arc-discharge, lithographic technique, spray pyrolysis, layer by layer growth, pulse laser desorption, and physical vapour condensation) and chemical (sol gel, chemical reduction, pyrolysis, photochemical method, CVD, and electrochemical) methods [12]. But these conventional techniques are associated with several disadvantages, such as toxic chemical reducing agents, high cost, high radiations,... etc. [13]. A safe, eco-friendly green synthesis approach is acquiring substantial importance to overcome these limitations. Till date, several methods have been reported for MNPs fabrication comprising thermal decomposition, co-precipitation, solvothermal, microemulsion, chemical vapor deposition, microwave-assisted, laser pyrolysis, combustion, carbon arc, electrochemical and green synthesis [16,17]. On the other hand, there are green processes involving using plant extracts (stem, roots, bark, flowers, fruits, leaves etc), microorganisms (fungi, bacteria, yeast), organic acids, enzymes, and vitamins [14,15].

MNPs exhibits marked advancement in oncology sectors over the past few years. MNPs are nanomagnetic material with a large area, finite size, economically facile synthesis, high mass transference, magnetic properties and superparamagnetic in state [18]. Compared to bulk magnetic materials, MNPs differ in magnetic parameters such as magnetic moment, magnetic anisotropy, coercivity field and Curie temperature. Because of their suitable physicochemical characteristics, MNPs have been synthesized for diverse application such as sensors, wastewater treatment, water purification, biosensors, cell labeling, tissue engineering, imaging, disease therapy, MRI, magnetic hyperthermia and drug delivery [21]. MNPs with a single magnetic domain and superparamagnetic behavior have potential medical applications because NPs with a size range of 10-50 nm are preferred in the biomedical field [19,20]. The potential in the biomedical application of MNPs is governed by the size, shape, morphology, chemical composition, and magnetic properties [3]. MNPs used in magnetic hyperthermia are first heated in magnetic field. MNPs are used in biomedical applications because they can bind with proteins, drugs, enzymes, nucleotides, and antibodies and be adsorbed successfully to an tumor, tissue or organ under the influence of an applied external magnetic field (EMF) [22]. MNPs acquired much attention among all nanoscale materials due to their absolute physical properties, stability, biocompability, magnetic susceptibility, and other pertinent characteristics. Moreover, MNPs can be influenced by EMF, which release itself at the targeted site at the proper rate and hence get a better of drawbacks that arise due to conformist techniques [23].

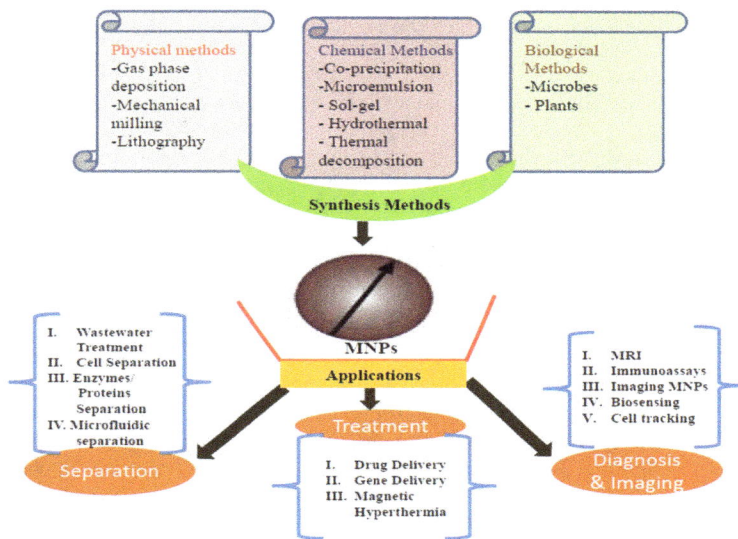

Fig. 1. Illustration of the fabrication routes and applications of multifunctional MNPs

2. Historical background

The first application of MNPs was in compass during the Qin dynasty (221-206 BC) in China. The compass was made of magnetite. In 1269, Petrus Peregrinus de Maricourt, a French scholar, discussed the idea of magnetic poles and the properties of magnets [24]. Another crucial technological and scientific advancement was made by Hans Christian Oersted, who studied the interaction between magnetic compass needles and electrical current [25]. This discovery provided the base for producing generators used for several purposes, including telecommunication, electromotors, electric light, etc. The magnetic tape recorder invented by Valdemar Poulsen was an additional principal milestone in the field of MNPs [26]. In 1949, Nobel Prize laureate, Louis Neel, gave intuitions about the magnetic behaviour of very small particles of magnetic material [24,27]. In the last half of the century, numerous applications have been dependent on nanoscale magnetic materials, with many superior properties different from bulk materials. Another breakthrough is the discovery of giant magnetoresistance of Fe/Cr superlattices synthesized by M. N. Baibich *et al.* using molecular beam epitaxy [28], leading to further development of giant magnetoresistance read heads used for reading data from hard disks [29].

Treatment of metastases and lymphatic nodes was carried out by Gilchrist et al. in the 1950s by administrating metallic particles by heating with a magnetic field. Soon after that, MNPs were more progressively implemented in enzyme immobilization, drug delivery, and empowered abundance of exhilarating biotechnological applications [30]. A significant development in the cancer nanomedicine sector was started with the evolution of liposome structure in 1964. Since 30 years long ago, the idea of NPs as targeting agents was introduced. In 1980, the first time targeted liposomes were tested for clinical trial against cancer (HER2, MM-302, and BIND-014) [31,32]. Earlier in 1986, tumor treatment was done by systematic administration of NPs. These NPs complies in the tumor by enhanced permeability and retention effect. From 1995 onwards, various active pharmaceutical ingredients viz. liposomal doxorubicin, nab-paclitaxel, micelle paclitaxel (Genexol-PM), CALAA-01, Ferumoxytol, BIND-014, etc. have been used as a therapeutic agent for cancer therapy [32]. Comprehensively, MNPs as a carrier of therapeutic molecules have been investigated over the last two decades, focusing on boosting the diagnostic effect of drugs and lessening their toxic side effects.

NPs play crucial functions in diagnosing cancer by envisioning cancer cells at a very early phase. Ralph Weissleder in 2006, in his literature, reported developing molecular imaging in cancer treatment [33]. Indeed, the progress in the field of molecular imaging is moderately slow due to substandard pharmacokinetics profile and expensive clinical development. Kim et al. [34] summarize the development of NPs as imaging modalities, specifically CT (computed tomography), MRI, PET (positron emission tomography) and Optical imaging.

3. Classification of MNPs

MNPs are broadly categorized based on their magnetic properties or the material type, as shown in fig 2. MNPs are grouped into weak and strongly magnetic materials based on magnetic properties. Weak magnetic materials are again classified into paramagnetic and diamagnetic materials. Similarly, strong magnetic materials are classified into ferromagnetic, antiferromagnetic, and ferromagnetic. Under the influence of an EMF, diamagnetic materials induce weak magnetic moment antiparallel to the applied field and have a negative susceptibility value (e.g. Cu, Au, Ag, etc.) [35]. However, paramagnetic materials show a magnetic field parallel to the applied field with positive susceptibility (e.g., Gd, Li, Mg, and Ta) [36,37]. Ferromagnetic materials (e.g., Ni, Fe, and Co) are composed of domains whose magnetic moment is produced by adding the magnetic moment value of an individual atom present in the domain. In the presence of an externally applied field, the magnetic moment of the ferromagnetic material domains aligns in the direction of the applied field, giving a net magnetic moment [38]. Antiferromagnetic and ferrimagnetic materials are composed of two different atoms present on various lattice sites. But antiferromagnetic materials have zero magnetic moments because the magnetic moment has equal magnitude with opposite direction and hence cancel each other. In contrast, ferrimagnetic materials pose magnetic moments with different magnitudes and cannot cancel out, resulting in non-zero magnetic moments. Both exhibit similar behaviour

Magnetic Nanoparticles for Biomedical Applications   Materials Research Forum LLC
Materials Research Foundations 143 (2023) 199-232   https://doi.org/10.21741/9781644902332-7

to that of ferromagnetic one [19]. The second classification of MNPs is based on material. MNPs are monometallic, multimetallic (metal alloy), ferrite, and metal oxides [39]. The monometallic MNPs are mainly composed of metals in elemental form like iron, cobalt, gold and nickel due to their immanent ferromagnetic properties. Metallic MNPs are often restricted by their low biocompatibility, chemical instability, high pyrophoricity and prone to easy oxidization. These features make metallic MNPs inadequate material for biomedical applications [40]. Intending to overcome these constraints, bimetallic MNPs composed of two metals for example iron-nickel, iron-platinum, cobalt-platinum and so forth come into effect. Bimetallic NPs reported opposition for oxidation without influencing their magnetic behavior. Metal nanoalloys containing dispersed NPs of metals have also been reported. The biomedical application of nanoalloys is due to their high surface-to-volume ratio [41]. Maghemite (γ-Fe_2O_3), magnetite (Fe_3O_4), and spinel ferrites have a large magnetic moment, are highly biocompatible, have high stability toward oxidation, excellent chemical stability with minimum toxicity [42]. These properties make them more relevant metal oxide MNPs for biomedical applications. Ferrites such as Mn-Fe_2O_4, Mn_3Zn_7 Fe_2O_4, Fe-MgO, Co-Fe_2O_4, $Mn_{0.6}Zn_{0.4}Fe_2O_4$, etc. own higher relaxation rate and magnetism, which grant an application in hyperthermia and MRI [40,39].

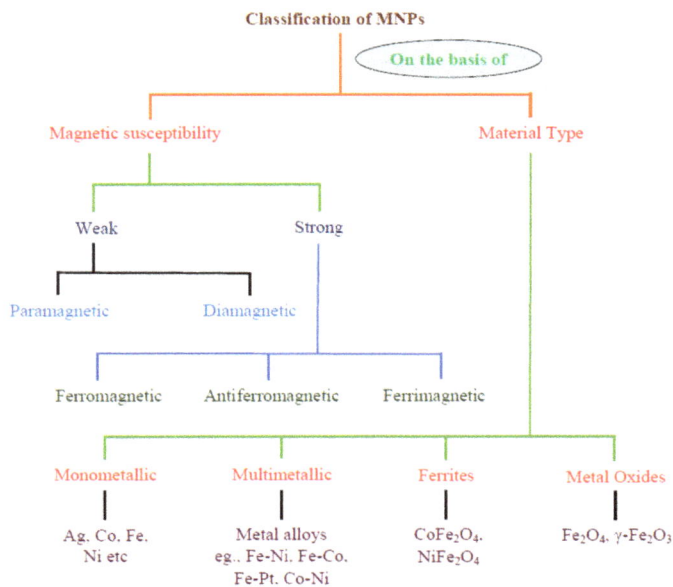

Fig. 2. MNPs classification on the basis of magnetic susceptibility and material type

4. Magnetism of MNPs

MNPs exhibit wide variety of surprising magnetic properties in contrast to bulk materials. Depending upon magnetic ordering and kind of magnetic material, magnetic properties of bulk magnetic materials get varied [43]. As discussed earlier ferromagnetic, ferrimagnetic are magnetically ordered materials. Meanwhile the size of bulk ferri or ferromagnetic materials is reduced to nanoscale, it was notice that the specific magnetic properties of bulk materials changes significantly, irrespective of kind of magnetic ordering [44]. Therefore, another special class acquired by Bean and Livingston in 1959 called superparamagnetism to distinguish magnetic nanomaterials (NMs) from bulk magnetic materials, i.e., ferro/ferromagnetic and paramagnetic ones [45]. Superparamagnetic material is itself magnetically ordered. It is noticed that at the nanoscale level magnetic domain holds numerous atoms with magnetic moments coupled with each other. Magnetically ordered ferromagnetic or ferrimagnetic are primarily used in biomedical applications, owing to their powerful magnetism. Superparamagnetic and ferromagnetic materials' shape and core size determine their magnetic behavior. Generally, magnetic properties largely depend on the temperature. As the temperature is elevated over specific Curie temperature, then the superparamagnetic and ferromagnetic behavior of material get changes into paramagnetic one [46].

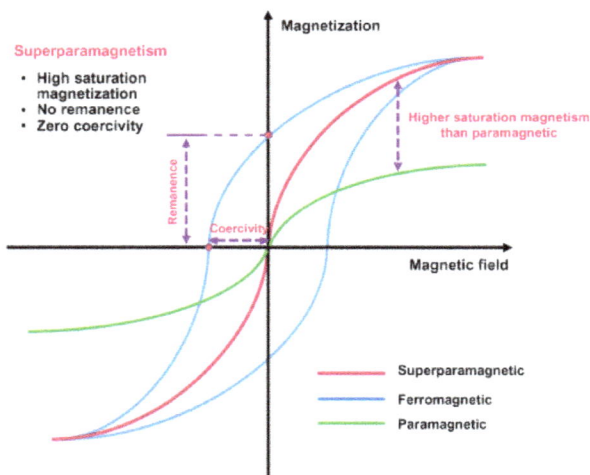

Fig. 3. Magnetization behaviour of ferromagnetic NPs [47]

Under an applied magnetic field, approximately wider magnetic hysteresis is observed for bulk ferri- and ferromagnetic materials. Bulk materials also have peculiar magnetization, saturation, and coercive fields. Nevertheless, MNPs, due to their small size, manifest significant changes even though the magnetic material characteristics are identical. The principal features of MNPs in the nano dimensional range are reported as follows:

1. When the size increases over critical diameter (D_c), (i) both bulk magnetic materials and MNPs poses similar magnetic domain structure, but with decreased domain number; (ii) depending upon the type of magnetic material (either hard or soft), the hysteresis phenomenon is more or less pronounced in an EMF; (iii) surface effects are not very pronounced. ,

2. At nanosize scale (below 10 nm) below threshold diameter (D_{th}), MNPs acts entirely different, (i) NPs being magnetized using Langevin function and superparamagnetism is evidenced (no hysteresis); (ii) as NPs own at most single domain impulsively magnetized to saturation; (iii) owing to strong surface effect, NPs exhibit marked drop in saturation magnetization under the influence of ordinary fields.

In between D_{th} and D_c, the phenomenon of Neel relaxation, i.e., magnetic relaxation (D), takes shape. When $D > D_{th}$, but nearer to $D_{th,}$ the single domain MNPs reveal more or less pronounced divergence from the Langevin function, depending on temperature condition, size, and nature of NPs [44]. For $D < D_c$, but nearer to D_c, the uni domain MNPs exhibits stable magnetization and magnetization of NPs done as reported by the Stoner-Wohlfarth model. Magnetic relaxation brings about the heating of NPs under the alternating magnetic field [48]. These heated NPs have biomedical applications, in cancer tumor therapy. NPs give an expeditious response to the EMF, making them more efficient in drug delivery and target medication applications. Moreover, because of two extraordinary properties of NPs, viz., small size and magnetism, they have applications as contrast agents in MRI [43].

The magnetism of MNPs is an immense development sector that encapsulates various disciplines such as condensed matter physics, planetary science, material science, medicine, biology, and others. Apart from synthesized MNPs, natural MNPs are also observed everywhere: in microbes (bacteria, algae), the human brain, bees, birds, ants, lacustrine, and soil sediments and meteorites [49]. Magnetism in MNPs is firmly governed by finite particle size, shape, and surface effect. The impact of the shape of particles on the magnetism of NPs is different. For instance, homogeneous magnetization can be accomplished only with a single-domain particle having an ellipsoidal shape. But particle shape distortion brings stabilization and anisotropy in single domain particles [50]. Corresponding dependence of the number of surface particles upon the shape of particles indicates the relation between surface effect and shape.

5.   Surface modification

MNPs exhibit a set of distinctive properties for various nanomedicine applications. Modification of surface ligand upgrades biocompatibility, stability, solubility, and other properties of MNPs; they, therefore, can be employed as a vector for gene transfer, drug delivery, thermotherapy, and MRI, respectively [51]. Thus, it is essential to take several factors into contemplation at each design step. MNPs surface modifications are categorized into encapsulation and ligand exchange. In encapsulation, one or more MNPs accommodate the liposaomal structure or inorganic shell or amphiphilic block copolymer. A coating of MNPs with organic moieties such as polymer, polysaccharides [52], small molecules like DNA [53,54], and inorganic materials such as silica, gold, and tantalum oxide is also be observed [55]. Even after encapsulation of MNPs and their native ligands, the physical properties and original metal atoms present on the surface of MNPs are conserved significantly. In ligand exchange, replacement of hydrophobic surfactants on MNPs surface with entering hydrophilic ligands. For successful substitution, anchoring groups of entering ligands must have greater affinity to MNPs surface than that of ligands present on MNPs surface reference. In iron oxide MNPs, substitution of ligand such as oleic acid by another small ligand with terminal connecting groups, namely sulfonates, carboxylic acid, thiols, phosphonates, silane and catechols. Ligand exchange is more advantageous over encapsulation because in encapsulation hydrodynamic size is increases which is avoided in ligand exchange [55,56].

Due to interface and surface effects, MNPs are highly unstable and aggregate easily. Coating MNPs with small organic particles, including dextran, PEG, polymers (polyethyleneimine and chitosan), is essential to overcome the agglomeration of MNPs [7]. In the biological system, MNPs are adsorbed by body proteins. It has been established that surface fabrications result in functionally modified surface NPs and stabilize NMs surfaces. Various surface-modified substances are currently available, such as folic acid, polyethylenimine, polyethylene glycol, liposome, inorganic materials, and noble metal [39]. Out of these, polyethylene glycol (PEG) has better water solubility, nontoxicity, and biocompatibility. PEG is anchored on the surface of MNPs by end group reactivity, which makes it an excellent binding agent for drugs, DNA, and other biological fragments, increasing the blood circulation rate [57,39].

6.   Methods for fabrication of MNPs

Many methods have been reported for MNPs synthesis for the past several decades. Different synthesis procedures are responsible for controlling the size, shape, dispersion ability, surface chemistry, and stability of MNPs. The methods comprising chemical vapor deposition, hydrothermal, laser pyrolysis, sonochemical, co-precipitation, thermal decomposition, microemulsion, combustion, biological synthesis, etc have been outlined for the fabrication of MNPs. It was unfeasible for us to describe all these methods, so we have endeavored to explain a few of them.

6.1   Hydrothermal synthesis

The hydrothermal synthesis, also termed the solvothermal method, is one of the best routes for MNPs and ultrafine powder synthesis [58]. This MNPs synthesis observed in aqueous media using higher pressure reactor or autoclaves accompanies at very high pressure over 13790 kPa and temperature above 200°C [59]. The morphology and the size of newly formed MNPs by the hydrothermal method directly relate to time and temperature. In hydrothermal conditions, ferrites synthesis can be achieved (i) neutralization of mixed metal oxides (ii) hydrolysis and oxidation. This method imparts preferable control over the geometry of NPs [60]. Accompanying superior morphological control and noble crystallization, many other advantages provided by the hydrothermal method make it a more fruitful way for MNPs fabrication. Cai et al. studied that surface-modified MNPs synthesized by hydrothermal technique pose promising application in MRI [61,23]. Manganese zinc ferrite powder $(Mn, Zn)Fe_2O_4$ [62] is synthesized by neutralizing mixed metal hydroxide resulting in MNPs with 11 nm size. Barium hexaferrite $(BaFe_{12}O_{19})$ NPs have been manufactured solvothermally under supercritical conditions [63]. Yadav et al. [64] reported magnetite NPs coated with (3- Aminopropyl) triethoxysilane (APTS) using hydrothermal approach (shown in Fig. 4) and its biomedical activity for in vivo delivery of drugs and bimolecular had been investigated.

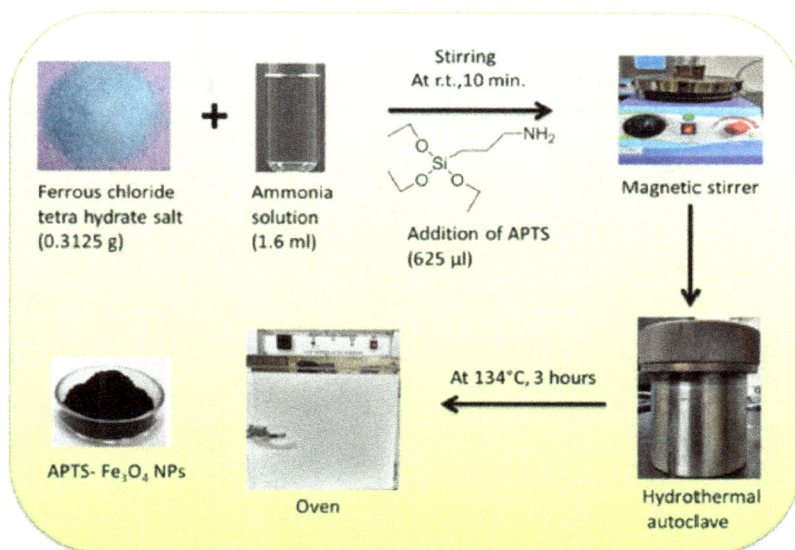

Fig. 4. Hydrothermal approach used to synthesize APTS functionalized Fe₃O₄ NPs [64].

6.2 Thermal decomposition

It is one of the easy techniques to synthesize MNPs by decomposing organometallic precursors. In anaerobic conditions, decomposition is achieved in organic solvent utilizing polymer or surfactants as a capping agent. As properties of newly synthesized MNPs is depended upon morphology, the polymer or surfactants are used to govern the size, morphology, crystallinity, and dispersion ability of NPs and protect particles from coalescing. It is widely used to synthesize magnetic iron oxide NPs [65,66]. The primary source organometallic precursor comprises $M^x (cup)_x$ (cup = N-nitrophenyl hydroxylamine) (M= Co, Fe, Ni, Mn, Cr), $[M^{n+} (acac)_n]$ (acac = acetylacetonate; n = 2 or 3) or carbonyls $[Fe(CO)_5]$ and surfactant used are hexadecyl amine, oleic acid and fatty acids [10]. Polyvinylpyrrolidone (PVP) polymer is used to limit particle size. Cobalt containing organometallic compound $(\eta^3\text{-}C_8H_{13})$ $(\eta^4\text{-} C_8H_{12})$ in the presence of H_2 and PVP polymer [67]. Ely et al. [68] synthesized bimetallic organometallic precursor Co_xPt_{1-x} NPs dispersed in PVP by thermal decomposition. The success rate of limiting small particle size can only be obtained by thermal decomposition, utilizing a high amount of expensive and toxic surfactant and precursor. Another drawback is the generation of organic soluble NPs, hence restricting its utilization in the medication field [69].

6.3 Sol-gel method

The generation of the sol-gel method for NPs is the most widely explored wet route technique. This method synthesizes films, glasses, fibers, monoliths, and powders. Traditionally, this method involves hydroxylation and condensing molecular precursors in an aqueous medium, commencing a 'sol' of NPs. Once sol is formed, condensation and polymerization occur, leading to the formation of metal oxide (wet gel) in the 3D network [70]. This method's synthesis depends upon several factors such as temperature, pH, solvent, precursors, additives, and catalyst, which influence growth, kinetics, hydroxylation, and condensation reaction [71]. Generally, the sol-gel method comprises five fundamental stages, hydrolysis, condensation, aging, drying, and finally, thermal decomposition, as shown in Fig 5 [72]. Sol-gel method has also been captivating for hexagonal ferrites preparation. For example, $Ba_{1-x}Sr_xFe_{12}O_{19}$, $BaZn_{2-x}Co_xFe_{16}O_{27}$, $La_{0.67}Ca_{0.33}MnO_3$ etc. [66]. This method is fascinating for the fabrication of binary and ternary compositions by using double alkoxide or mixed alkoxide. This route poses certain superiorities like low temperature, control reaction kinetics, etc. [73].

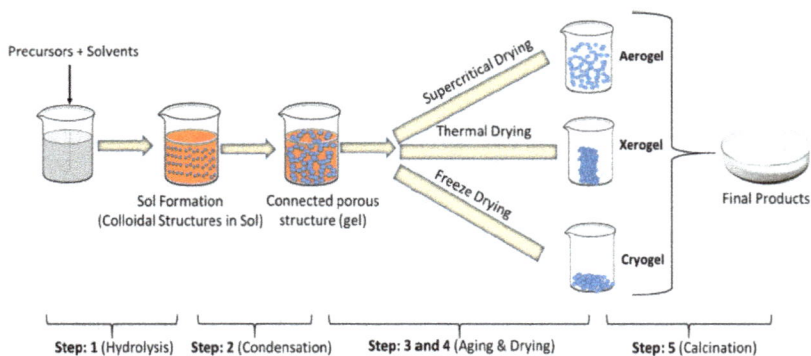

Fig. 5. Steps involve in synthesis of MNPs using sol-gel method [72].

6.4 Co-precipitation

Co-precipitation is the most proper and broadly used fabrication technique for MNPs synthesis. The synthesis process is accomplished under mild conditions. In the co-precipitation method, MNPs are synthesized from an aqueous monophasic salt solution by adding a base under anaerobic conditions at room or high temperatures [74]. It is known to be the most accurate and successful method for preparing superparamagnetic iron oxide MNPs. Nevertheless, there are some parameters like temperature and pH of the reaction medium, Fe (II): Fe (III) proportion, type of iron salts, and ionic strength of the solution that influences remarkably on composition, size and shape of iron oxide MNPs [75]. In the co-precipitation approach, after dissolving in acidic reagent, the nanosize magnetite prefers to decomposite into maghemite. Thus, magnetite dispersed in acidic medium subsequently the addition of Fe (III) nitrate, which gives substance with excellent stability at both acidic and basic conditions [76]. The size of MNPs can be controlled by adding polymer surface complexing agents or organic chelating agents throughout the synthesis of magnetite [77]. Freitas et al. reported variable size of iron oxide MNPs by using polyvinyl (alcohol) as a stabilizing agent in chainlike clusters revealing the significance of surfactant for stability purposes [78].

6.5 Micro-emulsion

The micro-emulsion thermodynamically stable solution is produced by mixing two immiscible solvents, generally the polar one water and non-polar one oil, together in the presence of surfactant molecule [79]. Micro-emulsion strategy is distinctive due to use of simple equipments, magnificent control over shape, size, crystalline structure, geometry, surface area and composition, thermodynamic stability, low interfacial tension, ambient pressure and temperature condition for synthesis [80,81]. The properties of MNPs synthesized using micro-emulsion route depends upon structure and type of surfactant [10]. In immiscible water-oil solution, when water microdroplets ($\approx$ 50 nm) is surrounded by surfactant molecules dispersed in hydrocarbon phase micro-emulsion of water-in-oil are formed [76]. Mixed metal-iron oxide MNPs, MFe_2O_4 (M = first row transition metal) synthesized by micro-emulsion is one the most interestingly studied material for electronic application [82]. Microemulsion synthesis of MNPs using ferrous dodecyl sulfate micellar solution produces nanosized MNPs whose size can be restrained by concentration and temperature of surfactant [83]. However this method is relatively difficult to apply at larger industrial scale because lower yield, additional methodology is needed to remove surfactant and massive amount of solvents are used in manufacturing of MNPs [84].

6.6 Green/Bio-inspired synthesis

Synthesis of MNPs using eco-friendly methods helps to understand better challenges advanced by chemical and physical processes. Recently, green nanotechnology has engrossed significant focus as it can eliminate toxic organic solvents and assist environmental restoration. Because of their genetic diversity and the presence of biomolecules and enzymes, both plants and microbes offer different routes for MNPs synthesis. The biocompatibility between different MNPs such as maghemite, magnetite, and zero-valent iron has been studied deliberately with vital scrutiny [85,86]. Moreover, utilization of plant extract, plants tissue, exudates, microorganisms, fungi, vitamins, enzymes, etc., has immerged as effectual substitutes for MNPs synthesis [87]. Occasionally, MNPs synthesis using plants is advantageous over biological entities because it eliminates extra work and time required for maintaining microbial culture. Also, plants act as a source of reducing agents like ascorbic acid, reductases, flavonoids, and dehydrogenase [88,89]. Chaudhary et al. copper NPs and ZnO NPs using green route [90,91]. Awwad et al. [92] synthesize magnetite NPs (Fe_3O_4) using carob leaf extract relatively at a lower temperature of 80-85°C. Newly formed MNPs have superior dispersion properties, and the average diameter is about 4-8 nm. Eatemadi et al. [69] aim to prepare magnetite NPs by using poly (ethylene glycol), PEO, as both surfactant as well as solvent simultaneously. PEO has been broadly employed as green solvent for number of organic synthesis because of its lower toxicity and higher boiling point. Studies have also reported that fungi are recommended to obtain huge quantity of MNPs extracellularly owing to presence of secretory parts [72]. Therefore, it can be assumed that more studies and work should be perform to recognize more efficient green synthesis methods. Nevertheless, toxicity and risks related with such methods must be investigates in detail manner.

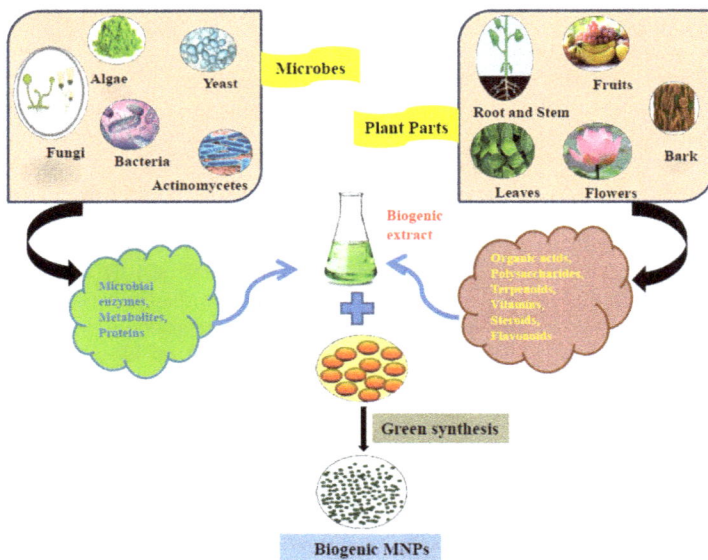

Fig. 6. Overview of green mediated synthesis of MNPs.

7. Cancer diagnosis with MNPs

MNPs overcome after the improvement in field of nanotechnology and have resulted in understanding of the NPs having leading edge application, endorsed with the suitability for cancer diagnosis via imaging and biosensor.

7.1 Imaging

Diagnosis of cancer by MNPs is an emerging domain. For example, non-invasive imaging could prevent the use of biopsy treatment and hence lessens the physical load on the patient and benefits the patient in crystal straightforward ways. Cancer imaging with higher susceptibility is pivotal and easily achievable imminent property of MNPs. In several pieces of literature, imaging can be combined with therapy by encapsulating drug molecules, hydrophobic or electrostatic interaction, and incorporating therapeutic drugs into polymer coating [93]. MNPs are also applied effectively in the cancer diagnosis by precisely imaging cancer resided areas, precisely and hence the application of MNPs with the MRI scans has significantly assisted in capturing the contrast image [94]. This can advance the treatment of small cancer metastases that were undetectable in the past. An *in-vitro* study of bladder cancer treatment by the Fe_3O_4 functional magnetic NMs (core/shell/crown) was performed and had been accomplished efficient therapeutic results [95].

Similarly, the gadolinium-doped iron oxide-NPs displayed the potentiality in magnetic resonance-based imaging and have attained good results in treating breast cancer at an *in-vitro* level. Further, the MRI and tumor magnetic fluid hyperthermia were also potentially achieved by the superparamagnetic PEG-modified La1-xSrxMnO3 (LSMO) NPs [96]. Furthermore, the wide application of MMPNs in imaging the breast cancer tumor via targeting the HER2/neu cancer marker (overexpressed) at an *in-vitro* and *in-vivo* level have been displayed, and the targeting of HER was approached by integrating the anti-HER antibody (Herceptin) on MPNs.

7.2 Biosensors

MNPs rooted biosensing policies have received significant attention owing to their distinctive significance over other approaches. For instance, physical and chemical stable, synthesis of MNPs are economically cheap, biocompatible, and eco-friendly. MNPs in sensing applications are applied via direct implementation of tagged support to the biosensor, then MNPs dispersion in the sample materials subsequently their attraction by applied EMF on biosensor active surface. In her literature review, Teresa explained MNPs based sensors and biosensors for detecting various analytes present in different samples by anticipating their parameters like LOD and linear range [97]. Issadore et al. [98] prepared a microfluidic chip-based micro-Hall detector (μ HD) which can detect targeted cell in the presence of millions of blood cells and do not need any additional washing treatment for purification. Additionally, they incorporated Manganese doped ferrite MNPs ($MnFe_2O_4$) with bio-orthogonal chemistry and distinctive magnetic properties. Synthesized MNPs simultaneously check for detection of the biomarkers EGFR, HER2/neu and EpCAM on individual cancer cell.

Lin et al. [99] developed a triple signal amplification scheme for immunosensing Carcinoembryonic antigen (CEA) cancer biomarker. For this, they used gold NPs modified ant-CEA antibodies. The strategy was achieved by immobilizing anti – CEA antibody on the graphene-chitosan surface coated with a glassy carbon electrode. Subsequently, BSA was added and finally incubated with CEA antigen. Poly (styrene-*co*-acrylic acid) micro-beads decorated with AuNPs were functionalized anti-CEA antibodies and eventually deposed to the biosensor. AuNPs persuade silver deposition resulting in the formation of AgNPs coated AuNPs. Voltammetry technique was used to analyze immunoreactions. It was observed that AgNPs oxidized preferably at a more negative potential with a reasonably sharp peak than AuNPs. The result indicates improvement in detection accuracy and sensitivity.

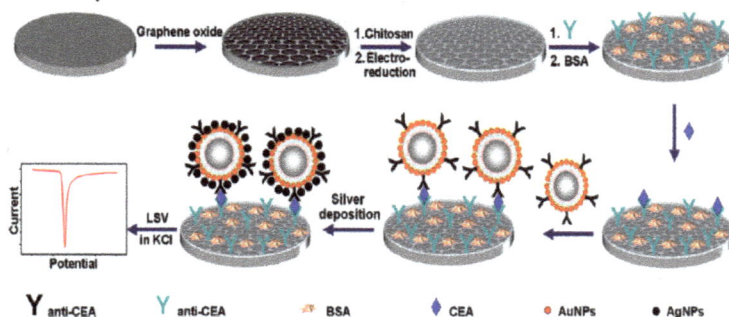

Fig. 7. Diagrammatic representation of silver-based immunosensor for CEA cancer biomarker detection [99].

8. Application in cancerous with MNPs

8.1 Drug delivery vehicle

MNPs as a vehicle for drug delivery purposes have been used to boost drugs' therapeutic activity and limit the secondary side effects accompanying traditional cancer therapies. Drug delivery is defined as delivering the therapeutic agent to the targeted site in the body. Drug delivery is an essential step toward medication of several ailments, like cardiovascular disease, microbial infected areas, and cancer. Among various NPs, MNPs are on the leading-edge for biomedical applications, including drug delivery. Lubbe et al. used MNPs in animal models. In 1996, the same group performed the first phase I clinical trial in patients using epirubicin-loaded MNPs [100]. The most interestingly studied MNPs is SIONs which act as drug delivery agent owing to their remarkable properties, as discussed earlier [47]. Frey et al., in their review, carefully summarize the fabrication and application of MNPs in drug delivery [101]. These MNPs bind together with chemotherapeutic, antibiotics, and other drugs and perform the function of drug vehicles. A scientist has illustrated that MNPs conjugated antibody and ovarian cancer detection also treat ovarian cancer because of their high level of ability to accumulate within cancer cells [18]. Rasaneh suggested it and his co-worker Dadras that use of MNPs with herceptin antibody may result in increasing therapeutic efficiency of antibody due to greater extent accumulation in tumor site [102].

The hydrophobic coating of MNPs limits their stability. To cope with this, SPIONs assembled with reducible copolymer were advanced to deliver chemotherapeutic drug DOX for cancer treatment. Additionally, it has also illustrated that the addition of iron oxide MNPs enhances the cell penetration efficiency of DOX NPs compared to free DOX and achieves better cellular response [103,104]. Nowadays, to improve therapeutic effectiveness, the combination of antibodies with medicinal drugs is captivating crucial importance. Aires et al. developed a dual-targeting therapy including novel iron oxide NPs with anti- CD44 antibody and gemcitabine derivatives. This newly formed structure is used for treatment of CD44-positive cancer cells [105].

Avedian et al. [106] prepared magnetic mesoporous silica NPs (MMSNs) composed of magnetite core coated with mesoporous silica and shell of poly (ethyleneimine) conjugated folic acid. The loaded Erlotinib named anticancer drug and its activity was assessed at two different pH 5.5 and 7.4 for the duration of 4 days. They observed that Erlotinib loaded MMSNs show toxicity towards HeLa cell lines. There is increased in release of Erlotinib of about 68% compared to 33% at normal pH. Further, folic acid labeled NPs exhibits higher cytotoxicity towards HeLa cells.

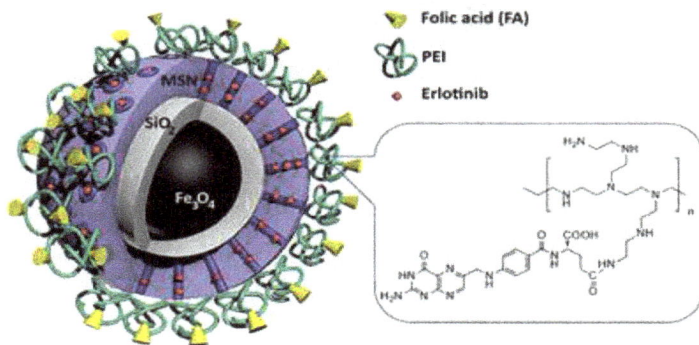

Biocompatible and folate sensitive MMSNs
with controllable and targeted Erlotinib delivery

Fig. 8. Targeted Erlotinib drug delivery with Fe₃O₄@MSN/PEI-FA [106].

The main objective of using MNPs for drug delivery stems from potentials such as a tiny amount of drug needed, capability to target particular organs, and reduction in the quantity of drug used minimizes various side effects. These advantages give ground for exponential growth in publications regarding MNPs for drug delivery applications. However, magnetic drug delivery poses limitations associated with a small size of NPs, the necessity of superparamagnetism, which is required to avoid agglomeration of MNPs after removing the magnetic field. Nanosized particles make it complicated to direct MNPs and retain them in surrounding of the targeted cells without influencing the drag of blood flows [107].

Table I: MNPs used in combination with various chemotherapeutic drugs to increase competency of cancer therapy.

Drug name	MNPs	Cancer cell line	Ref.
Docetaxel	MgNPs-Fe$_3$O$_4$	LNCaP, DU145, and PC-3	[108]
Docetaxel	β-Cyclodextrin Functionalized poly (5-amidoisophthalicacid) Grafted Fe$_3$O$_4$ MNPs	HEK293, HeLa and MDA-MB-231	[109]
Paclitaxel	IONPs coated with L-Aspartic acid	MCF-7	[110]
paclitaxel	Paclitaxel loaded core shell MNPs	A549 cells	[111]
Methotrexate	superparamagnetic iron oxide NPs	MCF-7	[112]
Gemcitabine	Carbopol/magnetite NPs with chitosan	PLCPRF-5, DLD-1, and MDA-231	[113]
Gemcitabine	Chitosan coated iron oxide NPs	SKBR and MCF-7	[114]
Bortezomib	chitosan coated superparamagnetic iron oxide NPs	HeLa and SiHa	[115]
Bortezomib	Iron oxide NPs	MDA-MB-468 and Caco-2	[116]
Erlotinib	Fe$_3$O$_4$ core coated with mesoporous silica	HeLa	[106]
Methotrexate/ Doxorubicin	Dendritic chitosan grafted mPEG coated (Fe$_3$O$_4$) MNPs	MCF-7	[117]

8.2 Hyperthermia

Several techniques, including microwave, ionizing radiations, and laser, are used as a healing tool for malignant tissues. Even though these approaches manage to increase intracellular temperature, they can cause harmful side effects like damage to neighboring healthy tissue and ionization of genetic material [118]. This motivates to explore another technique capable of rising temperature to damage cancerous cells without affecting healthy tissue. Magnetic hyperthermia is a novel and indigenous solution to all these issues. In 1957, Gilchrist et al. [119] introduced the concept of MNPs for hyperthermia applications. Nonmagnetic hyperthermia is a promising phenomenon for cancer treatment because; it is bound with hardly any side effects compared to chemotherapy and radiotherapy. Also, other conventional therapy modalities can use in combination with it [120]. The principal objective of hyperthermia is to increase the temperature between 41 and 45°C in tumor mass to kill cancerous cells, which are more sensitive to heat than normal cells.

The MNPs employed for hyperthermia are predominantly ferromagnetic NPs because they are highly potential in generating heat in the presence of an alternative magnetic field (AFM) [121]. The heat produced by MNPs is evaluated by specific absorption rate (SAR) or specific loss power (SLP) [122]. In the biological system, MNPs with magnetic fluid need to be administered directly into a tumor cell, followed by applying a magnetic field with radio wave frequency. The fluid contained MNPs must be tiny enough to be captured by cancer cells easily, innocuous, injectable, biocompatible, effective adsorption ability, and can function properly at physiological pH [123]. Fig. 8 shows the advantage of using MNPs in tumor cell treatment.

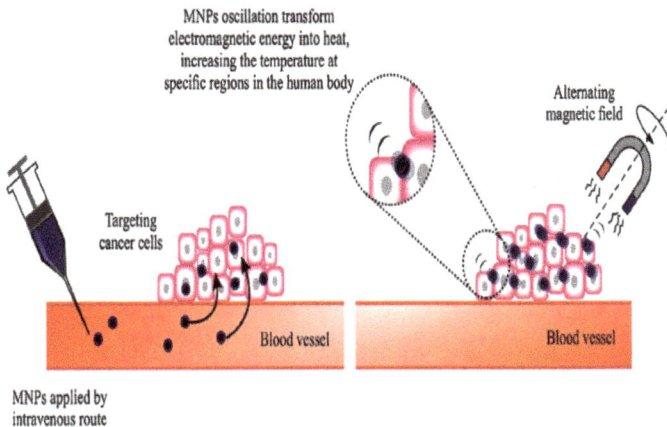

Fig. 9. Schematic illustration of hyperthermia therapy. MNPs accumulate in tumors, pointing to the cancer cells. Under the influence of an alternating magnetic field, MNPs align with the applied field and convert electromagnetic energy into heat, raising the temperature at this zone [23].

For hyperthermia treatment, the dose of heat produced is associated with the tumor volume and the applied procedure. For tumors with a diameter of 3 mm and MNPs with a concentration of 0.01 gcm^{-3}, the power required is more than 1 kWg^{-1}. The relationship between heat produced and clinical consequences parameter may be appraised by "cumulative equivalent minutes at 43°C". This parameter is interrelated with a heat exposure period and the stated temperature. Also, in the case of clinical trials, it is achieved at 90% of tumor-associated measure factors [6]. Furthermore, MNPs must be designed optimally to prevent unfavorable subordinate effects of high AFM strength on the patient.

A comprehensive review regarding hyperthermia with superparamagnetic iron oxide NPs (SPIONs) is elucidated by Laurent et al. [124]. Even though SPIONs manifest a moderate heating efficiency, it is more eminent in hyperthermia than other mentioned MNPs. Because of non-toxic, theragnostic potential, biodegradability, and biocompatibility, SPIONs represent a leading-edge tool in the nanomedicine sector [124,125]. To date, MNPs are made up of iron maghemite (γ-Fe$_2$O$_3$) and iron oxides magnetite (Fe$_3$O$_4$) owing to their low lethality and familiar metabolism pathway. Nevertheless, a major focus has been on magnetite because it is simpler than those necessary to produce maghemite. Hyperthermia can be explained by two loss mechanisms: hysteresis losses and Neel relaxation loss. In both loss mechanisms, different optimal particle sizes are observed. Specifically, both mechanisms exhibit monoatomic dependence of loss of particle size. Single domain ferromagnetic NPs produce more heat because single domain ferromagnetic NPs exhibit more hysteresis losses than multi-domain ferromagnetic NPs, Notwithstanding, superparamagnetic NPs own no hysteresis losses so they make heat as a consequence of Neel relaxation loss [126].

9.   Role of theranostic MNPs in cancer treatment

MNPs have been proposed as non-invasive NPs for several biomedical applications to introduce nanotechnology in the field of medicine. MNPs, in particular, have paved the way for theranostic agents, which allow for simultaneous diagnosis and treatment of disorders. The most significant factor in a good prognosis is early detection and treatment of disease. As a result, biomedical experts have put in much work to improve imaging techniques and therapy procedures. In the past decade, nanotechnology-derived concepts and technologies have solved difficulties with traditional advanced diagnosis and therapeutic strategies. Advances in NPs technology, in particular, have ushered in new paradigms for theranostics, or the combining of remedial and diagnostic agents on an individual platform.

Hyperthermia, or the process of elevating the temperature of tumor-loaded tissue to 40–43°C, is used in combination with several other cancer treatments such as chemotherapy and radiotherapy [127]. High temperatures (nearly 41-47° C) are applied to the tissue, damaging and destroying cancer cells. The idea of using hyperthermia as a cancer therapy is previously owned. Thermal therapy's therapeutic effects in cancer treatment were first discovered in the nineteenth century [128]. In the 1970s, the use of hyperthermia as a cancer treatment became increasingly severe, and controlled clinical trials on induced hyperthermia began. Then it was observed that cancer cells are more sensitive to hyperthermia than normal ones. On the other hand, healthy tissues can endure temperatures of 42–45 °C, whereas cancer cells go through apoptosis at those temperatures. Thermo-ablation is the treatment carried out above 46°C temperature. It causes necrosis of cancer cells and also affects surrounding healthy cells. Necrosis is another form of cellular damage which occurs due to premature death of cells found in body tissues and also commence an provocation response in nearby environment [129]. It is virtually every time dangerous, and it can even be fatal (i.e., it leads to septicemia or gangrene). As a result of the possibility for significant side effects on healthy tissues, thermoablation in this temperature range is not ideal, and hyperthermia-induced apoptosis is preferred.

Instead of positive results from increasing the intracellular temperature to induce death in cancer cells, specific detrimental side effects were detected in the surrounding healthy tissues. Some researchers have developed a new strategy to reduce this temperature increase due to this side effect [130]. The magnetic fluid hyperthermia (MFH), in which MNPs are used as heat mediators for local treatment, is a solution to this problem made possible by nanotechnology. MFH is based on the injection of a colloidal MNPs suspension that accumulates at the tumor either passively (due to the improved retention and permeability impact of MNPs) or actively (by using ligands at MNPs surface, specific for the surface receptors present on cancer cells) [131]. By increasing local temperature and modulating EMF, MNPs in tumor cells cause conversion of electromagnetic energy into heat. The use of MNPs can restrict the phenomenon of the heating of healthy tissue in the surrounding area.

Unlike traditional therapy methods, MNPs are recommended as a non-toxic and relatively non-invasive method for cancer therapy. Nonetheless, they have significant adverse effects such as damage to healthy organs and tissues and they are typically ineffective in treating metastatic cancer. With this frame of reference, in vivo, magnetic hyperthermia appears as a new cancer therapy with significant advantages over standard treatment.: i) tiny size particles can easily penetrate through the biological barrier, making injection less invasive; (ii) the use of specific agents can modulate the targeting properties of MNPs during the encapsulation activity; (iii) MNPs used in hyperthermia treatment can also be act as diagnostic representative for MRI; (iv) [131]

MNPs are less invasive and less harmful to the environment than traditional clinical treatments, including surgical resection, radiotherapy, and chemotherapy, and it has an effective therapeutic impact with few adverse effects. The Hayashi group [132] recently created superparamagnetic Fe_3O_4 clusters and improved their targeting ability by adding folic acid and PEG to the surface. The tail-vein-injected clusters generated hyperthermia for 20 minutes under an 8 kA m^{-1} field administered at 230 kHz (Hf = 1.8109 A m^{-1} s^{-1}) after 24 hours of enrichment. As a result, the temperature of the intratumor monitoring system was elevated to 65°C.

Peptide has received considerable attention as a powerful targeting ligand when coated with MNPs. A chemical reaction was used to make LTVSPWY peptide-modified PEGylated chitosan (LTVSPWY-PEG-CS), and 1H-NMR was used to confirm the chemical structure was reported by [133]. The solvent diffusion approach was used to prepare LTVSPWY-PEG-CS-modified MNPs successfully. The cellular uptake of the LTVSPWY-PEG-CS-modified MNPs was examined in a cocultured system of SKOV-3 cells that overexpress A549 and HER2 cells that are HER2-negative to investigate their selective targeting capacity. The MTT technique was used to assess the cytotoxicity of these NPas investigated using the SKOV-3-bearing nude mice model.

Conclusion

The chapter is focuses on recent advances in cancer therapy using multifunctional MNPs. In the mid-twentieth century, we observed a significant rise in our knowledge regarding the specification of MNPs properties. The breakthrough was the uncovering of superparamagnetic relaxation. In the present day, there has been an increase in the development of MNPs for cancer treatment such as hyperthermia, drug delivery, and cancer diagnosis methods like imaging and biosensors owing to their potentialities in surface functionality, high surface to volume ratio size, and shape. The current chapter discusses the construction and design of high-performance MNPs using chemical and green methods. But, considering the limitation posed by chemical methods, numerous researchers focus on a more convenient green synthesis method. However, surface modification of MNPs with proper ligand and coating material is essential to better lower biocompatibility, insufficient drug loading capacity, lower magnetic strength, etc.

In last couple of decades, the research in the use of MNPs in biomedical applications has been increasing expeditiously, resulting in the advancement of medical applications of NPs. However, it will be crucial to explore further how NPs and MNPs generally interact with biological systems in the human body. By knowing this, one can understand the toxicity of NPs and mapping. The role of MNPs in nature will gain extra attentiveness in the upcoming future. The significance of magnetic navigation for several animals has been recognized recently. Yet, understanding its biological system mechanism is an engrossing scientific challenge.

References

[1] Igor Pantic, Magnetic nanoparticles in cancer diagnosis and treatment : Novel approaches, Reviews on Advance Materials Science, 26 (2010) 67-73.

[2] S. Parvanian, S. Mojtaba, and M. Aghashiri, Multifunctional nanoparticle developments in cancer diagnosis and treatment, Sensing and Bio-Sensing Research, 13 (2016) 81-87. https://doi.org/10.1016/j.sbsr.2016.08.002

[3] G. Yeldag, A. Rice, and A. del Río Hernandez, Chemoresistance and the self-maintaining tumor microenvironment, Cancers, 10 (2018) 471. https://doi.org/10.3390/cancers10120471

Materials Research Forum LLC
https://doi.org/10.21741/9781644902332-7

[4] F. Cheng, C. Chan, B. Wang, Y. Yeh, and Y. Wang, The oxygen-generating calcium peroxide-modified magnetic nanoparticles attenuate hypoxia-induced chemoresistance in triple-negative breast cancer, Cancers, 13 (2021) 606. https://doi.org/10.3390/cancers13040606

[5] L. M. Colli, M. J. Machiela, H. Zhang, T. A. Myers, Landscape of combination immunotherapy and targeted therapy to improve cancer management, Cancer Research. 77 (2017) 3666–3672. https://doi.org/10.1158/0008-5472.can-16-3338

[6] P. I. P. Soares, I. M. M. Ferreira, R. A. G. B. N. Igreja, C. M. M. Novo, and P. M. R. Borges, Application of hyperthermia for cancer treatment : recent patents review, Recent Patents on Anti-Cancer Drug Discovery, 7 (2012) 64–73. https://doi.org/10.2174/157489212798358038

[7] N. Hinge, M. M. Pandey, G. Singhvi, G. Gupta, M. Mehta, S. Satija, M. Gulati, H. Dureja and K. Dua, Nanomedicine advances in cancer therapy, In Advanced 3D - Printed System and Nanosystems for Drug Delivery and Tissue Engineering, (2018) 219-253. https://doi.org/10.1016/B978-0-12-818471-4.00008-X

[8] H. Zhang, X. Li Liu, Yi F. Zhang, F. Gao, G. L. Li, Y. He, M. Li Peng and H. M. Fan, Magnetic nanoparticles based cancer therapy : current status and, Science China Life Sciences, 61 (2018) 400-414. https://doi.org/10.1007/s11427-017-9271-1

[9] A. Farzin, S. A. Etesami, J. Quint, A. Memic, and A. Tamayol, Magnetic Nanoparticles in Cancer Therapy and Diagnosis, Advanced Healthcare Materials, 1901058: (2020) 1–29. https://doi.org/10.1002/adhm.201901058

[10] S. Majidi, F. Z. Sehrig, S. M. Farkhani, and M. S. Goloujeh, Current methods for synthesis of magnetic nanoparticles, Artificial Cells, Nanomedicine, and Biotechnology, 44 (2014) 722-734. https://doi.org/10.3109/21691401.2014.982802

[11] M. Kouhi, A. Vahedi, A. Akbarzadeh, Y. Hanifehpour, and S. W. Joo, Investigation of quadratic electro-optic effects and electro-absorption process in GaN/AlGaN spherical quantum dot, Nanoscale Research Letters, 9 (2014) 131. https://doi.org/10.1186/1556-276X-9-131

[12] H. R. Ghorbani, A. A. Safekordi, H. Attar, and S. M. R. Sorkhabadi, Biological and Non-biological Methods for Silver Nanoparticles Synthesis, Chemical and Biochemical Engineering Quarterly. 25 (2011) 317–326.

[13] K. S. Kavitha, `S. Baker, D. Rakshith, H. U. Kavitha, H. C. Rao, B. P. Harini and S. Satish, Plants as Green Source towards Synthesis of Nanoparticles Plants as Green Source towards Synthesis of Nanoparticles, International Research Journal of Biological Sciences, 2 (2013) 66-76.

[14] M. Bin Ahmad, M. Y. Tay, K. Shameli, M. Z. Hussein, and J. J. Lim, Green synthesis and characterization of silver/chitosan/polyethylene glycol nanocomposites without any reducing agent, International Journal of Molecular Scienec, 12 (2011) 4872–4884. https://doi.org/10.3390/ijms12084872

[15] K. Parveen, V. Banse, and L. Ledwani, Green synthesis of nanoparticles : Their advantages and disadvantages, AIP Conference Proceeding, 1724 (2016) 20048. https://doi.org/10.1063/1.4945168

[16] R. G. Chaudhary, A. Mondal, T. Aziz, and A. Potbhare, Applications of metal / metal oxides nanoparticles in organic transformations, Material Research Foundation, 83 (2020) 134-156. https://doi.org/10.21741/9781644900970-6

[17] J. Kudr, Y. H. Id, L. Richtera, V. Adam, and O. Zitka, Magnetic nanoparticles : from design and synthesis to real world applications, Nanomaterials, 7 (2017) 243. https://doi.org/10.3390/nano7090243

[18] M. Wu, and S. Huang, Magnetic nanoparticles in cancer diagnosis , drug delivery and treatment (Review), Molecular and Clinical Oncology, 7 (2017) 738–746. https://doi.org/10.3892/mco.2017.1399

[19] B. Issa, I. M. Obaidat, B. A. Albiss, and Y. Haik, Magnetic nanoparticles : surface effects and properties related to biomedicine applications, International Journal of Molecular Sciences, 14 (2013) 21266–21305. https://doi.org/10.3390/ijms141121266

[20] I. M. Obaidat, B. Issa and Y. Haik, Magnetic properties of magnetic nanoparticles for efficient hyperthermia, Nanomaterials, 5 (2015) 63–89. https://doi.org/10.3390/nano5010063

[21] M. Javaid, A. Haleem, R. P. Singh, S. Rab, and R. Suman, Exploring the potential of nanosensors: A brief overvie, Sensors International, 2 (2021) 100130. https://doi.org/10.1016/j.sintl.2021.100130

[22] M. Mahdavi, M. B. Ahmad, M. J. Haron, F. Namvar, B. Nadi, M. Z. Ab Rahman, and J. Amin, Synthesis, surface modification and characterisation of biocompatible magnetic iron oxide nanoparticles for biomedical applications, Molecules, 18 (2013) 7533–7548. https://doi.org/10.3390/molecules18077533

[23] S. S. Khiabani, M. Farshbaf, S. Davaran and A. Akbarzadeh, Magnetic nanoparticles : preparation methods, applications in cancer diagnosis and cancer therapy, Artificial Cells, Nanomedicine, and Biotechnology, 45 (2016) 6-17. https://doi.org/10.3109/21691401.2016.1167704.

[24] S. Morup, M. F. Hansen, C. Frandsen, and K. Lyngby, Magnetic Nanoparticles, Department of Physics, Surface Physics and Catalysis, 1 (2011) 437-491.

[25] H. C. Orested, Experiments on the Effect of a current of electricity on the magnetic needle, Semantic Scholar, 16 (1820) 273-277. https://doi.org/10.1515/9781400864850.417

[26] J. M. Eargle, Analog Magnetic Recording, Handbook of Recording Engineering, (1996) 223–224.

[27] J. Alonso, M. Barandiarán, L. F. Barquín and A. G. Arribas, Magnetic nanoparticles, synthesis, properties, and applications, Magnetic Nanostructured Materials, (2018) 1-40. https://doi.org/10.1016/B978-0-12-813904-2.00001-2

[28] M. N. Baibich, J. M. Broto, A. Fert, F. Nguyen Van Dau and F. Petroff, Gaint Magnetoresistance of (001)Fe/(001)Cr Magnetic Superlattices, Physical Review Letters, 61 (1998) 2472. https://doi.org/10.1103/PhysRevLett.61.2472

[29] G. Binasch, P. Grunberg, F. Saurenbach, and W. Zinn, Enhanced magnetoresistance in layered magnetic structures antiferromagnetic interlayer exchange, Physical Review B, 39 (1989) 4828–4830. https://doi.org/10.1103/PhysRevB.39.4828

[30] E. M. Materon, C. M. Miyazaki, O. Carr, N. Joshi, P. H. S. Picciani, C. J. Dalmaschio, F. Davis and F. M. Shimizu, Magnetic nanoparticles in biomedical applications : A review, Applied Surface Science Advances, 6 (2021) 100163. https://doi.org/10.1016/j.apsadv.2021.100163

[31] M. Attia, N. Anton, J. Wallyn, Z. Omran, and T. Vandamme, An overview of active and passive targeting strategies to improve the nanocarriers efficiency to tumour sites, Journal of Pharmacy and Pharmacology, 71 (2019) 1185-1198. https://doi.org/10.1111/jphp.13098

[32] J. Shi, P. W. Kantoff, R. Wooster, and O. C. Farokhzad, Cancer nanomedicine : progress, challenges and opportunities, National Journal Publishers, 17 (2016) 20–37. https://doi.org/10.1038/nrc.2016.108

[33] R. Weissleder, Molecular Imaging in Cancer, Frontiers in Cancer Research, 312 (2016) 1168-1171. https://doi.org/10.1126/science.1125949

[34] J. Kim, N. Lee, and N. Lee, Recent development of nanoparticles for molecular imaging, Philosophical Transactions of the Royal Society A Journal, 375 (2017) 20170022. http://dx.doi.org/10.1098/rsta.2017.0022

[35] T. Hyeon, S. S. Lee, J. Park, Y. Chung, and H. Bin Na, Synthesis of highly crystalline and monodisperse maghemite nanocrystallites without a size-selection process," Journal of the American Chemical Society, 123 (2001) 12798–12801. https://doi.org/10.1021/ja016812s

[36] S. Sun, C. B. Murray, D. Weller, L. Folks, and A. Moser, Monodisperse FePt nanoparticles and ferromagnetic FePt nanocrystal superlattices, Science, 287 (2000) 1989-1992. https://doi.org/10.1126/science.287.5460.1989

[37] S. Sun, H. Zeng, D. B. Robinson, S. Raoux, P. M. Rice, Shan X. Wang, and G. Li (2004), "Monodisperse MFe_2O_4 (M = Fe , Co , Mn) Nanoparticles, Journal of the American Chemical Society, 126 (2004) 273-279. https://doi.org/10.1021/ja0380852

[38] A. Yadollahpour and S. Rashidi (2015), "Magnetic nanoparticles : A review of chemical and physical characteristics important in medical applications, Oriental Journal of Chemistry, 31 (2015). https://doi.org/10.13005/ojc/31.Special-Issue1.03

[39] J. Gellermann, W. Wlodarczyk, A. Feussner, H. Fahlings, J. Nadobny, B. Hildebrandt, R. Felix and P. Wust, Methods and potentials of magnetic resonance imaging for monitoring radiofrequency hyperthermia in a hybrid system, 21 (2005) 497-513. https://doi.org/10.1080/02656730500070102

[40] N. Senthilkumar, P. Kumar, N. Sood, and N. Bhalla, Designing magnetic nanoparticles for in vivo applications and understanding their fate inside human body, Coordination Chemistry Reviews, 445 (2021) 214082. https://doi.org/10.1016/j.ccr.2021.214082

[41] K. Mcnamara and S. A. M. Tofail, Nanosystem: the use of nanoalloys, metallic, bimetallic, and magnetic nanoparticles in biomedical applicatons, Physical Chemistry Chemical Physics, 17 (2015) 27981-27995. https://doi.org/10.1039/C5CP00831J

[42] A. Figuerola, R. Di, L. Manna, and T. Pellegrino, From iron oxide nanoparticles towards advanced iron-based inorganic materials designed for biomedical applications, Pharmacol, Res. 62 (2010) 126–143. https://doi.org/10.1016/j.phrs.2009.12.012

[43] D. Ho, X. Sun and S. Sun, Monodisperse Magnetic nanoparticles for theranostic applications, American Chemical Society, 44 (2011) 875-882. https://doi.org/10.1021/ar200090c

[44] C. Caizer, Nanoparticle Size Effect on Some Magnetic Properties, Springer Cham, (2016) 475-519. https://doi.org/10.1007/978-3-319-15338-4_24

[45] C. P. Bean and J. D. Livingston, Superparamagnetism, Journal of Applied Physics, 30 (1959) 120-129. https://doi.org/10.1063/1.2185850

[46] C. Rümenapp, B. Gleich, and A. Haase, Magnetic nanoparticles in magnetic resonance imaging and diagnostics, Pharmaceutical Research. 29 (2012) 1165-1179. https://doi.org/10.1007/s11095-012-0711-y

[47] S. Mirza, M. S. Ahmad, M. Ishaq, A. Shah, and M. Ateeq, Magnetic nanoparticles : drug delivery and bioimaging applications, In Metal nanoparticles for drug delivery and diagnostic applications, (2020) 189-208. https://doi.org/10.1016/B978-0-12-816960-5.00011-2

[48] Q. A. Pankhurst, J. Connolly, S. K. Jones, and J. Dobson, Applications of magnetic nanoparticles in biomedicine, Journal of Physics D: Applied Physics, 36 (2003) 167-181.

[49] Y. Koksharov and G. Khomutov, Organized ensembles of magnetic nanoparticles : preparation , structure , and properties, Magnetic Nanoparticles. 15 (2009) 117-95. https://doi.org/10.1002/9783527627561

[50] R. P. Cowburn, Property variation with shape in magnetic nanoelements, Journal of Physics D: Applied Physics, 33 (2000) 1–16.

[51] L. Sun, C. Huang, T. Gong, and S. Zhou, A biocompatible approach to surface modification : Biodegradable polymer functionalized super-paramagnetic iron oxide nanoparticles, Materials Science and Engineering C, 30 (2010) 583–589. https://doi.org/10.1016/j.msec.2010.02.009

[52] C. Ileana, C. Daniela, C. Matei, L. Diamandescu, and E. Vasile, Magnetic nanoparticles coated with polysaccharide polymers for potential biomedical applications, Journal of Nanoparticle Research, 13 (2011) 6169–6180. https://doi.org/10.1007/s11051-011-0452-6

[53] Y. Liu, T. Chen, C. Wu, L. Qui, R. Hu, J. Li, S. Cansiz, L. Zhang, C. Cui, G. Zhu, M. You, T. Zhang and W. Tan, Facile surface functionalization of hydrophobic magnetic nanoparticles, Journal of the American Chemical Society, 136 (2014) 12552-12555. https://doi.org/10.1021/ja5060324

[54] D. Ling, M. J. Hackett, and T. Hyeon, Surface ligands in synthesis, modification, assembly and biomedical applications of nanoparticles, Nano Today, 9 (2014) 457-477. https://doi.org/10.1016/j.nantod.2014.06.005

[55] T. Kang, F. Li, S. Baik, W. Shao, and D. Ling, Biomaterials surface design of magnetic nanoparticles for stimuli-responsive cancer imaging and therapy, Biomaterials, 136 (2017) 98–114. https://doi.org/10.1016/j.biomaterials.2017.05.013

[56] G. Palui, F. Aldeek, W. Wang, and H. Mattoussi, Strategies for interfacing inorganic nanocrystals with biological systems based on polymer-coating, Chemical Society Reviews, 44 (2015) 193-227. https://doi.org/10.1039/C4CS00124A

[57] E. C. Gryparis, M. Hatziapostolou, and E. Papadimitriou, Anticancer activity of cisplatin-loaded PLGA-mPEG nanoparticles on LNCaP prostate cancer cells, European Journal of Pharmaceutics and Biopharmaceutics, 67 (2007) 1–8. https://doi.org/10.1016/j.ejpb.2006.12.017

[58] P. M. De Molina, M. Zhang, A. V. Bayles, and M. E. Helgeson, Oil-in-water-in-oil multi-nanoemulsions for templating complex nanoparticles, Nano Letters, 16 (2016) 7325-7332. https://doi.org/10.1021/acs.nanolett.6b02073

[59] S. Natarajan, K. Harini, G. P. Gajula, B. Sarmento, M. T. Petersen, and V. Thiagarajan, Multifunctional magnetic iron oxide nanoparticles : diverse synthetic approaches, surface modifications, cytotoxicity towards biomedical and industrial applications, BMC Materials, 1 (2019) 1–22. https://doi.org/10.1186/s42833-019-0002-6

[60] A. Sobhani and M. Salavati-niasari, Synthesis and characterization of $FeSe_2$ nanoparticles and $FeSe_2/FeO(OH)$ nanocomposites by hydrothermal method, Journal of Alloys and Compounds, 625 (2015) 26–33. https://doi.org/10.1016/j.jallcom.2014.11.079

[61] H. Cai, X. An, J. Cui, J. Li, S. Wen, K. Li, M. Shen, L. Zheng, and X. Shi, Facile hydrothermal synthesis and surface functionalization of polyethyleneimine-coated iron oxide nanoparticles for biomedical applications, ACS Applied Materials & Interfaces, 5 (2013) 1722-1731. https://doi.org/10.1021/am302883m

[62] M. Rozman, M. Drofenik, I. Introduction, E. Procedure, P. Alto, and H. Wycombe, Hydrothermal synthesis of manganese zinc ferrites, Journal of the American Ceramic Society, 78 (1995) 2449-55. https://doi.org/10.1111/j.1151-2916.1995.tb08684.x

[63] Y. Hakuta, T. Adschiri, T. Suzuki, T. Chids, K. Seino, and K. Arai, Flow method for rapidly producing barium hexaferrite particles in supercritical water, Journal of the American Ceramic Society, 81 (1998) 2461–2464.

[64] N. Yadav, A. Singh, and M. Kaushik, Hydrothermal synthesis and characterization of magnetic Fe_3O_4 and APTS coated Fe_3O_4 nanoparticles : physicochemical investigations of interaction with DNA, The Journal of Materials Science: Materials in Medicine, 31 (2020) 1–11. https://doi.org/10.1007/s10856-020-06405-6

[65] S. Laurent D. Forge, M. Port, A. Roch, C. Robic, L. V. Elast and R. N. Muller Magnetic Iron Oxide Nanoparticles : Synthesis, Stabilization, Vectorization, Physicochemical Characterizations, and Biological Applications, Chemical Reviews, 108 (2008) 2064–2110. https://doi.org/10.1021/cr068445e

[66] M. A. Willard, L. K. Kurihara, E. E. Carpenter, S. Calvin, and V. G. Harris, Chemically prepared magnetic nanoparticles, International Materials Reviews, 49 (2004) 125–170. https://doi.org/10.1179/095066004225021882

[67] O. Margeat, M. Respaud, C. Amiens, P. Lecante, and B. Chaudret, Ultrafine metallic Fe nanoparticles : synthesis, structure and magnetism, Beilstein Journal of Nanotechnology, 1 (2010) 108–118. https://doi.org/10.3762/bjnano.1.13

[68] T. O. Ely, C. Pan, C. Amiens, B. Chaudret, F. Dassand, P. Lecante, M.-J. Casanove, A. Mosset, M. Respaud, and J.-M. Broto, Nanoscale Bimetallic Co_xPt_{1-x} Particles Dispersed in Poly (vinylpyrrolidone): Synthesis from Organometallic Precursors and Characterization, The Journal of Physical Chemistry B, 104 (2000) 695–702. https://doi.org/10.1021/jp9924427

[69] A. Eatemadi, H. Daraee, N. Zarghami, H. M. Yar, and A. Akbarzadeh, Nanofiber : Synthesis and biomedical applications, Artificial Cells, Nanomedicine and Biotechnology, 44 (2011) 111-121. https://doi.org/10.3109/21691401.2014.922568

[70] M. Alagiri, S. Ponnusamy and C. Muthamizhchelvan, Synthesis and characterization of NiO nanoparticles by sol – gel method, Journal of Materials Science: Materials in Electrons, 23 (2011) 728–732. https://doi.org/10.1007/s10854-011-0479-6

[71] C. Jeffrey Brinker, George W. Scherer, Sol-Gel Science: The physics and chemistry of sol-gel processing, Physical Sciences and Engineering, (1990) 1-17.

[72] M. Parashar, V. Kumar, and S. Ranbir, Metal oxides nanoparticles via sol – gel method : a review on synthesis, characterization and applications, Journal of Materials Science: Materials in Electrons, 31 (2020) 3729-3749. https://doi.org/10.1007/s10854-020-02994-8

[73] K. Gudikandula and S. C. Maringanti, Synthesis of silver nanoparticles by chemical and biological methods and their antimicrobial properties, Journal of Experimental Nanoscience, 11 (2016) 714–721. https://doi.org/10.1080/17458080.2016.1139196

[74] S. Majidi, F. Z. Sehrig, S. M. Fakhani, M. S. Goloujeh and A. Akbarzadeh, Current methods for synthesis of magnetic nanoparticles, Artificial Cells, Nanomedicine, and Biotechnology. 44 (2014) 722-734. https://doi.org/10.3109/21691401.2014.982802

[75] Z. Surowiec, M. Budzynski, K. Durak, and G. Czernel, Synthesis and characterization of iron oxide magnetic nanoparticles, Nukleonika, 62 (2017) 73–77. https://doi.org/10.1515/nuka-2017-0009

[76] S. Gul, S. B. Khan, I. U. Rehman, M. A. Khan, and M. I. Khan, A comprehensive review of magnetic nanomaterials modern day theranostics, Frontires in Materials. 6 (2019) 1–15. https://doi.org/10.3389/fmats.2019.00179

[77] M. C. Mascolo, Y. Pei, T. A. Ring, S. L. City, and S. Latium, Room temperature co-precipitation synthesis of Magnetite nanoparticles in a large pH window with different bases, Materials, 6 (2013) 5549–5567. https://doi.org/10.3390/ma6125549

[78] J. C. Freitas, R. M. Branco, I. G. Lisboa, T. P. da Costa, M. G. Campos, M. J. Junior and R. F. Marques, Magnetic nanoparticles obtained by homogeneous coprecipitation sonochemically assisted, Journal of Materials Research, 18 (2015) 220–224. https://doi.org/10.1590/1516-1439.366114

[79] D. T. Lucas, D. A. Sica, R. H. Cássia, and B. M. Luciano, Iron oxide magnetic nanoparticles as antimicrobial for therapeutics, Pharmaceutical Development and Technology, 23 (2013) 316-323. https://doi.org/10.1080/10837450.2017.1337793

[80] H. Zhao, R. Liu, Q. Zhang, and Q. Wang, Effect of surfactant amount on the morphology and magnetic properties of monodisperse $ZnFe_2O_4$ nanoparticles, Materials Research Bulletin, 75 (2016) 172-177. https://doi.org/10.1016/j.materresbull.2015.11.052

[81] A. Salabat and F. Mirhoseini, A novel and simple microemulsion method for synthesis of biocompatible functionalized gold nanoparticles, Journal of Molecular Liquids, 268 (2018) 849–853. https://doi.org/10.1016/j.molliq.2018.07.112

[82] L. Gutiérrez, R. Costo, C. Gruttner, F. Westphal, N. Gehrke, D. Heinke, A. Fornara, Q. A. Pankhurst, C. Johansson, S. Veintemillas-Verdaguer, and M. P. Morales, Synthesis methods to prepare single- and multi-core iron oxide nanoparticles for biomedical applications, Salton Transactions, 44 (2015) 2943-2952. https://doi.org/10.1039/c4dt03013c

[83] J. K. Yamchi, M. Mobasseri, A. Akbarzadeh, S. Davaran, A. Ostad-Rahimi, H. Hamishehkar, R. Salehi, Z. Bahmani, K. Nejati-Koshki, A. Darbin and M. R. Yamchi, Preparation of pH sensitive insulin-loaded nano hydrogels and evaluation of insulin releasing in different pH conditions, Molecular Biology Reports, 41 (2014) 6705-6712. https://doi.org/10.1007/s11033-014-3553-3

[84] M. J. Williams, E. Sanchez, E. R. Aluri, F. J. Douglas, D. A. MacLaren, O. M. Collins, E. J. Cussens, J. D. Budge, L. C. Sanders, M. Michaelis, C. M. Smales, J. D. Budge, L. C. Sanders, M. Michaelis, C. M. Smales, J. C. Jr., S. Lorrio, D. Krueger, Rafael T. M., and S. A. Corr, Microwave-assisted synthesis of highly crystalline, multifunctional iron oxide nanocomposites for imaging applications, RSC Advances, 6 (2016) 83520-83528. https://doi.org/10.1039/C6RA11819D

[85] N. Joshi, J. Filip, V. S. Coker, J. Sadhukhan, I. Safarik, H. Bagshaw and J. R. Lloyd, Microbial reduction of natural Fe (III) minerals ; toward the sustainable production of functional magnetic nanoparticles, Frontiers in Environmental Science, 6 (2018) 1–11. https://doi.org/10.3389/fenvs.2018.00127

[86] S. Shukla, R. Khan, and A. Daverey, Environmental Technology & Innovation Synthesis and characterization of magnetic nanoparticles , and their applications in wastewater treatment : A review, Environmental Technology and Innovation, 24 (2021) 101924. https://doi.org/10.1016/j.eti.2021.101924

[87] P. Singh, Y. Kim, D. Zhang, and D. Yang, Biological synthesis of nanoparticles from plants and microorganisms, Trends in Biotechnology, 34 (2016) 588–599. https://doi.org/10.1016/j.tibtech.2016.02.006

[88] R. G. Chaudhary, A. K. Potbhare, P. B. Chouke, A. R. Rai, R. Mishra, M. F. Desimone, and A. A. Abdala, Graphene-based materials and their nanocomposites with metal oxides : biosynthesis, electrochemical, photocatalytic and antimicrobial applications, Material Research Forum. 83 (2020) 79-116.

[89] P. B. Chouke, K. M. Dadure, A. K. Potbhare, G. S. Bhusari, A. Mondal, K. Chaudhary, V. Singh, M.F. Desimone, R.G. Chaudhary, D.T. Masram, Biosynthesized δ-Bi_2O_3 Nanoparticles from Crinum viviparum flower extract for photocatalytic dye degradation and molecular docking, ACS Omega, 2022, 7 (24),20983–20993. https://doi.org/10.1021/acsomega.2c01745

[90] P. B. Chouke, A. Potbhare, G. Bhusari, S. Somkuwar, D. P. Shaik, R. Mishra and R. G. Chaudhary, Green fabrication of zinc oxide nanospheres by aspidopterys cordata for effective antioxidant and antibacterial activity, Advanced Materials Letter, 10 (2019) 355–360. https://doi.org/10.5185/amlett.2019.2235

[91] J. A. Tanna, R. G. Chaudhary, N. V. Gandhare, A. R. Rai, S. Yerpude and H. D. Juneja, Copper nanoparticles catalysed an efficient one-pot multicomponent synthesis of chromenes derivatives and its antibacterial activity, Journal of Experimental Nanoscience, 11 (2016) 884-900. http://dx.doi.org/10.1080/17458080.2016.1177216

[92] A. M. Awwad and N. M. Salem, A Green and Facile Approach for Synthesis of Magnetite Nanoparticles, *nanoscience and nanotechnology*, 2 (2012) 208-213. https://doi.org/10.5923/j.nn.201206.09

[93] M. V Yigit, A. Moore, and Z. Medarova, Magnetic Nanoparticles for Cancer Diagnosis and Therapy, Pharmaceutical Research, 29 (2012) 1180–1188. https://doi.org/10.1007/s11095-012-0679-7

[94] S. Motaali, M. Pashaeiasl, S. Davaran, and A. Akbarzadeh, Synthesis and characterization of smart N-isopropylacrylamide-based magnetic nanocomposites containing doxorubicin anti-cancer drug, Artificial Cells, Nanomedicine, and Biotechnology, 43 (2017) 560-567. https://doi.org/10.3109/21691401.2016.1161640

[95] Y. Wang, Y. Liao, C. Liu, J. Yu, Y. Yamauchi, S. A. Hossain, and K. C. Wu, Tri-functional Fe_3O_4/CaP/Alginate core-shell- corona nanoparticles for magnetically guiding, pH- responsive, and chemically targeting chemotherapy, ACS Biomaterials Science & Engineering, 3 (2017) 2366-2374. https://doi.org/10.1021/acsbiomaterials.7b00230

[96] N. D. Thorat, R. A. Bohara, S. M. Tofail, Z. A. Alothman, M. A. Shiddiky, S. A. Hossain, Y. Yamauchi, and K. Wu, Superparamagnetic gadolinium ferrite nanoparticles with controllable curie temperature – cancer theranostics for MR-imaging-guided magneto-chemotherapy, European Journal of Inorganic Chemistry, 28 (2016) 4586-4597. https://doi.org/10.1002/ejic.201600706

[97] T. A. P. Rocha-santos, Sensors and biosensors based on magnetic nanoparticles, Trends in Analytical Chemistry, 62 (2014) 28–36. https://doi.org/10.1016/j.trac.2014.06.016

[98] D. Issadore, J. Chung, H. Shao, M. Liong, A. A. Ghazani, C. M. Castro, R. Weissleder, and H. Lee, Ultrasensitive clinical enumeration of rare cells ex vivo using a micro-hall detector, Science Translational Medicine, 4 (2012) 141. https://doi.org/10.1126/scitranslmed.3003747

[99] D. Lin, J. Wu, M. Wang, F. Yan, and H. Ju, Triple signal amplification of graphene film, polybead carried gold nanoparticles as tracing tag and silver deposition for ultrasensitive electrochemical immunosensing, Analytical Chemistry, 84 (2012) 3662-3668. https://doi.org/10.1021/ac3001435

[100] A. S. Lãbbe, C. Bergemann, H Riess, F. Schriever, P. Reichrdt, K. Possinger, M. Matthias, B. Dorken, F. Herrmann, R. Gurtler, P. Hohenberger, N. Haas, R. Sohr, B. Sander, A J Lemke, D. Ohlendorf, W. Huhnt, and D. Huhn, Clinical experiences with magnetic drug targeting : a phase I study with 4'- epidoxorubicin in 14 patients with advanced solid tumors, Cancer Research, 56 (1996) 4686-4693.

[101] N. A. Frey, S. Peng, K. Cheng, and S. Sun, Magnetic nanoparticles : synthesis, functionalization, and applications in bioimaging and magnetic energy storage, Chemical Society Reviews, 38 (2009) 2532–2542. https://doi.org/10.1039/b815548h

[102] S. Rasaneh and M. Dadras, The possibility of using magnetic nanoparticles to increase the therapeutic efficiency of Herceptin antibody, Biomedical Engineering / Biomedizinisch. Technik, 60 (2015). https://doi.org/10.1515/bmt-2014-0192

[103] J. Chen, M. Shi, P. Liu, A. Ko, W. Zhong, W. Liao, and M. Xing, Reducible polyamidoamine-magnetic iron oxide self-assembled nanoparticles for doxorubicin delivery, Biomaterials, 35 (2014) 1240–1248. https://doi.org/10.1016/j.biomaterials.2013.10.057

[104] E. Augustin, B. Czubek, A. M. Nowicka, A. Kowalczyk, Z. Stojek, and Z. Mazerska, Improved cytotoxicity and preserved level of cell death induced in colon cancer cells by doxorubicin after its conjugation with iron-oxide magnetic nanoparticles, Toxicology in Vitro, 33 (2016) 45-53. https://doi.org/10.1016/j.tiv.2016.02.009

[105] A. Aires, S. M. Ocampo, B. M. Simoes, M. J. Rodriguez, J. J. Cadenas, P. Coulead, K. Spence, A. Latorre, R. Miranda, A. Somoza, R. B. Clarke, J. L. Carrascosa, and A. L. Cortajarena, Multifunctionalized iron oxide nanoparticles for selective drug

delivery to CD44-positive cancer cells, Nanotechnology, 27 (2016) 065103. https://doi.org/10.1088/0957-4484/27/6/065103

[106] N. Avedian, F. Zaaeri, M. P. Daryasari, H. A. Javar, and M. Khoobi, pH-Sensetive biocompatible mesoporous magnetic nanoparticles labeled with folic acid as an efficient carrier for controlled anticancer drug delivery, Journal of Drug Delivery Science and Technology, 44 (2018) 323-332. https://doi.org/10.1016/j.jddst.2018.01.006

[107] M. Namdeo, S. Saxena, R. Tankhiwale, M. Bajpai, Y. M. Mohan and S. K. Bajpai, Magnetic nanoparticles for drug delivery applications, Journal of Nanoscience and Nanotechnology, 8 (2018) 3247-3271. https://doi.org/10.1166/jnn.2008.399

[108] A. Sato, N. Itcho, H. Ishiguro, D. Okamoto, N. Kobayashi, K. Kawai, H. Kasai, D. Kurioka, H. Uemura, Y. Kubota, and M. Watanabe, Magnetic nanoparticles of Fe_3O_4 enhance docetaxel-induced prostate cancer cell death, International Journal of Nanomedicine, 8 (2013) 3151–3160. https://dx.doi.org/10.2147%2FIJN.S40766

[109] R. Tarasi, M. Khoobi, H. Niknejad, A. Ramazani, L. Ma'mani, S. Bahadorikhalili, and A. Shafiee, β -Cyclodextrin functionalized poly (5-amidoisophthalicacid) grafted Fe_3O_4 magnetic nanoparticles: A novel biocompatible nonocomposite for targeted docetaxel delivery, Journal of Magnetism and Magnetic Materials, 417 (2016) 451-459. https://doi.org/ https://doi.org/10.1016/j.jmmm.2016.05.080

[110] T. Mahsa, N. Hamed, K. M. Hamidreza, and B. K. Ali, Preparation, characterization and *in vitro* anticancer activity of paclitaxel conjugated magnetic nanoparticles, Drug Development and Industrial Pharmacy, (2018) 1520-5762. https://doi.org/10.1080/03639045.2018.1508222

[111] H. Yu, Y. Wang, S. Wang, X. Li, W. Li, D. Ding, X. Gong, M. Keidar, and W. Zhang, Paclitaxel-Loaded Core-shell Magnetic Nanoparticles and Cold Atmospheric Plasma Inhibit Non-small Cell Lung Cancer Growth, ACS Applied Materials & Interfaces, 10 (2018) 43462-43471. https://doi.org/10.1021/acsami.8b16487

[112] L. Zhao, M. Huo, J. Liu, Z. Yao, D. Li, Z. Zhao, and J. Tang, In vitro investigation on the magnetic thermochemotherapy mediated by magnetic nanoparticles combined with methotrexate for breast cancer treatment, Journal of Nanoscience and Nanotechnology, 13 (2013) 741–745. https://doi.org/10.1166/jnn.2013.6080

[113] J. L. Viota, A. Carazo, J. A. Munoz-gamez, K. Rudzka, R. Gómez-sotomayor, A. Ruiz-extremera, J. Salmeron and A. V. Delgado, Functionalized magnetic nanoparticles as vehicles for the delivery of the antitumor drug gemcitabine to tumor cells. Physicochemical in vitro evaluation, Material Science and Engineering: C Materials for Biological Applications, 33 (2013) 1183–1192. https://doi.org/10.1016/j.msec.2012.12.009

[114] M. Parsian, G. Unsoy, P. Mutlu, S. Yalcin, A. Tezcaner, and U. Gunduz, Loading of Gemcitabine on chitosan magnetic nanoparticles increases the anti-cancer efficacy

of the drug, European Journal of Pharmacology, 784 (2016) 121-128.
https://doi.org/10.1016/j.ejphar.2016.05.016

[115] G. Unsoy, S. Yalcin, R. Khodadust, P. Mutul, O. Onguru, and U. Gunduz, Chitosan magnetic nanoparticles for pH responsive Bortezomib release in cancer therapy, Biomedicine & Pharmacotherapy. 68 (2014) 641-648.
https://doi.org/10.1016/j.biopha.2014.04.003

[116] M. P. Alvarez-Berriios, A. Castillo, C. Rinaldi, and M. Torres-Lugo, Magnetic fluid hyperthermia enhances cytotoxicity of bortezomib in sensitive and resistant cancer cell lines, International Journal of Nanomedicine, 9 (2014) 145–153.
https://dx.doi.org/10.2147%2FIJN.S51435

[117] M. Rahimi, K. D. Safa, and R. Salchi, Co-delivery of dexorubicin and methotrexate by dendritic chitosan-g-mPEG as a magnetic nanocarrier for multi-drugs delivery in combination chemotherapy, Polymer Chemistry, 8 (2017) 7333-7350.
https://doi.org/10.1039/C7PY01701D

[118] I. Zaman, F. M. Nor, B. Manshoor, A. Khalid, and S. Araby, Influence of interface on epoxy/clay nanocomposites: 2. mechanical and thermal dynamic properties, Procedia Manufacturing, 2 (2015) 23-27.
http://dx.doi.org/10.1016/j.promfg.2015.07.005

[119] K. McNamara and Syes A. M. Tofail, Nanosystems: the use of nanoalloys, metallic, biometallic, and magnetic nanoparticles in biomedical applications, Physical Chemistry Chemical Physics, 17 (2015) 27981-27995.
https://doi.org/10.1039/c5cp00831j

[120] B. Thiesen and A. Jordan, Clinical applications of magnetic nanoparticles for hyperthermia, International Journal of Hyperthermia, 26 (2009) 467-474.
https://doi.org/10.1080/02656730802104757

[121] S. Hatamie, Z. Malaie, M. Mahdi, and T. Mortezazadeh, Hyperthermia of breast cancer tumor using graphene oxide-cobalt ferrite magnetic nanoparticles in mice, Journal of Drug Delivery Sciences and Technology, 65 (2021) 102680.
https://doi.org/10.1016/j.jddst.2021.102680

[122] S. K. Sharma, N. Shrivastava, F. Rossi, L. D. Tung, and N. T. Thanh, Nanoparticles-based magnetic and photo induced hyperthermia for cancer treatment, Nanotoday, 29 (2019) 100795. https://doi.org/10.1016/j.nantod.2019.100795

[123] S. Hatamie, M. Ahadian, M. Ghiass, A. Zad, R. Saber, B. Pareh, M. Oghabian, and S. Zadeh, Graphene/Cobalt nanocarrier for hyperthermia therapy and MRI diagnosis, Colloids Surfaces B Biointerfaces, 146 (2016) 271-279.
https://doi.org/10.1016/j.colsurfb.2016.06.018

[124] S. Laurent, S. Dutz, U. O. Häfeli, and M. Mahmoudi, Magnetic fluid hyperthermia : Focus on superparamagnetic iron oxide nanoparticles, Advances in Colloid and Interface Science, 166 (2021) 8–23. https://doi.org/10.1016/j.cis.2011.04.003

Materials Research Forum LLC
https://doi.org/10.21741/9781644902332-7

[125] C. Grüttner, K. Müller, J. Teller, and F. Westphal, Synthesis and functionalisation of magnetic nanoparticles for hyperthermia applications, International Journal of Hyperthermia. 29 (2013) 777-789. https://doi.org/10.3109/02656736.2013.835876

[126] T. Kobayashi, Cancer hyperthermia using magnetic nanoparticles, Biotechnology Journal, 6 (2011) 1342–1347. https://doi.org/10.1002/biot.201100045

[127] Harvey B. Simon, Hyperthermia, The New England Journal of Medicine, 329 (1993) 483-487. https://doi.org/10.1056/NEJM199308123290708

[128] M. Ba, A. Teijeiro, and J. Rivas, Magnetic nanoparticle-based hyperthermia for cancer treatment, Reports of Practical Oncology & Radiotherapy, 18 (2013) 397–400. https://doi.org/10.1016/j.rpor.2013.09.011

[129] P. Golstein and G. Kroemer, Cell death by necrosis : towards a molecular definition, Trends in Biochemical Sciences, 32 (2017) 37-43. https://doi.org/10.1016/j.tibs.2006.11.001

[130] A. Jordan, R. Scholz, P. Wust, H. Fahling, and R. Felix, Magnetic fluid hyperthermia (MFH): Cancer treatment with AC magnetic field induced excitation of biocompatible superparamagnetic nanoparticles, Journal of Magnetism and Magnetic Materials. 201 (1999) 413–419.

[131] A. Hervault, and N. K. Thanh, Magnetic nanoparticle-based therapeutic agents for thermo-chemotherapy treatment of cancer, Nanoscale, 6 (2014) 11553-11573. https://doi.org/10.1039/C4NR03482A

[132] K. Hayashi, M. Nakamura, W. Sakamoto, T. Yogo, H. Miki, S. Ozaki, M. Abe, T. Matsumoto, and K. Ishimura, Superparamagnetic nanoparticle clusters for cancer theranostics combining magnetic resonance imaging and hyperthermia treatment, Theranostics, 3 (2013) 366-376. https://doi.org/10.7150/thno.5860

[133] L. Jie, L. Cai, L. Wang, X. Ying, R. Yu, M. Zhang, and Y. Du, Actively-targeted LTVSPWY peptide-modified magnetic nanoparticles for tumor imaging, International Journal of Nanomedicine, 7 (2012) 3981–3989. https://doi.org/10.2147/IJN.S33593

Materials Research Forum LLC
https://doi.org/10.21741/9781644902332-8

Chapter 8

Magnetic Nanoparticles for Drug Delivery Applications

Ayushi G. Patel[1], Rajshree B. Jotania[1,a*], Martin F. Desimone[2,b**]

[1]Department of Physics, Electronics and Space science, University School of Sciences, Gujarat University, Ahmedabad 380 009, India

[2]Universidad de Buenos Aires. IQUIMEFA-CONICET, Facultad de Farmaciay Bioquímica, (1113) Junin 956 Piso 3. Buenos Aires. Argentina

[a*]rbjotania@gujaratuniversity.ac.in,rajshree_jotania@yahoo.co.in
[b**]martinfdesimone@gmail.com, desimone@ffyb.uba.ar

Abstract

Magnetic nanoparticles (MNPs) possess different structural, magnetic and dielectric properties compared to bulk magnetic particles and they have found potential applications in the biomedical field (due to their response to applied magnetic field) including MRI, targeted drug delivery, bio sensors, tissue engineering, cancer therapy, and diagnosis, biomaterial coating devices, hyperthermia, cell selection and separation, magnetorelaximetry, antibacterial agent, biomolecules extraction, immunoassay, lab-on-a-chip etc. The synthesis of MNPs can be done by various methods pertaining to the application one is interested in. In present chapter we discuss basic properties (surface and magnetic) of magnetic nano particles and their biomedical applications.

Keywords

Magnetic Nano Particles, Spinel Ferrites, Surface Morphology, Magnetic Properties, Biomedical Applications

Contents

1.   Introduction

Magnetic nanoparticles (MNPs) are being exceedingly considered lately because of their unique chemical, structural, and magnetic properties. MNPs are an important class of functional materials for applications in biomedical research due to their biocompatibility, non-toxicity and inducible magnetic moments [1–5]. The applications are not limited to only in vitro diagnosis, but in vivo diagnosis/imaging, drug delivery, surgery, tissue engineering, device coating, implanting materials [6]. Researchers have developed MNPs with specific chemical composition, size, shape and surface coating (core-shell nanostructure) for various biomedical applications. Mostly iron oxide based nanoparticles (< 100 nm), coated with organic or nonorganic shell are used. For example, $Fe_3O_4@SiO_2$ NPs were prepared by Z. Sharafi *et al.* using the Stober process, it was reported that prepared NPs show good physiochemical properties and can be used for genomic DNA extraction [7]. NPs with transition metal ions (Co, Fe), alloys (FeCo, FePt, CoPt), and metal oxides ($CoFe_2O_4$, $\gamma\text{-}Fe_2O_3$, Fe_3O_4) have been prepared by many researchers to improve sensitivity of MRI [8–10]. As a matter of fact, iron oxide NPs are used in vitro diagnosis since 1957 [11]. It was reported that $\gamma\text{-}Fe_2O_3$ and Fe_3O_4 are very promising MNPs because of their biocompatibility and they can be easily functionalized with polymers like PVA [12–14], PEG [15–17], dextrant [18–20].

2.   Synthesis of magnetic nanoparticles

The functionalization of MNPs is what renders them useful for various biomedical applications. Coating or combining MNPs with biofunctional molecules such as ligands, antibodies, drugs, dyes, receptors results in biomolecules capable of specific site attaching along with the response to magnetic fields, which helps in bacteria detection, bimodal imaging with MRI as well as optical imaging [21]. Another way is to use MNPs by combining them with other nanostructures like different biocompatible metallic nanoparticles and encapsulation by polymer that makes them useful in drug delivery as well as multimodal imaging [21]. The synthesis of metallic nanoparticles can be carried out by physical, chemical or biological means. The common methods implemented by various researchers till now are ball-milling technique, coprecipitation, thermal decomposition, hydrothermal, microemulsion, sol-gel synthesis and biological methods. MNPs having different morphologies such as solid nanospheres, hollow nanospheres, porous nanospheres, mesoporous nanospheres, nanocubes, nanopyramids, triangular nanoplates, hexagonal nanoplates, nanowires, nanotubes, nanorods, nanoflowers, elliptical nanorings are explored by various researchers through the above mentioned synthesis methods [22].

2.1   Ball-milling technique

Developed by Benjamin J. in 1970 [23], it is a physical grinding method that uses a hollow enclosure containing raw materials as well as steel balls that aid in the breaking down of bulk materials. The bulk material undergoes kinetic abrasions and forms particles of smaller size by continuous milling inside the enclosure. Various versions of the ball milling such as planetary ball mill, high energy ball mill, vibrating ball mill have been developed where the number of steel balls, milling time, rotation speeds, dry or wet milling, ball to bulk powder ratio [24] are all factors that are crucial in getting crystalline uniform size nanoparticles.

2.2   Co-precipitation method

This method is by far the extensively used method for synthesis of iron nanoparticles, Fe^{+2} and Fe^{+3} containing salts react with reducing agents and give iron oxide nanoparticles through nucleation and growth phenomena. Various modifications to the synthesis route have been implemented for obtaining MNPs helpful in biomedical applications. Synthesis of PEG integrated iron nanoparticles of size 33 ± 8 nm was done with the help of ultrasonic assisted co-precipitation technique [26]. MNPs prepared by co-precipitation method and directly coated with Oleic acid, SiO_2 and PEG with average crystal size 6-7 nm, 10 nm and 10 nm respectively, show suitability for MRI application [27]. Fe_3O_4 nanoparticles prepared using citic acid modified co-precipitation technique resulted in pure magnetite nanoparticles in the range 11- 15 nm [28]. Polystyrene sulfonic acid (PSS) coated MNPs prepared by co-precipitation having iron oxide core of 8-18 nm were obtained. The particles were monodisperse, superparamagnetic with good biocompatibility having potential application as magnetic carriers [29]. Changing magnetic field and ultrasonic assisted co-precipitation synthesis produced Fe_3O_4 nanoparticles of various sizes from 35 nm to 6 nm [30]. The reaction temperature, metal ions concentration, pH value all these factors affect in composition of nanoparticles, size and shape [31].

2.3   Microemulsion method

An isotropic, thermodynamically stable liquid system of oil, water and amphiphile is defined as a microemulsion [32]. It is possible to integrate amphiphilic drugs in microemulsion, tailor the viscosity for a given application through formulation changes and thus use this method for a variety of drug delivery applications [33]. P.A. Trzaskowska, *et al.* prepared oil in water (o/w) microemulsion with methyl methacrylate (MMA) and polymerized it by using lecithin as surfactant for the application of biocompatible coating. The effect of microemulsion polymerization time on the coating properties showed that 3h of polymerization corresponding SS-PMMALec coating presented higher cell viability (mean 91.5%) than bare SS (mean 87.5%) and was more resistant to washing [34]. The effect of Z. bungeanum essential oil and microemulsion (ZO-ME) on skin permeation of drugs was investigated by Liu et al., ZO-ME was prepared by extracting essential oil and using labrasol as surfactant, propylene glycol as co-surfactant. The essential oil in microemulsion loaded with model drugs show enhancement in permeation on complex components with reduced irritation and improved stability of Z. bungeanum oil [35]. Insulin loaded oil-in-water (o/w) emulsions were prepared using light liquid paraffin as the oily phase and various combinations of Tween® 80 and snail mucin powder as a surfactant. Improved encapsulation efficiency and stabilised insulin complexes resulting from such formulations could be a potential for oral insulin delivery application [36]. A study demonstrates smaller particle size and controlled morphology can be achieved when synthesizing Hydroxyapatite (HAP), bioceramic material using polyoxyethylene lauryl ether as surfactant using microemulsion method [37].

2.4   Thermal decomposition technique

As the name suggests, this method involves decomposition of the precursors at high temperature. A reaction between organometallic precursors and organic solvents under high temperature and high pressure yields monodispersed magnetic nanoparticles. Although this method is highly suitable for preparing nanoparticles with good crystallinity and desired size, safety issues with high temperature and pressure as well as toxic organic solvents are the demerits of this method [38]. Oleic acid (OA)-coated $Mg_{1-x}Ca_xFe_2O_4$ ($x =$ 0.0-0.9) MNPs were synthesized using thermal decomposition and then coated with Pluronic F127 for magnetic hyperthermia applications. All the prepared MNPs were in the range of 8-19 nm with saturation magnetization ranging from 20.29 emu/g to 52.5 emu/g. The authors suggest that these MNPs could be thermoseeds for magnetic hyperthermia treatment due to the estimated size of F127-coated nanoparticles, heating ability, and hemo-compatibility [39].

2.5   Sol-gel technique

The precursors containing metal salts are dissolved in water and a sol formation occurs through polycondensation or hydrolysis [25], which is then converted to gel by constant heating and stirring followed by drying to obtain the final product. The nanoparticles obtained from this method have good crystallinity and can be made in large quantities as the method is fairly simple. S. Bhullar *et al.* synthesized TiO_2 NPs for their possible future use in drug delivery applications. The intensive study with varying molarity, pH value and calcination temperature gave optimized values for spherical rutile TiO_2 NPs of size > 20 nm, but < 30 nm [40].

2.6   Hydrothermal

A reaction between aqueous solution vapours with solid material [41] occurring at high temperature and high pressure results in nanoparticles formed through crystal growth [42]. The distribution and size of nanoparticles in this method is affected by reaction temperature, the amount of precursors used and time duration of reaction [2]. ^{223}Ra-doped BaFe nanoparticles were synthesized for targeted drug application using hydrothermal method. The magnetic properties of MNPs with trastuzumab monoclonal antibody show that under the effect of an external magnetic field, the bioconjugate can be used against cancerous cells [43].

2.7   Biological synthesis

Biological or the green synthesis method of preparing nanoparticles for corresponding application in biomedical areas involves mainly two biological substrates such as fungi, plant and bacteria. The use of plant based material such as extract of leaves, fruit, seed, flower in synthesis produces magnetic nanoparticles that are biocompatible and non-toxic [2]. The amino acids, polyphenols, nitrogenous bases, and reducing sugars present in the plant extracts act as reducing and stabilizing agents [44].

2.8   Other methods

Apart from these methods, laser ablation, ion sputtering, laser pyrolysis, flow injection technique, microwave-assisted [45], solvo-thermal [46], sonochemical, polyol [47], electro-deposition [48], combustion are methods which are being implemented recently for better quality nanoparticles [49].

In retrospect, no single method is considered the only suitable method for preparation of MNPs which have further use in drug delivery applications. Certain broader drawbacks like contamination by impurities, longer synthesis duration, involvement of toxic solvents, expensive technology entails with implementation of all the methods. Thus, a method is chosen according to research needs and abilities. The co-precipitation, microemulsion, hydrothermal methods are the least complicated methods that provide good crystallinity and required size of nanoparticles for specific drug applications. While thermal decomposition, sonochemical, microwave methods also provide uniform shape and size nanoparticles but require complex setups. Thus, these points can be considered as challenges that need to be overcome in order to prepare large quantities and easily reproducible MNPs.

3.   Drug delivery with magnetic nanoparticles

Magnetic nanoparticles are being developed for the localized drug delivery in patients with different pathologies. These drug delivery systems can be directed through the bloodstream to the site of action with the aid of a magnetic field. Once they reach with the drug to the patient's site of action, the magnetic field also contribute to keep the magnetic particles in the targeted site and favour the local delivery of the drugs [50]. For example, the treatment

Magnetic Nanoparticles for Biomedical Applications   Materials Research Forum LLC
Materials Research Foundations 143 (2023) 233-252   https://doi.org/10.21741/9781644902332-8

of neurological diseases is challenging because of the ineffective targeting techniques [51]. Thus, magnetic nanoparticles have been proposed as true candidates to reach the brain after an intra-nasal administration. These NPs can be driven across the cribriform plate using remote pulsed magnetic fields [52–54]. Though, they contribute to overcome the difficulties in the treatment of several diseases with low permeation of therapeutic agents across biological barriers [55]. Indeed, the targeted or localized drug delivery achieved with magnetic nanoparticles would reduce the risk associated with systemic administration, hinder side effects and diminish the dose employed for treatment [56]. Moreover, magnetic nanoparticles can be coated with different stabilizing agents including natural or synthetic polymers which would also contribute to drug delivery [57]. Great efforts are being made to improve chemical aspects of the design and synthesis of magnetic drug delivery systems that favours the delivery of drugs and to understand their pharmacodynamic behaviour [58,59].

Chen *et al.*, employed a solvothermal method to produce different spinel ferrites particles ($M_xFe_{3-x}O_4$, M= Co, Cu, Fe, Mg, Mn, Ni, Zn) with an average diameter between 200 and 350 nm [60]. These ferrites possess a mesoporous or hollow-mesoporous structure which enables the direct loading of drugs. The authors evaluated the drug loading and release profiles employing the chemotherapeutic drug VP16. VP-16, also known as Etoposide, is a semi-synthetic podophyllin derived from the podophyllum plant *Podophyllum notatum*. It is used as a medicine to treat some types of cancer due to its ability to inhibit the multiplication of tumour cells [61]. Concerning the drug loading evaluation, an ethanol solution of the drug was directly mixed with the magnetic particles and stirred until the equilibrium state was reached. Afterwards, the drug loaded particles were resuspended in a saline solution to evaluate the release profiles [62]. The hollow-mesoporous nanoparticles could store the drug both in the mesopores and in the cavities providing a higher loading of drug molecules. Interestingly, more drug was released under microwave irradiation than with only stirring. The increased drug release was attributed to increase in temperature provoked by microwave irradiation of the ferrite nanocarriers [60].

Doxorubicin is a widely used drug in cancer chemotherapy. It is an antibiotic of the anthracycline family being a DNA intercalator. Wang *et al.*, developed a multifunctional microsphere with the simultaneous ability to work as drug delivery system and magnetic resonance imaging agent [63]. The authors employed an oil-in-water emulsion solvent-evaporation method, in which poly (ε-caprolactone) was employed as the capping agent of Fe_3O_4 nanoparticles. Doxorubicin loading and encapsulation efficiency were 36.7% and 25.8%, respectively. The release exhibits a burst initial release followed by a sustained release reaching 62% of the total drug loaded after 30 days. Moreover, a time and dose dependent cytotoxicity effect on HeLa cells was observed. Similarly, a dry powder drug delivery vehicle with superparamagnetic iron oxide nanoparticles and loaded with doxorubicin was successfully targeted in the presence of a magnetic field to specific regions of the lung. In addition, an increased therapeutic toxicity in A549 adenocarcinoma cells resulted from the synergistic effect of reactive oxygen species production and increased cellular uptake of nanoparticles, which improved the cytotoxic effects of doxorubicin [64].

An interesting strategy for mitochondrial targeting therapy was developed by Wang *et al.* After the accumulation of nanoparticles at the tumour site and entrance in the tumour cells, the second stage of mitochondrial targeting involves detachment of the nanoparticles shell through a redox reaction. Subsequently, a decrease in mitochondrial membrane potential is achieved with near-infrared light irradiation, thus the doxorubicin loaded in the nanoparticles can penetrate the mitochondria rapidly and provoke damage to the mitochondrial DNA, leading to apoptosis of the cancerous cells and favouring the treatment [65]. In this sense, various drug delivery systems loaded with doxorubicin were reported. Most of them were effectively transferred to cells with the help of an external magnet [66,67]. Cobalt ferrite nanoparticles coated with hydrophilic polymers gum arabic, and guar gum poly (methacrylic acid) were loaded with doxorubicin. The hydrodynamic size of the nanoparticles was *ca.* 35 nm and with the polymer coating it reaches *ca.* 80 nm. The saturation magnetization was 77.2 emu/g for uncoated nanoparticles and a lower saturation magnetization was observed for coated nanoparticles due to the non-magnetic nature of the polymer layer. The release of the drug was lower than 12% in all cases when no magnetic field was applied. While the release of the drug under the effect of external applied magnetic field is enhanced up to 77% and this value is highly affected by the polymer employed [68]. The polymer coatings did not significantly affect the magnetic properties while presenting cytocompatibility, thus they were proposed as promising carriers for the delivery of doxorubicin. Wang *et al.*, successfully developed magnetic nanostructures with the ability to develop on-demand heat therapy, and doxorubicin release in the presence of a magnetic field, leading to an efficient synergistic effect in inhibiting tumour growth without any side effect during in vivo animal experiments [69].

The drug 5-Fluorouracil, also known as 5-FU, is a potent antimetabolite used in cancer treatment. Fe_3O_4 nanoparticles coated with polyvinyl alcohol and Zn/Al-layered double hydroxide loaded with a 5-fluorouracil possessed higher effect towards liver cancer cells (HepG2) when compared to the free 5-FU drug. Moreover the absence of cytotoxicity towards 3T3 fibroblast cell lines highlights their potential for cancer treatment [70]. Similarly, Mg/Al-layered double hydroxide were employed for the same purpose with similar results in terms of 5-fluorouracil controlled release and superior anticancer effect than the drug alone [71]. Interestingly, the saturation magnetization value for magnetite nanoparticles was *ca.* 80 emu/g, while for magnetite coated with polyvinyl alcohol (FPVA) the value decreases to 49 emu/g. The lowest value was obtained with magnetite coated with PVA, and co-coated with Mg–Al-layered double hydroxide loaded with 5FUas it was *ca.* 11 emu/g (figure 1).These differences can be attributed to the presence of successive grafted layers. The authors also reported similar behaviour employing iron oxide nanoparticles coated with polyethylene glycol, polyethylene glycol co-coated with Mg/Al-LDH and polyethylene glycol co-coated with Zn/Al-LDH with magnetic saturation values decreasing from 56 emu/g to 40 emu/g and finally 27 emu/g [72].

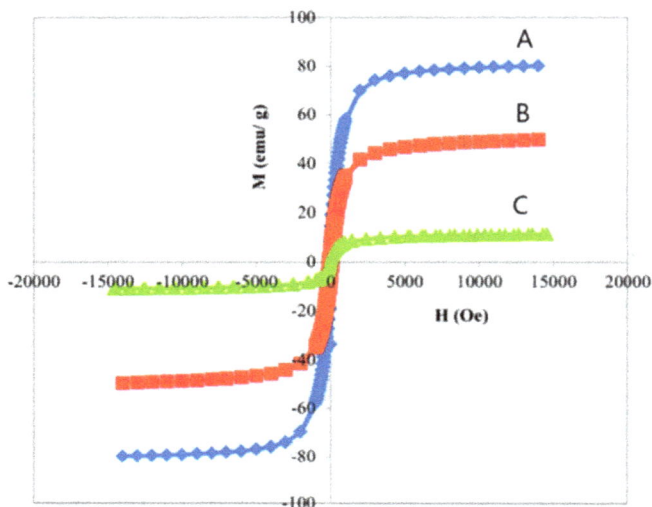

Fig. 1. Magnetization curves of (A) magnetite nanoparticles; (B) magnetite-coated polyvinyl alcohol, (C) FPVA-FU-MLDH nanoparticles. Notes: The data are presented in terms of Ms, mass magnetization (emu/g), versus H, applied magnetic field (Oe). Reproduced from [71] under the Creative Commons Attribution license (http://creativecommons.org/licenses/by/3.0/).

Chlorambucil is an alkylating agent that has been used to treat different types of cancer. Currently, its main indication is for the treatment of chronic lymphatic leukaemia. Iron oxide nanoparticles were coated with chitosan and loaded with chlorambucil. The release profile showed a pseudo-second-order kinetic model with 90% release within about 83 h. The IC50 against WEHI cancer cells was 11.12 µg/mL, showing good biocompatibility with 3T3 normal cells [73].

Antracyclinic antibiotics provokes modifications in DNA helix shape, avoiding the replication of DNA and the transcription of RNA. They possess recognized antitumor activity in solid tumors, especially for the treatment of breast cancer, malignant lymphomas and leukemia. The antitumor effect of magnetic nanoparticles loaded with Violamycine B1was evaluated on Michigan Cancer Foundation-7 (MCF-7) cells. Violamycine B1 was adsorbed on NPs by electrostatic interactions between amino sugars present in the drug and iron oxide groups present on the NPs. The results revealed a higher cell uptake and superior anticancer activity by the magnetic complexes, in comparison with the drug alone [74].

Amoxicillin/clavulanic acid, is an antibiotic employed for the treatment of various diseases. It is a combination consisting of amoxicillin, a β-lactam antibiotic, and potassium clavulanate, a β-lactamase inhibitor. The binding between magnetic nanoparticles and the antibiotic amoxicillin clavulanate suggest the participation of either hydrogen bonds or weak induction forces [75]. The parenchymatous organs were not affected after an intravenous injection into the rat's tail vein of these magnetic nanoparticles. Moreover, after 24 h the nanoparticles were completely discarded from the body [75].

Gentamicin is an antibiotic of the aminoglycoside group used as a broad-spectrum antibiotic and bactericidal action for the treatment of infections. Interestingly, magnetic nanoparticles loaded with gentamicin exposed to a magnetic field experienced an improved penetration into mature biofilm of S. aureus. Consequently, the delivery of the antibiotic is more effective for the successful eradication of biofilms frequently associated with infections [76].

In a different approach, Chorny et al., produced magnetic nanocarriers loaded with catalase and superoxide dismutase to protect endothelial cells from oxidative stress mediated damage and could be an alternative to treat cardiovascular diseases [77]. Under magnetic guidance enzyme-loaded nanocarriers were rapidly taken up by endothelial cells providing increased resistance to oxidative stress. It is worth to mention that the enzyme superoxide dismutase catalyses the dismutation of superoxide into oxygen and hydrogen peroxide. While catalase is an enzyme belonging to the oxidoreductase category that catalyses the breakdown of hydrogen peroxide into oxygen and water. Both are considered important enzymes involved in the antioxidant defence in most cells exposed to oxygen. With same aim, Aryan et al., numerically simulated magnetic drug delivery to the carotid artery bifurcation and concluded that both magnet configuration and nanoparticle size are the main parameters that must be considered to improve the efficiency of the magnetic drug delivery system [78].

Magnetic nanoparticles conjugated with therapeutic or bioactive molecules, such as drugs, enzymes and growth factors can contribute to enhance tissue regeneration [79]. Nerve growth factor (NGF) is a neurotrophic factor and neuropeptide primarily involved in the regulation of maintenance, growth and proliferation of neurons. Recently, nerve growth factor was conjugated with iron oxide nanoparticles and controlled magnetic fields to deliver the complexes to target sites. On one hand, the nanosystem was injected and directed to the sciatic nerve. On the other hand, the nanosystem was injected intravenously and using an external magnet placed next to one of the eyes, it was possible to accumulate them in the retina of a mouse. These results, highlight the possibility of delivering drugs specifically to injured tissues [80].

Kinetic and spatial control of ibuprofen release was achieved employing magnetite nanoparticles and poly(lactic acid) [81]. Ibuprofen is a nonsteroidal anti-inflammatory, frequently used as antipyretic, analgesic and anti-inflammatory. Poly(lactic acid) is a biocompatible and biodegradable polymer which is employed in various biomedical applications. The authors successfully developed a nanocomposite with these components and conclude that the kinetic control provided by the polymer and the decrease of toxicity promoted by the spatial control achieved with magnetite nanoparticles, would promote patient comfort and compliance making a more efficient treatment [81].

Prindopril, is an inhibitor of the angiotensin-converting enzyme (ACE) with antihypertensive effect. Magnetic nanoparticles with a mean diameter of 6 nm were coated with chitosan and the prindopril erbumine drug resulting in nanoparticles with a mean diameter of 15 nm (figure 2). The release study revealed first order kinetic with ca. 89% of the drug released within 93 h. The saturation magnetization values of magnetic iron oxide nanoparticles, chitosan coated nanoparticles and iron oxide coated with chitosan with the drug were 44.65 emu/g, 38.57 emu/g and 27.66 emu/g, respectively. Cell viability tests performed with 3T3 cells revealed that these nanomaterials were biocompatible [82].

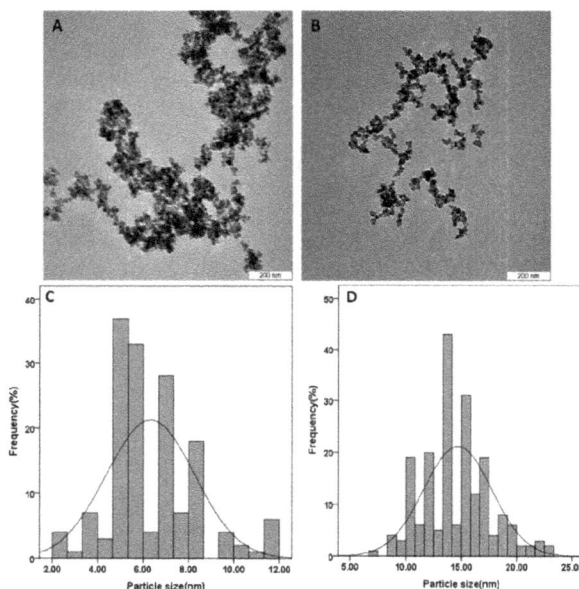

Fig. 2. TEM micrographs (A) iron oxide magnetic nanoparticles with 200 nm microbar; (B) iron oxide nanoparticles coated with chitosan-prindopril erbumine (FCPE) with 200 nm microbar; (C) particle size distribution of iron oxide nanoparticles; and (D) particle size distribution of FCPE nanocomposite. Reproduced from [82] with permission under the Creative Commons Attribution license (http://creativecommons.org/licenses/by/3.0/).

Octreotide is a drug analogous to natural somatostatin and, therefore, with pharmacological effects similar to it, with neuroprotective and anti-angiogenic properties. The main difference with natural somatostatin is marked by its longer duration of action. Drug delivery systems resulted from the octreotide (OCT) bound to magnetic nanoparticles possess reduced biodegradation and long-term drug release and consequently reduce the injection frequency. These nanoparticles were nontoxic with both mouse retinal explants and in human retinal endothelial cells (HRECs). The ultrastructural analysis confirmed the presence of the nanoparticles in the retinal pigment epithelium (RPE). Indeed, nanoparticles were internalized in various endocytic vesicles (figure 3) [83].

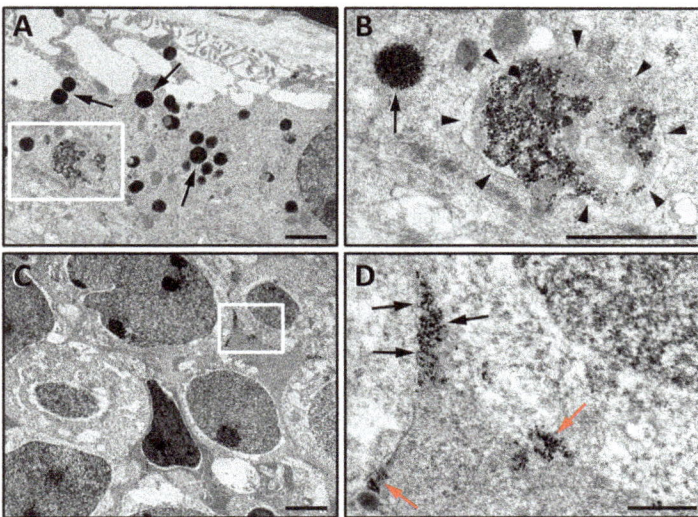

Fig. 3. High resolution TEM images of ultrathin retinal sections at the level of the RPE (A,B) and of the INL (C,D). Images in (A,B) are from a Balb/c mouse retina, in which RPE cells do not contain pigment and therefore MNP-OCT can be easily detected. (B) is a higher power of the boxed area in (A). In the RPE, MNP-OCT were found to be internalized in several endocytic vesicles. The scission phase of endocytic process releasing primary endocytic vesicle is highlighted by arrowheads (B), while other MNP-containing vesicles are likely to represent different stages of the following intracellular trafficking (A,B, arrows). (D) is a higher power of the boxed area in (C). The nanoparticles in the INL were densely packed into narrow intercellular spaces (D, black arrows) or distributed in dense spots adjacent to the extracellular side of the plasma membrane of INL cells (D, red arrows). Scale bars, 1 µm in (A,B,D); 2 µm in (C). Reproduced from [83] with permission under the Creative Commons Attribution license (https://creativecommons.org/licenses/by/4.0/)

Conclusion

Until now, a great variety of drug delivery systems employing magnetic nanoparticles has been developed. Numerous methods of synthesis have been implemented according to their merits in preparation of MNPs for specific applications [84, 85]. The size and shape of nanoparticles deeply affect the use as drug vehicles as well as other biomedical applications. In this sense, different drugs including antibiotics, anti-angiogenic, antihypertensive, among others, have been successfully delivered to the site of action. Undoubtedly, antitumorigenic drugs are a clear example of the potentialities of using magnetic carriers to diminish the side effects associated with systemic administration. In conclusion, while new magnetic nanoparticles with their associated synthesis procedures are continuously being developed, the potentialities to achieve better and safer therapies are enormous.

References

[1] V.F. Cardoso, A. Francesko, C. Ribeiro, M. Bañobre-López, P. Martins, S. Lanceros-Mendez, Advances in Magnetic Nanoparticles for Biomedical Applications, Adv. Healthc. Mater. 7 (2018) 1700845. https://doi.org/10.1002/adhm.201700845

[2] M.I. Anik, M.K. Hossain, I. Hossain, A.M.U.B. Mahfuz, M.T. Rahman, I. Ahmed, Recent progress of magnetic nanoparticles in biomedical applications: A review, Nano Sel. 2 (2021) 1146–1186. https://doi.org/10.1002/nano.202000162

[3] N. Tran, T.J. Webster, Magnetic nanoparticles: biomedical applications and challenges, J. Mater. Chem. 20 (2010) 8760. https://doi.org/10.1039/c0jm00994f

[4] C. Xu, S. Sun, Superparamagnetic nanoparticles as targeted probes for diagnostic and therapeutic applications, Dalt. Trans. (2009) 5583. https://doi.org/10.1039/b900272n

[5] C. Xu, S. Sun, New forms of superparamagnetic nanoparticles for biomedical applications, Adv. Drug Deliv. Rev. 65 (2013) 732–743. https://doi.org/10.1016/j.addr.2012.10.008

[6] W. Cai, Engineering in Translational Medicine, Springer London, London, 2014. https://doi.org/10.1007/978-1-4471-4372-7

[7] Z. Sharafi, B. Bakhshi, J. Javidi, S. Adrangi, Synthesis of Silica-coated Iron Oxide Nanoparticles: Preventing Aggregation without Using Additives or Seed Pretreatment., Iran. J. Pharm. Res. IJPR. 17 (2018) 386–395. http://www.ncbi.nlm.nih.gov/pubmed/29755569

[8] R. Hao, R. Xing, Z. Xu, Y. Hou, S. Gao, S. Sun, Synthesis, Functionalization, and Biomedical Applications of Multifunctional Magnetic Nanoparticles, Adv. Mater. 22 (2010) 2729–2742. https://doi.org/10.1002/adma.201000260

[9] Y. Jun, J. Choi, J. Cheon, Shape Control of Semiconductor and Metal Oxide Nanocrystals through Nonhydrolytic Colloidal Routes, Angew. Chemie Int. Ed. 45 (2006) 3414–3439. https://doi.org/10.1002/anie.200503821

[10] Y. Jun, J. Lee, J. Cheon, Chemical Design of Nanoparticle Probes for High-Performance Magnetic Resonance Imaging, Angew. Chemie Int. Ed. 47 (2008) 5122–5135. https://doi.org/10.1002/anie.200701674

[11] R.K. Gilchrist, R. Medal, W.D. Shorey, R.C. Hanselman, J.C. Parrott, C.B. Taylor, Selective Inductive Heating of Lymph Nodes, Ann. Surg. 146 (1957) 596–606. https://doi.org/10.1097/00000658-195710000-00007

[12] H. Pardoe, W. Chua-anusorn, T.G. St. Pierre, J. Dobson, Structural and magnetic properties of nanoscale iron oxide particles synthesized in the presence of dextran or polyvinyl alcohol, J. Magn. Magn. Mater. 225 (2001) 41–46. https://doi.org/10.1016/S0304-8853(00)01226-9

[13] A.A. Novakova, V.Y. Lanchinskaya, A.V. Volkov, T.S. Gendler, T.Y. Kiseleva, M.A. Moskvina, S.B. Zezin, Magnetic properties of polymer nanocomposites containing iron oxide nanoparticles, J. Magn. Magn. Mater. 258–259 (2003) 354–357. https://doi.org/10.1016/S0304-8853(02)01062-4

[14] A. Petri-Fink, B. Steitz, A. Finka, J. Salaklang, H. Hofmann, Effect of cell media on polymer coated superparamagnetic iron oxide nanoparticles (SPIONs): Colloidal stability, cytotoxicity, and cellular uptake studies, Eur. J. Pharm. Biopharm. 68 (2008) 129–137. https://doi.org/10.1016/j.ejpb.2007.02.024

[15] L.R. Hirsch, R.J. Stafford, J.A. Bankson, S.R. Sershen, B. Rivera, R.E. Price, J.D. Hazle, N.J. Halas, J.L. West, Nanoshell-mediated near-infrared thermal therapy of tumors under magnetic resonance guidance, Proc. Natl. Acad. Sci. 100 (2003) 13549–13554. https://doi.org/10.1073/pnas.2232479100

[16] M. Li, M.J. Mondrinos, X. Chen, M.R. Gandhi, F.K. Ko, P.I. Lelkes, Elastin Blends for Tissue Engineering Scaffolds, J. Biomed. Mater. Res. Part A. 79 (2006) 963–73. https://doi.org/10.1002/jbm.a

[17] B.A. Moffat, G.R. Reddy, P. McConville, D.E. Hall, T.L. Chenevert, R.R. Kopelman, M. Philbert, R. Weissleder, A. Rehemtulla, B.D. Ross, A Novel Polyacrylamide Magnetic Nanoparticle Contrast Agent for Molecular Imaging using MRI, Mol. Imaging. 2 (2003) 324–332. https://doi.org/10.1162/153535003322750664

[18] A. Moore, E. Marecos, A. Bogdanov, R. Weissleder, Tumoral Distribution of Long-circulating Dextran-coated Iron Oxide Nanoparticles in a Rodent Model, Radiology. 214 (2000) 568–574. https://doi.org/10.1148/radiology.214.2.r00fe19568

[19] A.S. Arbab, L.A. Bashaw, B.R. Miller, E.K. Jordan, B.K. Lewis, H. Kalish, J.A. Frank, Characterization of Biophysical and Metabolic Properties of Cells Labeled with Superparamagnetic Iron Oxide Nanoparticles and Transfection Agent for Cellular MR Imaging, Radiology. 229 (2003) 838–846. https://doi.org/10.1148/radiol.2293021215

[20] C.C. Berry, S. Wells, S. Charles, A.S.G. Curtis, Dextran and albumin derivatised iron oxide nanoparticles: influence on fibroblasts in vitro, Biomaterials. 24 (2003) 4551–4557. https://doi.org/10.1016/S0142-9612(03)00237-0

Materials Research Forum LLC
https://doi.org/10.21741/9781644902332-8

[21] J. Gao, H. Gu, B. Xu, Multifunctional Magnetic Nanoparticles: Design, Synthesis, and Biomedical Applications, Acc. Chem. Res. 42 (2009) 1097–1107. https://doi.org/10.1021/ar9000026

[22] A.G. Niculescu, C. Chircov, A.M. Grumezescu, Magnetite nanoparticles: Synthesis methods – A comparative review, Methods. 199 (2022) 16–27. https://doi.org/10.1016/j.ymeth.2021.04.018

[23] J.S. Benjamin, Dispersion strengthened superalloys by mechanical alloying, Metall. Trans. 1 (1970) 2943–2951. https://doi.org/10.1007/BF03037835

[24] A. Ali, T. Shah, R. Ullah, P. Zhou, M. Guo, M. Ovais, Z. Tan, Y.K. Rui, Review on Recent Progress in Magnetic Nanoparticles: Synthesis, Characterization, and Diverse Applications, Front. Chem. 9 (2021) 1–25. https://doi.org/10.3389/fchem.2021.629054

[25] S. Liu, B. Yu, S. Wang, Y. Shen, H. Cong, Preparation, surface functionalization and application of Fe_3O_4 magnetic nanoparticles, Adv. Colloid Interface Sci. 281 (2020) 102165. https://doi.org/10.1016/j.cis.2020.102165

[26] L.M. AL-Harbi, M.S.A. Darwish, Functionalized iron oxide nanoparticles: synthesis through ultrasonic-assisted co-precipitation and performance as hyperthermic agents for biomedical applications, Heliyon. 8 (2022) e09654. https://doi.org/10.1016/j.heliyon.2022.e09654

[27] H. Mohammadi, E. Nekobahr, J. Akhtari, M. Saeedi, J. Akbari, F. Fathi, Synthesis and characterization of magnetite nanoparticles by co-precipitation method coated with biocompatible compounds and evaluation of in-vitro cytotoxicity, Toxicol. Reports. 8 (2021) 331–336. https://doi.org/10.1016/j.toxrep.2021.01.012

[28] A. Bahadur, A. Saeed, M. Shoaib, S. Iqbal, M.I. Bashir, M. Waqas, M.N. Hussain, N. Abbas, Eco-friendly synthesis of magnetite (Fe_3O_4) nanoparticles with tunable size: Dielectric, magnetic, thermal and optical studies, Mater. Chem. Phys. 198 (2017) 229–235. https://doi.org/10.1016/j.matchemphys.2017.05.061

[29] B.W. Chen, Y.C. He, S.Y. Sung, T.T.H. Le, C.L. Hsieh, J.Y. Chen, Z.H. Wei, D.J. Yao, Synthesis and characterization of magnetic nanoparticles coated with polystyrene sulfonic acid for biomedical applications, Sci. Technol. Adv. Mater. 21 (2020) 471–481. https://doi.org/10.1080/14686996.2020.1790032

[30] M. Jafari Eskandari, I. Hasanzadeh, Size-controlled synthesis of Fe_3O_4 magnetic nanoparticles via an alternating magnetic field and ultrasonic-assisted chemical co-precipitation, Mater. Sci. Eng. B Solid-State Mater. Adv. Technol. 266 (2021) 115050. https://doi.org/10.1016/j.mseb.2021.115050

[31] P. Tartaj, M. Del Puerto Morales, S. Veintemillas-Verdaguer, T. González-Carreño, C.J. Serna, The preparation of magnetic nanoparticles for applications in biomedicine, J. Phys. D. Appl. Phys. 36 (2003). https://doi.org/10.1088/0022-3727/36/13/202

[32] I. Danielsson, B. Lindman, The definition of microemulsion, Colloids and Surfaces. 3 (1981) 391–392. https://doi.org/10.1016/0166-6622(81)80064-9

[33] M.J. Lawrence, G.D. Rees, Microemulsion-based media as novel drug delivery systems, Adv. Drug Deliv. Rev. 64 (2012) 175–193. https://doi.org/10.1016/j.addr.2012.09.018

[34] P.A. Trzaskowska, A. Poniatowska, K. Tokarska, C. Wiśniewski, T. Ciach, E. Malinowska, Promising electrodeposited biocompatible coatings for steel obtained from polymerized microemulsions, Colloids Surfaces A Physicochem. Eng. Asp. 591 (2020) 124555. https://doi.org/10.1016/j.colsurfa.2020.124555

[35] X. Liu, L. Xu, X. Liu, Y. Wang, Y. Zhao, Q. Kang, J. Liu, H. Lan, L. Yu, Q. Wu, Combination of essential oil from Zanthoxylum bungeanum Maxim. and a microemulsion system: Permeation enhancement effect on drugs with different lipophilicity and its mechanism, J. Drug Deliv. Sci. Technol. 55 (2020) 101309. https://doi.org/10.1016/j.jddst.2019.101309

[36] M.A. Momoh, K.C. Franklin, C.P. Agbo, C.E. Ugwu, M.O. Adedokun, O.C. Anthony, O.E. Chidozie, A.N. Okorie, Microemulsion-based approach for oral delivery of insulin: formulation design and characterization, Heliyon. 6 (2020) e03650. https://doi.org/10.1016/j.heliyon.2020.e03650

[37] V. Collins Arun Prakash, I. Venda, V. Thamizharasi, Synthesis and characterization of surfactant assisted hydroxyapatite powder using microemulsion method, Mater. Today Proc. 51 (2021) 1788–1792. https://doi.org/10.1016/j.matpr.2021.05.059

[38] M. Faraji, Y. Yamini, M. Rezaee, Magnetic nanoparticles: Synthesis, stabilization, functionalization, characterization, and applications, J. Iran. Chem. Soc. 7 (2010) 1–37. https://doi.org/10.1007/BF03245856

[39] P.Y. Reyes-Rodríguez, D.A. Cortés-Hernández, C.A. Ávila-Orta, J. Sánchez, M. Andrade-Guel, A. Herrera-Guerrero, C. Cabello-Alvarado, V.H. Ramos-Martínez, Synthesis of Pluronic F127-coated magnesium/calcium ($Mg_{1-x}Ca_xFe_2O_4$) magnetic nanoparticles for biomedical applications, J. Magn. Magn. Mater. 521 (2021). https://doi.org/10.1016/j.jmmm.2020.167518

[40] S. Bhullar, N. Goyal, S. Gupta, A recipe for optimizing TiO_2 nanoparticles for drug delivery applications, OpenNano. 8 (2022) 100096. https://doi.org/10.1016/j.onano.2022.100096

[41] P.G. Jamkhande, N.W. Ghule, A.H. Bamer, M.G. Kalaskar, Metal nanoparticles synthesis: An overview on methods of preparation, advantages and disadvantages, and applications, J. Drug Deliv. Sci. Technol. 53 (2019) 101174. https://doi.org/10.1016/j.jddst.2019.101174

[42] A. V. Samrot, C.S. Sahithya, J. Selvarani A, S.K. Purayil, P. Ponnaiah, A review on synthesis, characterization and potential biological applications of superparamagnetic

iron oxide nanoparticles, Curr. Res. Green Sustain. Chem. 4 (2021) 100042.
https://doi.org/10.1016/j.crgsc.2020.100042

[43] W. Gawęda, M. Pruszyński, E. Cędrowska, M. Rodak, A. Majkowska-Pilip, D. Gaweł, F. Bruchertseifer, A. Morgenstern, A. Bilewicz, Trastuzumab Modified Barium Ferrite Magnetic Nanoparticles Labeled with Radium-223: A New Potential Radiobioconjugate for Alpha Radioimmunotherapy, Nanomaterials. 10 (2020) 2067. https://doi.org/10.3390/nano10102067

[44] Y.C. López, M. Antuch, Morphology control in the plant-mediated synthesis of magnetite nanoparticles, Curr. Opin. Green Sustain. Chem. 24 (2020) 32–37. https://doi.org/10.1016/j.cogsc.2020.02.001

[45] M. Masuku, L. Ouma, A. Pholosi, Microwave assisted synthesis of oleic acid modified magnetite nanoparticles for benzene adsorption, Environ. Nanotechnology, Monit. Manag. 15 (2021) 100429. https://doi.org/10.1016/j.enmm.2021.100429

[46] Z. Shaoqiang, T. Dong, Z. Geng, H. Lin, Z. Hua, H. Jun, L. Yi, L. Minxia, H. Yaohua, Z. Wei, The influence of grain size on the magnetic properties of Fe_3O_4 nanocrystals synthesized by solvothermal method, J. Sol-Gel Sci. Technol. 98 (2021) 422–429. https://doi.org/10.1007/s10971-018-4909-2

[47] S. Cabana, A. Curcio, A. Michel, C. Wilhelm, A. Abou-Hassan, Iron Oxide Mediated Photothermal Therapy in the Second Biological Window: A Comparative Study between Magnetite/Maghemite Nanospheres and Nanoflowers, Nanomaterials. 10 (2020) 1548. https://doi.org/10.3390/nano10081548

[48] J.F. Mir, S. Rubab, M. Shah, Photo-electrochemical ability of iron oxide nanoflowers fabricated via electrochemical anodization, Chem. Phys. Lett. 741 (2020) 137088. https://doi.org/10.1016/j.cplett.2020.137088

[49] L. Mohammed, H.G. Gomaa, D. Ragab, J. Zhu, Magnetic nanoparticles for environmental and biomedical applications: A review, Particuology. 30 (2017) 1–14. https://doi.org/10.1016/j.partic.2016.06.001

[50] P.M. Price, W.E. Mahmoud, A.A. Al-Ghamdi, L.M. Bronstein, Magnetic Drug Delivery: Where the Field Is Going, Front. Chem. 6 (2018). https://doi.org/10.3389/fchem.2018.00619

[51] I. Venugopal, N. Habib, A. Linninger, Intrathecal magnetic drug targeting for localized delivery of therapeutics in the CNS, Nanomedicine. 12 (2017) 865–877. https://doi.org/10.2217/nnm-2016-0418

[52] S. Jafari, L.O. Mair, I.N. Weinberg, J. Baker-McKee, O. Hale, J. Watson-Daniels, B. English, P.Y. Stepanov, C. Ropp, O.F. Atoyebi, D. Sun, Magnetic drilling enhances intra-nasal transport of particles into rodent brain, J. Magn. Magn. Mater. 469 (2019) 302–305. https://doi.org/https://doi.org/10.1016/j.jmmm.2018.08.048

[53] A. Nacev, I.N. Weinberg, P.Y. Stepanov, S. Kupfer, L.O. Mair, M.G. Urdaneta, M. Shimoji, S.T. Fricke, B. Shapiro, Dynamic Inversion Enables External Magnets To Concentrate Ferromagnetic Rods to a Central Target, Nano Lett. 15 (2015) 359–364. https://doi.org/10.1021/nl503654t

[54] L.B. Thomsen, T. Linemann, K.M. Pondman, J. Lichota, K.S. Kim, R.J. Pieters, G.M. Visser, T. Moos, Uptake and Transport of Superparamagnetic Iron Oxide Nanoparticles through Human Brain Capillary Endothelial Cells, ACS Chem. Neurosci. 4 (2013) 1352–1360. https://doi.org/10.1021/cn400093z

[55] M.L. Formica, D.A. Real, M.L. Picchio, E. Catlin, R.F. Donnelly, A.J. Paredes, On a highway to the brain: A review on nose-to-brain drug delivery using nanoparticles, Appl. Mater. Today. 29 (2022) 101631. https://doi.org/https://doi.org/10.1016/j.apmt.2022.101631

[56] M. Arruebo, R. Fernández-Pacheco, M.R. Ibarra, J. Santamaría, Magnetic nanoparticles for drug delivery, Nano Today. 2 (2007) 22–32. https://doi.org/10.1016/S1748-0132(07)70084-1

[57] M.V. Tuttolomondo, M.E. Villanueva, G.S. Alvarez, M.F. Desimone, L.E. Díaz, Preparation of submicrometer monodispersed magnetic silica particles using a novel water in oil microemulsion: Properties and application for enzyme immobilization, Biotechnol. Lett. 35 (2013). https://doi.org/10.1007/s10529-013-1259-6

[58] K. Ulbrich, K. Holá, V. Šubr, A. Bakandritsos, J. Tuček, R. Zbořil, Targeted Drug Delivery with Polymers and Magnetic Nanoparticles: Covalent and Noncovalent Approaches, Release Control, and Clinical Studies, Chem. Rev. 116 (2016) 5338–5431. https://doi.org/10.1021/acs.chemrev.5b00589

[59] A. Mittal, I. Roy, S. Gandhi, Magnetic Nanoparticles: An Overview for Biomedical Applications, Magnetochemistry. 8 (2022). https://doi.org/10.3390/magnetochemistry8090107

[60] P. Chen, B. Cui, Y. Bu, Z. Yang, Y. Wang, Synthesis and characterization of mesoporous and hollow-mesoporous $M_xFe_{3-x}O_4$ (M=Mg, Mn, Fe, Co, Ni, Cu, Zn) microspheres for microwave-triggered controllable drug delivery, J. Nanoparticle Res. 19 (2017) 398. https://doi.org/10.1007/s11051-017-4096-z

[61] K.R. Hande, Etoposide: four decades of development of a topoisomerase II inhibitor, Eur. J. Cancer. 34 (1998) 1514–1521. https://doi.org/10.1016/S0959-8049(98)00228-7

[62] P. Chen, B. Cui, X. Cui, W. Zhao, Y. Bu, Y. Wang, A microwave-triggered controllable drug delivery system based on hollow-mesoporous cobalt ferrite magnetic nanoparticles, J. Alloys Compd. 699 (2017) 526–533. https://doi.org/https://doi.org/10.1016/j.jallcom.2016.12.304

[63] G. Wang, D. Zhao, N. Li, X. Wang, Y. Ma, Drug-loaded poly (ε-caprolactone)/Fe_3O_4 composite microspheres for magnetic resonance imaging and

controlled drug delivery, J. Magn. Magn. Mater. 456 (2018) 316–323.
https://doi.org/https://doi.org/10.1016/j.jmmm.2018.02.053

[64] D.N. Price, L.R. Stromberg, N.K. Kunda, P. Muttil, In Vivo Pulmonary Delivery and
Magnetic-Targeting of Dry Powder Nano-in-Microparticles, Mol. Pharm. 14 (2017)
4741–4750. https://doi.org/10.1021/acs.molpharmaceut.7b00532

[65] Y. Wang, G. Wei, X. Zhang, X. Huang, J. Zhao, X. Guo, S. Zhou, Multistage
Targeting Strategy Using Magnetic Composite Nanoparticles for Synergism of
Photothermal Therapy and Chemotherapy, Small. 14 (2018) 1702994.
https://doi.org/https://doi.org/10.1002/smll.201702994

[66] M.-L. Chen, Y.-J. He, X.-W. Chen, J.-H. Wang, Quantum Dots Conjugated with
Fe_3O_4-Filled Carbon Nanotubes for Cancer-Targeted Imaging and Magnetically
Guided Drug Delivery, Langmuir. 28 (2012) 16469–16476.
https://doi.org/10.1021/la303957y

[67] W.-H. Chiang, V.T. Ho, H.-H. Chen, W.-C. Huang, Y.-F. Huang, S.-C. Lin, C.-S.
Chern, H.-C. Chiu, Superparamagnetic Hollow Hybrid Nanogels as a Potential
Guidable Vehicle System of Stimuli-Mediated MR Imaging and Multiple Cancer
Therapeutics, Langmuir. 29 (2013) 6434–6443. https://doi.org/10.1021/la4001957

[68] M.W. Mushtaq, F. Kanwal, A. Batool, T. Jamil, M. Zia-ul-Haq, B. Ijaz, Q. Huang,
Z. Ullah, Polymer-coated $CoFe_2O_4$ nanoassemblies as biocompatible magnetic
nanocarriers for anticancer drug delivery, J. Mater. Sci. 52 (2017) 9282–9293.
https://doi.org/10.1007/s10853-017-1141-3

[69] X. Wang, Y. Qi, Z. Hu, L. Jiang, F. Pan, Z. Xiang, Z. Xiong, W. Jia, J. Hu, W. Lu,
Fe_3O_4@PVP@DOX magnetic vortex hybrid nanostructures with magnetic-responsive
heating and controlled drug delivery functions for precise medicine of cancers, Adv.
Compos. Hybrid Mater. 5 (2022) 1786–1798. https://doi.org/10.1007/s42114-022-
00433-2

[70] M. Ebadi, S. Bullo, K. Buskaran, M.Z. Hussein, S. Fakurazi, G. Pastorin, Dual-
Functional Iron Oxide Nanoparticles Coated with Polyvinyl Alcohol/5-
Fluorouracil/Zinc-Aluminium-Layered Double Hydroxide for a Simultaneous Drug
and Target Delivery System, Polymers (Basel). 13 (2021).
https://doi.org/10.3390/polym13060855

[71] M. Ebadi, K. Buskaran, B. Saifullah, S. Fakurazi, M.Z. Hussein, The Impact of
Magnesium–Aluminum-Layered Double Hydroxide-Based Polyvinyl Alcohol Coated
on Magnetite on the Preparation of Core-Shell Nanoparticles as a Drug Delivery
Agent, Int. J. Mol. Sci. 20 (2019). https://doi.org/10.3390/ijms20153764

[72] M. Ebadi, B. Saifullah, K. Buskaran, M. Hussein, S. Fakurazi, Synthesis and
properties of magnetic nanotheranostics coated with polyethylene glycol/5-
fluorouracil/layered double hydroxide, Int J Nanomedicine. 14 (2019) 6661–6678.
https://doi.org/https://doi.org/10.2147/IJN.S214923

[73] H.-A.-A. SH, H. MZ, B. S, A. P., Chlorambucil-Iron Oxide Nanoparticles as a Drug Delivery System for Leukemia Cancer Cells, Int J Nanomedicine. 16 (2021) 6205–6216. https://doi.org/doi:10.2147/IJN.S312752

[74] A. Marcu, S. Pop, F. Dumitrache, M. Mocanu, C.M. Niculite, M. Gherghiceanu, C.P. Lungu, C. Fleaca, R. Ianchis, A. Barbut, C. Grigoriu, I. Morjan, Magnetic iron oxide nanoparticles as drug delivery system in breast cancer, Appl. Surf. Sci. 281 (2013) 60–65. https://doi.org/https://doi.org/10.1016/j.apsusc.2013.02.072

[75] K. Dobretsov, S. Stolyar, A. Lopatin, Magnetic nanoparticles: a new tool for antibiotic delivery to sinonasal tissues. Results of preliminary studies, Acta Otorhinolaryngol. Ital. 35 (2015) 97–102

[76] X. Wang, A. Deng, W. Cao, Q. Li, L. Wang, J. Zhou, B. Hu, X. Xing, Synthesis of chitosan/poly (ethylene glycol)-modified magnetic nanoparticles for antibiotic delivery and their enhanced anti-biofilm activity in the presence of magnetic field, J. Mater. Sci. 53 (2018) 6433–6449. https://doi.org/10.1007/s10853-018-1998-9

[77] M. Chorny, E. Hood, R.J. Levy, V.R. Muzykantov, Endothelial delivery of antioxidant enzymes loaded into non-polymeric magnetic nanoparticles, J. Control. Release. 146 (2010) 144–151. https://doi.org/10.1016/j.jconrel.2010.05.003

[78] H. Aryan, B. Beigzadeh, M. Siavashi, Euler-Lagrange numerical simulation of improved magnetic drug delivery in a three-dimensional CT-based carotid artery bifurcation, Comput. Methods Programs Biomed. 219 (2022) 106778. https://doi.org/https://doi.org/10.1016/j.cmpb.2022.106778

[79] R.P. Friedrich, I. Cicha, C. Alexiou, Iron Oxide Nanoparticles in Regenerative Medicine and Tissue Engineering, Nanomaterials. 11 (2021). https://doi.org/10.3390/nano11092337

[80] M. Marcus, A. Smith, A. Maswadeh, Z. Shemesh, I. Zak, M. Motiei, H. Schori, S. Margel, A. Sharoni, O. Shefi, Magnetic Targeting of Growth Factors Using Iron Oxide Nanoparticles, Nanomaterials. 8 (2018). https://doi.org/10.3390/nano8090707

[81] E. Daher Pereira, S. Thomas, F. Gomes de Souza Junior, J. da Silva Cardoso, S. Thode Filho, V. Corrêa da Costa, F. da Silveira Maranhão, N. Ricardo Barbosa de Lima, F. Veloso de Carvalho, M. Galal Aboelkheir, Study of controlled release of ibuprofen magnetic nanocomposites, J. Mol. Struct. 1232 (2021) 130067. https://doi.org/https://doi.org/10.1016/j.molstruc.2021.130067

[82] D. Dorniani, M.Z. Bin Hussein, A.U. Kura, S. Fakurazi, A.H. Shaari, Z. Ahmad, Sustained release of prindopril erbumine from its chitosan-coated magnetic nanoparticles for biomedical applications, Int. J. Mol. Sci. 14 (2013) 23639–23653. https://doi.org/10.3390/ijms141223639

[83] R. Amato, M. Giannaccini, M. Dal Monte, M. Cammalleri, A. Pini, V. Raffa, M. Lulli, G. Casini, Association of the Somatostatin Analog Octreotide With Magnetic Nanoparticles for Intraocular Delivery: A Possible Approach for the Treatment of

Materials Research Forum LLC
https://doi.org/10.21741/9781644902332-8

[84] C. C. Chauhan, T. Gupta, S. S. Meena, M. F. Desimone, A. Das, C. Singh Sandhu, K. R. Jotania, R. B. Jotania, Tailoring magnetic and dielectric properties of $SrFe_{12}O_{19}/NiFe_2O_4$ ferrite nanocomposites synthesized in presence of Calotropis gigantea (crown) flower extract. J. Alloys Comp, 900, 163415, 2022.https://doi.org/10.1016/j.jallcom.2021.163415

[85] C. C.Chauhan, A. A.Gor, T. Gupta, M. F. Desimone, N. Patni, R. B. Jotania. Investigation on structural, optical, magnetic, and dielectric properties of calcium hexaferrite synthesized in presence of Azadirachta indica and Murraya koenigii leaves extract. Ceramics Int., 48, 14, 20134-20145, 2022. https://doi.org/10.1016/j.ceramint.2022.03.292

Materials Research Forum LLC
https://doi.org/10.21741/9781644902332-9

Chapter 9

Magnetic Nanoparticles for Immune System Related Diseases Treatment

Federico G. Baudou[1,2,#], Exequiel Giorgi[1,2,#], María Eugenia Diaz[1,2], Gabriela F. Rocha[1,2], Florencia S. Conti[1,2], Sofia Genovés[1,2], Gabriel A. De Diego[1,2], Liliana N. Guerra[1,2,3], Mauricio C.De Marzi[1,2]*

[1]Universidad Nacional de Luján (UNLu), Depto. de Ciencias Básicas, Luján, Buenos Aires, Argentina

[2]Laboratorio de Inmunología, Instituto de Ecología y Desarrollo Sustentable (INEDES), UNLu-CONICET, Luján, Buenos Aires, Argentina

[3]Universidad de Buenos Aires. Facultad de Ciencias Exactas y Naturales. Departamento de Química Biológica, Buenos Aires, Argentina

[#] both authors contribute equally

*mdemarzi@unlu.edu.ar

Abstract

Nanotechnology is a continuously rising field with different and varied purposes. Nanoparticles (NPs) are defined as particles between 1 and 100 nanometers in diameter, but within this classification larger particles can be incorporated up to 500 nm. Magnetic NPs (MNPs) are generally composed with magnetic elements such as nickel, cobalt and iron, and have the ability to be manipulated through a magnetic field. Recently, these have been the subject of numerous studies because their attractive properties for several potential applications. Among them, we can include its use for diagnosis and medical treatment (theragnosis). Due to their properties, MNPs could be used with the aim of developing immunomodulatory therapies and deal with infectious diseases, cancer, allergies, autoimmune diseases, etc. The use of MNPs provides a novel tool to manipulate the immune response towards a profile according to the proposed objectives. That is why this chapter describes the interaction of MNPs with the immune system as well as its possible applications in immunomodulatory therapies.

Keywords

Magnetic Nanoparticles, Immune System, Infectious Diseases, Cancer, Allergies, Autoimmunity

Contents

1.   Introduction

Nanotechnology is a research area in continuous expansion that involves materials at the molecular and atomic scale. By definition, it refers to structures of 1-100 nm in size, but commonly can be used to those that are up to 500 nm [1]. The development of nanotechnologies contributes to most fields of science, including biology and specifically its interaction with the immune system. Its application in human health has demonstrated promising results, especially in the treatment of cancer [2].

Nanoparticles (NPs) possess unique physicochemical properties making them ideal for novel applications. Specially, magnetic NPs (MNPs) can be manipulated or detected by magnetic fields presenting several opportunities to *in vivo* uses. Among this, MNPs are widely studied for their potential applications in nanomedicine, such us hyperthermia therapy, drug and gene delivery, magnetic separation, magnetic imaging, diagnostics, etc. The development of MNPs with surface functionalization has also been described [3]. They generally consist of magnetic elements such as cobalt, nickel, iron, etc. Iron oxide NPs (IONPs) constitute robust platforms for drug delivery system development because they can be loaded with large quantities of drugs, as well as targeting them for the efficient delivery of these compounds to selected locations [4]. Beside this, MNPs present a surface-area / volume ratios and high magnetic moments for what they are postulated as potent tool for cancer therapy by hyperthermia. Additionally, they can be used to improve the sensitivity of diagnostic tools and biosensors by acting as contrast agents for magnetic resonance imaging (MRI) [5].

On the other way, immunomodulators are substances with the ability to increase or decrease the immune response. This capacity has wide beneficial uses in infectious, allergic, neoplastic and autoimmune diseases therapies. The immune system is made up of a set of tissues, cells and molecules whose objective is to defend the body from foreign agents. This is based in a complex organization used to fight pathogens as well as to maintain a dynamic equilibrium among immune / inflammatory processes, levels of homeostatic activity and regulatory / suppressive functions. On the other hand, the immune response involves the coordinated action of both the innate immune response and the adaptive immune response [6,7]. Innate immunity is considered to be the first line of defense, involving both chemical and physical responses as well as cellular responses against pathogens (natural killer cells, dendritic cells, basophils, eosinophils, mast cells, neutrophils, macrophages, among others). These cells are responsible for cytokines secretion, producing local inflammation. This inflammatory process is frequently followed by the adaptive immune response. Adaptive immune response involves B and T lymphocytes activation, antigen-specific cells against instigators of the immune system and pathogens. Both responses are intimately connected [8].

Due to their properties, MNPs could be used with the aim of developing immunomodulatory therapies and challenge allergies, infectious diseases, etc (Fig. 1). NPs provide a potential tool with the aim of inducing a certain profile of the immune response. That is why herein we describe the interaction of MNPs with the immune system as well as its possible applications in immunomodulatory therapies.

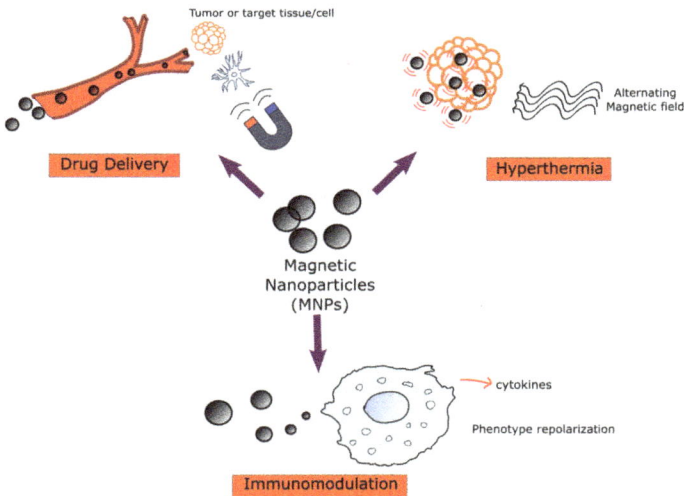

Fig. 1. Schematic representation of different potential applications of MNPs for disease treatment.

2. MNPs for infectious disease treatment

A "disease" can be defined as any condition that disturbs the regular function of an organ and / or system of the body, of the mind, or of the complete organism, and it can be linked with specific symptoms and signs. An infectious disease is referred as a disorder triggered by the entry of a foreign agent to a host, affecting its normal function of organs and / or systems and, also could be transmitted to other individuals [9]. Despite of the human immune system ability to protect our body from infection, this mechanism is able to fail, leading to disease after the infection [10]. Therefore, people that have their immune system compromised, children and elder are affected more frequently by pathogenic agents, such as bacteria, viruses, fungi, protozoa, and helminthes than the rest of the people. A common classification of this type of diseases is dividing them between emerging and re-emerging infectious diseases. The term emerging infectious diseases is referred to new diseases, in contrast with re-emerging infectious diseases that reappear after the improvement of the pathogen drug resistance. Due to this, any treatment or control of these diseases turns to be difficult [11]. Microorganisms' drug-resistance change occurs after exposition to antimicrobial medications (such as antibiotics, antivirals, antifungals, antimalarials, and antihelmintics). This results in a reduction of medication efficiency and in a persistence of infections in the body, increasing their potential spreading among the population. Thus, drug resistance hampers infectious disease treatments and therefore represents a serious threat to global public health, affecting animals, human and environmental health. With the consequent increasing of mortality and morbidity, there are high economic costs because of its burden of medical care [12]. Furthermore, the exacerbate use of antibiotics during the COVID-19 pandemic will augment microbial resistance, leading to a higher mortality [13]. That is the reason for improving the developing of new therapeutics and novel strategies looking for a way of overcoming drug resistance.

In this sense, the expansion of NPs as a strategy to treat infectious diseases is currently booming, specifically MNPs appear to be an effective alternative. MNPs are formed by a magnetic core stabilized with a biocompatible coating that protects its surface, allowing a reduction of their potential toxicity and also providing a protection to the magnetic core from corrosion [14]. The properties of these nanosized magnetic particles are different from those of the corresponding conventional magnetic materials [15]. In particular, MNPs that have a core diameter of 10-100 nm, exhibit special magnetic properties called "superparamagnetism" [16,17]. Superparamagnetic iron oxide nanocrystals (SPIONs) are formed by iron oxide. Nevertheless, due to their small dimensions, there is not any magnetization exhibited by them with the exception of the cases that it is applied an external magnetic field [18]. Due to this, there is less probability of aggregation in absence of magnetic fields, making them a very recommendable platform to biological applications [3]. In addition to this, MNPs can trap lipophilic, hydrophilic, lipophobic and hydrophobic drugs; which allows that these NPs could remain protected against the body's immune system and, as a consequence, they could prolong their blood circulation time [19].

Currently most studies carried out with MNPs focus on two types of strategies to treat infectious diseases:

2.1 Drug delivery

Only a small proportion of delivered medicine (orally or intravenously) reaches the target organ, and even less reaches cellular targets, making precise delivery of an agent to the site of infectious illness a major pharmacological problem [20]. To ensure a sufficient concentration of the drug at the target site, the administered dose is usually increased [21,22]. This inefficiency leads to unwanted effects and a possible toxicity, which can limit the clinical use of promising treatments. Furthermore, non-selective administration can also lead to negative immune responses at the site of administration [23]. In this sense, MNPs can be used to control drug delivery in the body by directing the medicine to where it is supposed to act by applying external magnetic fields. Furthermore, the surface of MNPs can be modified to respond to a specific exogenous (e.g., temperature fluctuations, ultrasound intensity) or endogenous (e.g., pH, enzyme concentration, redox gradients) trigger, allowing them to exploit specific changes in the microenvironment associated with pathological situations [24]. In this way, the drug deliverability and therapeutic efficacy at the disease site can be improved and the toxicity on human cells can be reduced [25,26]. Cai *et al.* (2021) demonstrated that the green synthesized superparamagnetic NPs exhibit the ability to adsorption and drug release in a pH-dependent manner, using ofloxacin and pefloxacin as pharmacological models, while Mohapatra *et al.* (2017) demonstrated that the antibiotic vancomycin could be released from chitosan microbeads embedded with MNPs by frequency alternating magnetic field [27,28].

Various nanostructures of MNPs have been reported as antibacterial agents to kill a range spectrum of bacteria species, including multidrug-resistant bacteria and bacterial biofilms, without harming human host cells [29]. In several works it has been shown that SPIONs by themselves or with certain chemical modifications in their structure have the ability to inhibit the development of certain bacteria, such as gentamicin-resistant staphylococci [30,31]. On the other hand, MNPs conjugated with antibiotics, and their derivatives have been extensively investigated for their potential to penetrate into bacterial cells and biofilm mass, which inactivates antibiotic resistant bacteria [32]. Geilich *et al.* (2017) showed that an antibiotic delivery built with MNP produces drug penetration and generates high concentrations of antibiotics in the different biofilm layers, while the antibiotic can only control planktonic bacteria since it does not have the ability to penetrate biofilms [33]. Maleki *et al.* (2016) studied the antimicrobial activity of superparamagnetic iron oxide NPs with a gold shell conjugated to the antimicrobial peptide cecropin mellithin (NP-CM). The authors found that the nanosystem had a lower minimum inhibitory concentration for *Escherichia coli* than the soluble peptide. This could be due to the synergistic effect of multiple peptide molecules on the surface of the NPs. In addition, the authors showed that NP-CM was highly internalized by endothelial cells and macrophages and accumulated mainly in the endolysosomal compartment. In this case, it was demonstrated that nanosystems can have desirable properties, such as attacking microorganisms without harming the cells of the human body [34].

Viral infections such as AIDS, hepatitis, herpes keratitis, and cold sores have also become resistant to drugs, making it difficult to develop effective treatments. Furthermore, there have been recent outbreaks of illnesses caused by numerous viruses, such as distinct strains of influenza virus (H5N1, H1N1 and H3N2), dengue, Ebola, Zika, and the latest coronavirus [35], with devastating consequences and demanding the development of novel antiviral therapies. Studies are currently focused on designing nano delivery systems to control viral diseases and overcome drug resistance. These systems are based on broad-spectrum antivirals that target the viral membrane or cell receptor to compete with the binding of the virus to the target cell and prevent its internalization. One example is silver NPs, which are considered promising antiviral agents against human norovirus [36]. Park *et al.* (2014) found that a new magnetic colloid of NPs and silver produced a 2-log reduction in murine norovirus after 1 h of treatment. The study also found that these particles damaged viral nucleic acids and envelope proteins [37].

In the same way, Jayant *et al.* (2014) designed magnetic NPs system for the sustained release of tenofovir, which showed a stronger antiviral activity in infected astrocytes and greater ability to cross the Blood-Brain Barrier (BBB) compared with the free drug. In addition, the authors packaged HIV latency-breaking agents such as vorinostat into the nanosystem to simultaneously reactivate and kill HIV in a sustained manner for 5 to 7 days throughout the BBB. The nanoformulation showed good BBB transmigration capacity with marked *in vitro* antiviral efficacy in primary human astrocytes, with good cell viability after HIV infection [38,39].

Other studies have revealed the potential inactivation and virucidal effects of various types of magnetic NPs on certain viruses with structural similarities to SARS-CoV-2 [40], such as the suppression of herpes simplex virus by aziridine-terminated polyethyleneimine-functionalized core-shell superparamagnetic NPs [41].

Although the World Health Organization has not designated fungal and parasite diseases as a public health emergency, it is worth emphasizing that fungi, which are single or multicellular eukaryotic microorganisms, can cause infections ranging from superficial to systemic, which can result in a increased mortality.

Candida is one of the fungus species that is widely recognized for producing both superficial and systemic illnesses. Using two types of medicines, amphotericin B and nystatin, Niemirowicz *et al.* (2016) investigated the antifungal and anti-biofilm potential of functionalized MNPs against clinical isolates of Candida species. When tested against Candida strains in the presence of pus the authors discovered that nanosystems were more effective than free agents, suppressing Candida biofilm formation. Furthermore, nanosystems had a higher biocompatibility than free medications, as the lytic activity of antibiotics against host cells was significantly reduced when they were attached to MNPs surfaces [42].

Drug toxicity and resistance make it difficult to treat parasite infections; hence a combination of two or more antiparasitic medications is used to improve therapeutic results. Drug resistance in parasitic illnesses like malaria is a result of poor treatment compliance [43]. As a new technique to treat malaria infection, Wang *et al.* (2020) created MNPs based on hollow mesoporous ferrite NPs coupled with artemisinin in the internal part of magnetite and heparin in the outside mesoporous layer. The interaction between the malaria parasite's hemozoin (paramagnetic) and the MNPs produces the targeting of the delivery mechanism. Magnetic adsorption, according to the authors, not only prevented the discharge of merozoites but also improved the antimalarial action by raising the local concentration of artemisinin around infected red blood cells [44].

2.2   Hyperthermia

Cells are susceptible to heat damage because it disrupts cellular pathways, inhibits DNA damage repair, etc. When MNPs are influenced by an alternating magnetic field (AMF), lose energy as heat, achieving an increase of temperature in a specific point. In this way, if MNPs binds to specific cells, it could destroy them by transferring heat. Due to this, hyperthermia is investigated as a way to attack tumor cells (as reported later), but it can also attack specific pathogens as bacteria, protozoa, etc.

Despite the great advantages that MNPs have to be used in hyperthermia procedures, there is one big limitation: a tendency of these MNPs to aggregate if they are not stabilized with other substances, reducing heating efficiency. Conjugation of antibodies over MNPs surface is another way to improve the hyperthermia treatment. Because of this, Kim *et al.* (2013) used a MNP composed by magnetite and conjugated to its surface an anti-protein-A antibody to specific binding of NPs to *S. aureus* surface. These MNPs showed an effective inactivation of bacteria *in vitro* after a brief AMF exposure and were also tested in a model of *S. Aureus* cutaneous wound infection in mice. A thermal inactivation was demonstrated *in vivo*, maintaining a low level of wound bacterial burden at day 2 post-infection, and also improving the wound healing. In spite of the temperature increase in the wound while AMF was active, it was observed a fast return to normal temperature for the tissue surrounding the wound edge. Although, the effect of MNP hyperthermia on the innate immune response was not evaluated [45].

In the same way, Chen *et al.* (2016) used antibodies to direct particles to *S. aureus*, using magnetic crystals highly homogenous, which do not aggregate. These MNPs can swim forward along magnetic field lines and, with the adsorption of rabbit polyclonal antibody (anti-MO-1), they can attach to *S. aureus*. Thanks to this, when an alternating magnetic field is applied, hyperthermia occurs near *S. aureus*, reducing their viability. As *S. aureus* tends to form biofilms, currently available topical antibiotic treatment cannot be efficient enough. For that, this strategy using antibody-coated magnetotatic bacteria has a huge potential for treatment of infected wounds by *S. aureus* and allow immune system to act more efficiently [46].

Not only an AMF can be used as inducer of heating in MNPs, but also it is possible to irradiate them with an NIR laser. Yu *et al.* (2011) synthesized alumina (Al) coated IONPs to be used as photothermal agents to selectively kill bacteria. It seems to be an important interaction of these particles with phosphorylated proteins of most bacteria through Al-phosphate chelation. MNPs were attached to different bacteria, an aggregation was generated by magnet, and after that, a treatment with an 808 nm laser during short time produced a photothermal killing of these bacteria. In addition to these results, they also found a weak interaction with A549 cells (carcinomic human alveolar basal epithelial cell) and a low reduction in their cell viability, thus they concluded that the alumina-coated magnetic IONPs were not only efficient to kill bacteria by photothermal activity, but also has good biocompatibility [47].

Most of the research that apply hyperthermia for infections are directed on bacterial infections, but Chudzik *et al.* (2016) tested the effectiveness of this treatment against *Candida albicans*. This pathogen is less sensitive to higher temperatures than human cells, establishing a huge limitation in the application of hyperthermia. They used MNPs coated with meso-2,3-dimercaptosuccinic acid (DMSA) and grafted with anti-*C. albicans* type A antibody attached by the linker EDC (1-ethyl-3-[3-dimethylaminopropyl] carboiimide hydrochloride). These NPs were called IMNPs (immunogenic magnetic NPs). It showed an ability to bind to pathogen surface and, after a short exposure to an AMF, an increase of temperature at 43°C, reducing cell viability and proliferation activity of the pathogen. Using no more than 43-45°C temperature range assures a safe application of the hyperthermia treatment [48].

Some MNPs can be more complex, such as those generated by Singh *et al.* (2015), mesoporous ZnO microparticles (200-600 nm), with Fe_3O_4 NPs (10 nm) embedded in the mesopores. It was tested in *S. aureus* and *E. coli*, causing oxidative stress by ROS generated. Bacterial growth inhibition was dependent of NPs concentration and incubation time. When an AMF was applied, bacteria viability was also reduced, probably as result of hyperthermia and ROS generation. In addition, NPs were tested in HeLa cell culture, which had a 90 % of cell viability after an incubation at high concentrations of NPs, demonstrating a reasonable biocompatibility [49].

Other work reported initial approaches made by Grazú *et al.* (2012) that used IONPs in presence of an AMF to kill *Crithidia fasciculate*. They also found a high reduction of their viability [50].

Berry *et al.* (2019) evaluated magnetic hyperthermia, using IONPs to kill protozoan parasite *Leishmania mexicana in vitro*. This parasite invades mononuclear phagocytic cells and it generates a disease that can be presented in two forms. The most common form is cutaneous leishmaniasis, which provokes skin lesions susceptible to posterior bacterial infections. The presence of the IONPs alone had no effect on amastigote (a stage of *L. Mexicana* reached after infect macrophages) viability, but it produced a 70 % reduction of viability after a magnetic hyperthermia treatment. The next challenge is to target the amastigote when is located intracellular and to test it in animal models [51].

Recently amino acids functionalized MNPs were tested as antiplasmodial tools against *Plasmodium falciparum*, an important pathogen that has been acquiring resistance to the usual drugs used for treatment [52]. Natural amino acids attached to NPs surface stabilize MNPs in physiological conditions, providing a solution for the MNPs aggregation, which, as mentioned above, is one of the main problems of these particles with a therapeutic potential. These NPs showed an important heat capacity by hyperthermia, but the authors demonstrated that functionalized MNPs *per se* had a significant antiplasmodial activity in comparison with MNPs without functionalization. In addition, these MNPs compatibility was demonstrated in cytotoxicity analysis in VERO cells. This report proves that amino acids functionalized MNPs have potential as a possible malaria treatment, so it is necessary to add evidence to contrast their effect *in vivo*.

Although MNPs are commonly considered as a potential antitumor tool, based on the evidence presented here, it is clear that these NPs can be used in different infections since their ability to reduce or eliminate various bacteria and parasites has been demonstrated through drug delivery and hyperthermia (Fig. 2). Most of the experiments published so far have been done *in vitro* and only a few have presented initial animal model approaches. Because of treatments based on the use of MNPs for drug-resistant infections have a promising future; it is necessary to advance in models that allow their use in humans.

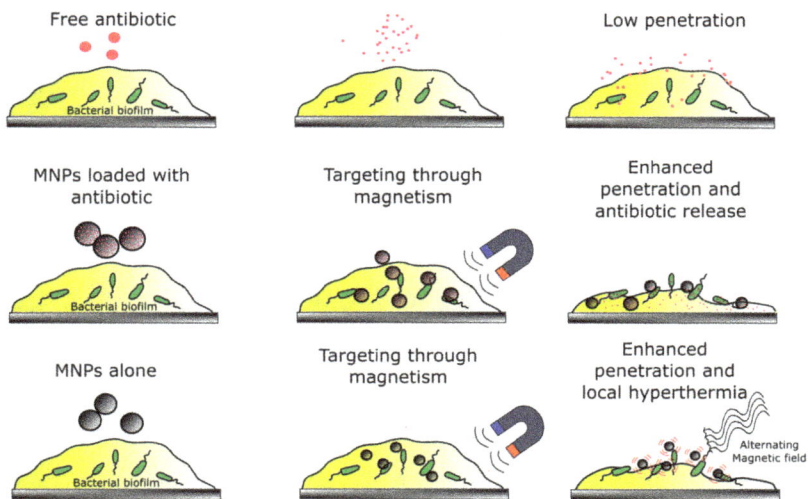

Fig. 2. Strategies currently under study and that involve the use of MNPs with the aim of treating infectious diseases.

3. Applications of MNPs against cancer

As we previously mentioned, the use of nanotechnology in medicine (nanomedicine) is a hopeful strategy for the design of new efficacious tools that overcome the limitations of traditional treatments. In recent times, special interest has been shown in MNPs in order to use them in theranostic applications. This is due to its ability to be used in both therapeutic and diagnostic applications [53–55]. MNPs offer great advantages, mainly due to their controllable size and functionalization capability and also because they may be controlled remotely by external methods like magnetic fields [56].

Drug´s administrations that can be directed towards the target site, as well as hyperthermia, are some of the different therapeutic applications that they can offer us, as mentioned for infectious diseases. They can be located in the tumor area thanks to manipulation through the magnetic field, a property that also allows them to be used as drug carriers [54]. When responding to an external stimulus, they can be guided towards the target site. On the other hand, as most drugs have hydrophobic properties, being encapsulated in a polymeric structure would improve their bioavailability in the body [5]. As for diagnostic applications, MNPs can be used in magnetic resonance images and can serve as new tracers due to its potential as a contrasting agent [5,53].

Currently there are two major strategies for the manufacture of multifunctional MNPs (Fig. 3). They can be molecularly functionalized by binding antibodies, proteins or dyes, and also as metallic NPs or MNPs with functional nanocomponents such as quantum dots (QD). Thanks to these types of multifunctional characteristics simultaneously with synergistic properties, they could present unique advantages in the biomedicine field [56].

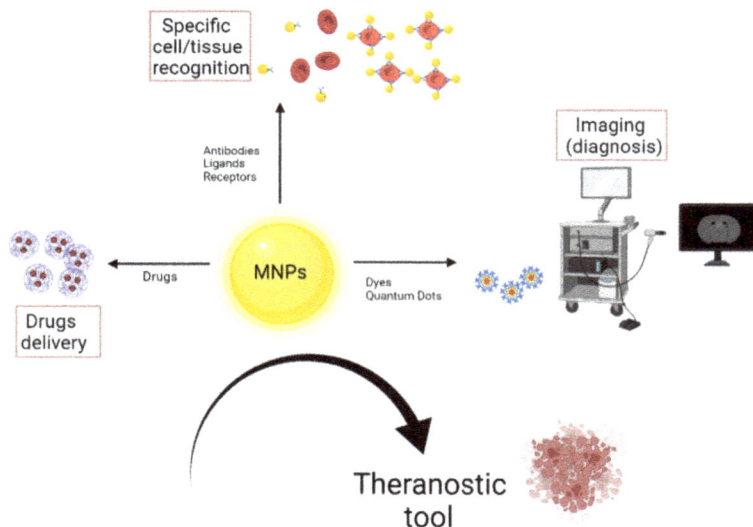

Fig. 3. Schematization of the possible uses of multifunctional MNPs for treatment, diagnosis and as a theranostic tool, against cancer.

Relating to the application in cancer, MNPs can interact with the immune system, so immunotherapy is currently presented as the therapy option with potential best results. This can be carried out systemically, as is the case with the checkpoint blocking antibodies, but also directly in an *in situ* vaccination strategy on identified tumors. In this case, the aim is to stimulate the immune system in order to respond against the tumor, generating a more favorable microenvironment. This can include attenuated microorganisms, physical disruptors, recombinant proteins, and also NPs, among others [57]. As has been said, NPs can load different drugs that can activate immune and cancer cells, generating an immune stimulated tumor microenvironment that allows the immune system to attack and kill them.

In this sense, the immunomodulation of macrophages against cancer mediated by MNPs, emerges as a promising therapeutic strategy [58]. Using a model mice in which melanoma and / or breast cancer are generated, authors reported a novel work in which they used genetically modified cell membrane-coated MNPs (gCM-MNPs) that can deactivate macrophages set by cancer cells. This MNPs inactivate the regulatory protein SIRPα, generating that tumor cells were recognized and destroyed by cytotoxic T lymphocytes and their remnants phagocyted by macrophages (in addition to being a powerful activator of the T lymphocyte response). In addition, they inactivate the polarization to a tumorigenic M2 phenotype of macrophages (cancer cells induce M2 phenotype by secreting colony-stimulating factors). gCM can over express variants of the SIRPα signal protein that, with high specificity, competently blocks the CD47-SIRPα pathway; on the other hand, the nucleus formed by MNPs generates the repolarization of M2 to M1 macrophages and induce the response of antitumor T cells. Both effects behave synergistically, triggering a powerful macrophage immune response. The protective layer of gCM that covers the core of MNPs does not allow MNPs immune clearance, and at the same time, under magnetic navigation, it targets the tumor area, delivering the gCM, where it accumulates. gCM-MNPs significantly showed an increase in the overall survival of the mice, controlling distant tumor metastasis, as well as local tumor growth.

Although cancer immunotherapy is currently emerging as a promising therapeutic method, it is effective only for certain types of cancer. Photoimmunotherapy, it is photothermal and photodynamic therapy that would radically improve the immune response of immunotherapy, but the immunogenicity caused could induce severe inflammatory effects. Hyperthermia is a therapeutic technique that consists of increasing the temperature above the normal physiological temperature. Unlike normal cells, cancer cells are sensitive to hyperthermia because in the cancer microenvironment there is a pH decrease, causing these cells to be less thermotolerant. In this way, MNPs can be magnetically manipulated by an external field and produce heat due to hysteresis loss during a magnetization cycle [5].

Also based on biomimetic characteristics, Jiang *et al.* (2020) camouflaged MNPs with platelet membranes (called Fe_3O_4-SAS@PLT) that increase ferroptosis in cancer immunotherapies. Fe_3O_4-SAS@PLT are mesoporic Fe_3O_4 MNPs loaded with sulfasalazine and camouflaged with a platelet-derived membrane that trigger ferroptotic cell death, considerably increasing the efficiency of programmed cell death immune checkpoint 1 blocking therapy and removing the tumor in murine 4T1 metastatic tumors models. These results revealed that the use of Fe_3O_4-SAS @ PLT it would induce a specific immune response against the tumor and could also repolarize immunosuppressive macrophages to antitumor inflammatory ones (M2 to M1 phenotype) [59].

Another type of research against cancer using immunotherapy, but trying to exploit toll like receptors (TLR), was carried out by Bocanegra *et al.* (2018) in mice model. They used zinc-doped IONPs vaccines together with phospholipid micelles in combination with ovalbumin as antigen. Due to this complex formed, a synergistic activation between the immune response and the direct death of cancer cells was produced by the combination of the TLR using a polyIC tail as a ligand for TLR3 and imiquimod (R837) as a TLR7 agonist. A large innate immune response in the lymph nodes was triggered by the binding of agonists to the TLR, which could be monitored by magnetic resonance imaging and tracking migration from the site of application to the lymph nodes and tumor. They also observed that when the complex with MNPs had an immunosuppressive ligand of the PDL-1 checkpoint, the antitumor results improved. As we can see, vaccines for cancer immunotherapy could have potential clinical use due to the combined use of synergistic TLR agonists together with MNPs and immune checkpoint blockers [60].

Similarly, the use of nanoparticulate material composed of ferumoxytol together with micelles, in complexes formed together with anti-PDL-1 antibodies and inhibitors of the protein serine / threonine kinase (BRAF) has also been used in mouse models for the study of cancer of anaplastic thyroid with good results [61]. The quantification of iron in ferumoxytol by magnetic resonance imaging allows marking tumor-associated macrophages (TAMs) in murine models. In this way, researchers were able to study TAM levels, which were found to be increased in the thyroid, in addition to observing a high vascularization thanks to the combined use of particulate material together with the inhibitor for BRAF and the antibody against PDL-1checkpoint.

Other studies in mice allow us to see the versatility of this type of NPs, MNPs called IO@FuDex3 based on fucoidan / dextran conjugated with inhibitors against checkpoints such as PDL-1 and activators of T lymphocytes (anti CD3 and anti CD28) were generated [62]. By commanding the IO@FuDex3 from external magnetic fields that drive them towards the tumor, its secondary effects are minimized, while the complex can repair the tumor-generated immunosuppressive microenvironment due to the revitalization of lymphocytes infiltrated in the zone. Interestingly, the mean survival of mice treated with this immunotherapy, in addition to the considerably reduced adverse effects, increases approximately twice as long with a single dose of less than 1% compared to the use of anti-PDL-1 only.

On the other way, the lack of biomarkers that predict the response to a specific antitumor treatment limits the development of vaccines that can generate a response against tumors. To this end, a work has recently been published using MNPs that activate dendritic cells (DCs) and also early predict an antitumor immune response through magnetic resonance imaging [63]. In studies carried out with a mice model as pre-clinical tumor supports, MRI indicates that responding animals after the vaccines had been applied, had significantly smaller tumors and their survival increased by more than 70% compared to non-responders. Thus, the design of multifunctional iron magnetic liposomes loaded with RNAs that induce a strong immune response against the tumor that function as an anticipatory bioindicator of the reaction to the treatment, seems to be an alternative that allows an early prediction of the antitumor response by means of MRI. The inclusion of the iron oxide that makes up the MNPs improves DCs transfection and allows monitoring their migration by obtaining images, with a correlation between the intensity of the MRI in the lymph nodes and the DCs trafficking.

In this sense, the use of hybrid vesicles composed of an ascorbic acid nucleus and covered by poly (lactic-co-glycolicacid) that incorporate iron oxide nanocubes is also presented as a strategy in immunotherapy. The Fenton reaction triggered by a magnetic force produced from the outside causes the release of ascorbic acid from the nucleus, generating an increase in ferrous ions by reacting with the iron-containing shell, improving the immune response mediated by cell death similar to ferroptosis. The oxidative stress generated in this reaction causes the release of the protein calreticulin from the interior of tumor cells to the cell surface and external environment. This causes the maturation of DCs and the entry of cytotoxic T lymphocytes into the tumor environment, achieving a significant suppression of tumor size. Furthermore, the change in ferric ion concentration can also be monitored during the treatment [64].

Regarding the use of different types of magnetic fields that direct MNPs, there is great interest in understanding these effects on cells, in the immune system and in living organisms [65]. In general, the cells that are part of the immune system are slightly affected by low magnetic fields, but they are strongly affected by moderate-high intensities. Such is the case of macrophages, whose cytoskeleton, membrane receptors and ion channels are altered, as well as iron metabolism, by affecting the expression of genes involved. However, MNPs combined with magnetic fields can affect T cells and play a favorable role in immunity. These involve a regrouping of their TCRs, which increases cell-cell contact and communication and consequently a greater capacity for tumor destruction.

In summary, the use of MNPs, for therapy, diagnosis or theranostic, allows the rapid identification of the phenotype of cancer cells, thus facilitating the correct selection of drugs and their administration. Navigation through magnetic fields also allows a better observation of tumors and how immunotherapy is conducted using this type of complex, since the negative consequences of the medicament used to combat certain types of tumors can be reduced.

4.   MNPs and allergy

Allergic diseases consist in IgE-mediated immune hypersensitivity reactions in response to innocuous antigens. They include allergic rhinitis, asthma, food allergy and atopic dermatitis, and its prevalence is increasing worldwide. DCs and macrophages fagocitate and present the allergens, to that were exposure, on major histocompatibility complex, causing the activation of immune cells. This produce the polarization of T helper (Th) cells to a Th2 phenotype and therefore the stimulation of plasma cells and IgE production. IgE binds to basophils and mast cells and after a second exposure to the allergen, inflammatory mediators as histamine, heparin, prostaglandins and cytokines are released [66]. As consequence, different symptom appears that could be aggravated with further exposure to the same antigen, like runny nose, watery eyes, bronchoconstriction and sometimes anaphylaxis.

Due to the increasing prevalence of this kind of diseases, new therapeutic tools like NPs are being studied. However, there is evidence showing that different types of NPs exacerbate the hypersensitivity observed in allergy and this effects inversely correlate with particle size [67–70]. In particular, a carbohydrate-coated SPION-based drug that is administered to patients with chronic kidney disease, caused hypersensitivity including severe anaphylactic reactions [71]. In addition, SPIONs-based medications used for MRI as contrast agent have been associated with allergic-like reactions [72,73]. Different mechanisms have been proposed as the responsible of these reactions, however, it still remain unknown (Fig. 4) [74].

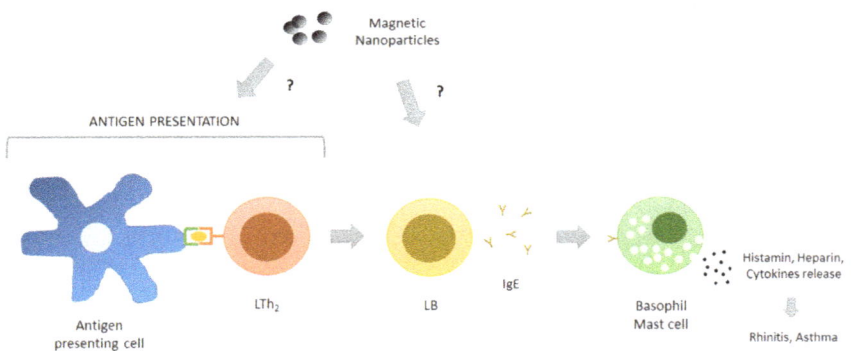

Fig. 4. Diagram of the immune response induction process in the triggering of an allergic reaction in which MNPs could interfere. The mechanisms responsible for these effects remain unknown.

Despite the evidence showing that MNPs can cause allergic reactions, there is a study showing that they could suppress the immune response. Blank *et al.* (2010) showed that the exposure of human monocyte-derived DCs to poly vinylalcohol-coated SPIONs (PVA-SPIONs) did not produce changes in surface marker expression or antigen-uptake, but the capacity to process antigen, to stimulate of CD4+ T cells and to induce cytokines were diminished. Due to this result some authors proposed to use this type of NPs in vaccines for allergen-specific immunotherapy [75,76]. Therefore, taking into account the available information, more studies are needed to assess whether MNPs are a useful tool for the treatment of allergies.

Further, MNPs have been very useful to design methods more sensitive to detect allergen-specific IgE, which is required for diagnosis of allergic diseases. These NPs can be conjugated with different molecules and they could be easily separated from other components. Teste *et al.* (2010) developed a magnetic-core-shell NPs based immunoassay capable of detecting specific milk allergen-IgE [77]. Han *et al.* (2020) developed a method based on MNPs with surface modification of nickel (II) nitrilotriaceticacid (Ni-NTA) moieties (Fe3O4@SiO2-NTA), this allows the quantitative detection of specific IgE in serum samples due to the immobilization of allergenic recombinants proteins. One of the advantages of this method is that, under an external magnetic field, the solid-liquid separation can be performed and therefore, the sensitivity is improved compared to traditional methods [78]. In another work, $Fe_3O_4@SiO_2$-NTA was employed to improve the detection of IgE by using gold NPs with HRP-labeled anti-IgE that enhanced the signal [79,80]. Other authors developed an IgE-detection method that correlated the magnetophoretic deflection velocity with allergen-specific IgE concentration in serum [81].

5. Use of MNPs in autoimmune disease diagnosis and treatment

Generally, the immune system can discriminate between foreign cells and self-cells. That is why it has the ability to recognize foreign entities and attack pathogens such as bacteria, parasites and viruses. However, autoimmune diseases develop when the immune system erroneously attacks the body itself. In an autoimmune disease, the immune system kicks in and attacks healthy cells themselves (as in type 1 diabetes, systemic lupus erythematosus (-SLE-), etc.). That is why a large part of the treatments against them lies in reducing immune activation and the general inflammatory process. However, this strategy can cause immunosuppression, so the most current therapies under developing are focused on trying to reduce directly the specific autoimmune response.

Different metal oxide NPs (MONPs) developed with Zinc, Iron, Nickel, Copper, Cobalt, Gold, have been used in therapeutic applications as drug nanocarriers, alone and as diagnostic tools for the last decade.

IONPs have been employed in biological requests, because their minimum toxicity and excellent biocompatibility, as MRI contrast agents. They are usual in MRI diagnosis of rheumatoid arthritis and are the only metal oxide NPs approved by the Food and Drug Administration (FDA) for this use. They show an important biodegradability, since IONPs lower than 10 nm can be removed by renal clearance and large IONPs are dissolved in acidic lysosomes, while the released iron is incorporated into ferritin protein [82].

Moreover, SPION are widely used as contrast agents for MRI [83]. SPION are non-cytotoxic, biodegradable and easy to develop. Therefore, they are a good choice for drug delivery and diagnostic tools. These NPs are accepted, by FDA and European Council, for clinical use. SPION are negative contrast agent in MRI, they are quickly captured by macrophages that infiltrate the damaged organs in autoimmune disease, even in the absence of important vascular permeability, and in some stages of the autoimmune disease multiple sclerosis. SPION were also widely used to evaluate progression of transplants rejection response [84].

Renal injury caused during lupus disease (other autoimmune disease) results in glomerulonephritis, which is generally treated with immunosuppressive drugs. Nowadays MRI with SPION protocols display information about renal parenchyma and vessels as well as functional data. Because these NPs are small, they stay in colloidal solution and they could penetrate tissues, maintaining their physical characteristics allowing them measurable by radiological methods. These agents enable evaluate renal inflammation, which is generally diagnosed by a biopsy [85].

These NPs used as imaging tools improve the ability to optimize treatment of the patients using non-invasive methods to evaluate the disease. Recent advance in development of nano-sized contrast agents is a new perspective for diagnostic.

Regarding the therapeutic applications, IONPs have been approved by FDA as a stand-along drug, since they are capable of inducing an increase of IL-1 in the THP-1 macrophage cell line. Others, such as Zn oxide NPs, present anti-inflammatory effects in view of the fact that reduce IL-1 and IL-6 in RAW264.7 cells line. In contrast, cobalt oxide NPs increases the production of TNF-α in peripheral macrophages in an animal model [86].

In fact, Nickel NPs have been reported as activators through TLR to trigger inflammation responses in innate immunity. Moreover, Zn oxide NPs modulate immune response via TLRs, inducing proinflammatory cytokines in human macrophage U-937 cell line [87]. So, these kinds of MNPs must be used with precaution because they can generate adverse effects on patients with autoimmune disease.

Finally, Gold NPs have been evaluated for different autoimmune disease; including SLE, rheumatoid arthritis, multiple sclerosis and celiac disease. There are contradictory reports about their possible pro-inflammatory or anti-inflammatory effects. Some of them demonstrated that these NPs could transient increase TNF-α, IL-6 and IL-1 gene expression (pro-inflammatory molecules) in animal models, independent of their size. Nevertheless, Gold complexes have been usually used as an important tool for rheumatoid arthritis since them have immune-suppressive effect inhibiting the production of the pro-inflammatory cytokines [88].

Definitively, MNPs are a good approach for rheumatoid arthritis disease, since they are an option for delivery therapeutic agents using as a magnetic probe. They are able to targeting the local lesion more accurately without side effects [55] but much remains to be studied of these NPs regarding their possible adverse effects in patients with autoimmune diseases and even more so in their potential use for treating them.

Conclusion

NPMs are promising tools, mainly for the treatment of cancer, infectious diseases among others, in addition to the diagnosis of numerous pathologies. But there is still much to study to understand the real impact of these NPs on human health. The effects that the different MNPs cause in macrophages or the mechanisms that are triggered have not yet been fully clarified [89]. MNPs, and specifically IONPs, have been described as useful tools for biomedical purposes given their magnetic properties, stability and biocompatibility [90]. MNPs have the capacity to accumulate fundamentally in macrophages, which is why they are considered as a potential therapeutic tool against cancer [91]. But, it was found that after phagocytosis, IONPs are degraded in lysosomes participating in different activation pathways and entering the recycling route [92]. Among MNPs, only IONPs have been approved by the FDA for the treatment of iron deficiency. Finally, the biological effects of NPs (absorption, biodistribution, cellular internalization and degradation) are related to their size, surface charge, shape, chemical composition so, much remains to be learned about the interaction of these NPs with the immune system and to fully understand their potential as a therapeutic tool [93–95].

References

[1] O.C. Farokhzad, R. Langer, Impact of nanotechnology on drug delivery, ACS Nano. 3 (2009) 16–20. https://doi.org/10.1021/nn900002m

[2] S. Bayda, M. Adeel, T. Tuccinardi, M. Cordani, F. Rizzolio, The history of nanoscience and nanotechnology: From chemical-physical applications to nanomedicine, Molecules. 25 (2020) 1–15. https://doi.org/10.3390/molecules25010112

[3] K. Wu, D. Su, J. Liu, R. Saha, J.P. Wang, Magnetic nanoparticles in nanomedicine: A review of recent advances, Nanotechnology. 30 (2019). https://doi.org/10.1088/1361-6528/ab4241

[4] T. Vangijzegem, D. Stanicki, S. Laurent, Magnetic iron oxide nanoparticles for drug delivery: applications and characteristics, Expert Opin. Drug Deliv. 16 (2019) 69–78. https://doi.org/10.1080/17425247.2019.1554647

[5] A. Farzin, S.A. Etesami, J. Quint, A. Memic, A. Tamayol, Magnetic Nanoparticles in Cancer Therapy and Diagnosis, Adv. Healthc. Mater. 9 (2020) 1–29. https://doi.org/10.1002/adhm.201901058

[6] R. Medzhitov, C.A. Janeway, Innate immunity: Impact on the adaptive immune response, Curr. Opin. Immunol. 9 (1997) 4–9. https://doi.org/10.1016/S0952-7915(97)80152-5

[7] L.B. Nicholson, The immune system, Essays Biochem. 60 (2016) 275–301. https://doi.org/10.1042/EBC20160017

[8] M.F. Flajnik, L. Du Pasquier, Evolution of innate and adaptive immunity: can we draw a line?, Trends Immunol. 25 (2004) 640–644. https://doi.org/https://doi.org/10.1016/j.it.2004.10.001

[9] N.I. Nii-Trebi, Emerging and Neglected Infectious Diseases: Insights, Advances, and Challenges, Biomed Res. Int. 2017 (2017). https://doi.org/10.1155/2017/5245021

[10] National Institutes of Health (US), Biological Sciences Curriculum Study, Understanding Emerging and Re-emerging Infectious Diseases, in: NIH Curric. Suppl. Ser., 2007. https://www.ncbi.nlm.nih.gov/books/NBK20370/

[11] B.A. Aderibigbe, Metal-based nanoparticles for the treatment of infectious diseases, Molecules. 22 (2017). https://doi.org/10.3390/molecules22081370

[12] World Health Organization, Antibiotic resistance, (2020). https://www.who.int/news-room/fact-sheets/detail/antibiotic-resistance

[13] M.A.A. Majumder, S. Rahman, D. Cohall, A. Bharatha, K. Singh, M. Haque, M. Gittens-St Hilaire, Antimicrobial stewardship: Fighting antimicrobial resistance and protecting global public health, Infect. Drug Resist. 13 (2020) 4713–4738. https://doi.org/10.2147/IDR.S290835

[14] R. Costo, M.P. Morales, S. Veintemillas-Verdaguer, Improving magnetic properties of ultrasmall magnetic nanoparticles by biocompatible coatings, J. Appl. Phys. 117 (2015) 1–8. https://doi.org/10.1063/1.4908132

[15] P. Dutta, M.S. Seehra, S. Thota, J. Kumar, A comparative study of the magnetic properties of bulk and nanocrystalline Co3O4, J. Phys. Condens. Matter. 20 (2008). https://doi.org/10.1088/0953-8984/20/01/015218

[16] S.M. Devi, A. Nivetha, I. Prabha, Superparamagnetic Properties and Significant Applications of Iron Oxide Nanoparticles for Astonishing Efficacy—a Review, J. Supercond. Nov. Magn. 32 (2019) 127–144. https://doi.org/10.1007/s10948-018-4929-8

[17] L. Xiang, O.U. Akakuru, C. Xu, A. Wu, Harnessing the intriguing properties of magnetic nanoparticles to detect and treat bacterial infections, Magnetochemistry. 7 (2021) 1–13. https://doi.org/10.3390/magnetochemistry7080112

[18] Z. Xiao, Q. Zhang, X. Guo, J. Villanova, Y. Hu, I. Külaots, D. Garcia-Rojas, W. Guo, V.L. Colvin, Libraries of Uniform Magnetic Multicore Nanoparticles with Tunable Dimensions for Biomedical and Photonic Applications, ACS Appl. Mater. Interfaces. 12 (2020) 41932–41941. https://doi.org/10.1021/acsami.0c09778

[19] S. Maiti, K.K. Sen, Introductory Chapter: Drug Delivery Concepts, Adv. Technol. Deliv. Ther. (2017) 1–12. https://doi.org/10.5772/65245

[20] M. Torrice, Does nanomedicine have a delivery problem?, ACS Cent. Sci. 2 (2016) 434–437. https://doi.org/10.1021/acscentsci.6b00190

[21] M.W. Freeman, A. Arrott, J.H.L. Watson, Magnetism in medicine, J. Appl. Phys. 404 (1960) 127–129. https://doi.org/10.1063/1.1984765

[22] H. Wen, H. Jung, X. Li, Drug Delivery Approaches in Addressing Clinical Pharmacology-Related Issues: Opportunities and Challenges, AAPS J. 17 (2015) 1327–1340. https://doi.org/10.1208/s12248-015-9814-9

[23] D.D. Stueber, J. Villanova, I. Aponte, Z. Xiao, V.L. Colvin, Magnetic nanoparticles in biology and medicine: Past, present, and future trends, Pharmaceutics. 13 (2021). https://doi.org/10.3390/pharmaceutics13070943

[24] S. Mura, J. Nicolas, P. Couvreur, Stimuli-responsive nanocarriers for drug delivery, Nat. Mater. 12 (2013) 991–1003. https://doi.org/10.1038/nmat3776

[25] J. Chen, S.M. Andler, J.M. Goddard, S.R. Nugen, V.M. Rotello, Integrating recognition elements with nanomaterials for bacteria sensing, Chem. Soc. Rev. 46 (2017) 1272–1283. https://doi.org/10.1039/c6cs00313c

[26] K.E. Albinali, M.M. Zagho, Y. Deng, A.A. Elzatahry, A perspective on magnetic core–shell carriers for responsive and targeted drug delivery systems, Int. J. Nanomedicine. 14 (2019) 1707–1723. https://doi.org/10.2147/IJN.S193981

[27] W. Cai, X. Weng, W. Zhang, Z. Chen, Green magnetic nanomaterial as antibiotic release vehicle: The release of pefloxacin and ofloxacin, Mater. Sci. Eng. C. 118 (2021) 111439. https://doi.org/10.1016/j.msec.2020.111439

[28] A. Mohapatra, M.A. Harris, D. LeVine, M. Ghimire, J.A. Jennings, B.I. Morshed, W.O. Haggard, J.D. Bumgardner, S.R. Mishra, T. Fujiwara, Magnetic stimulus responsive vancomycin drug delivery system based on chitosan microbeads embedded with magnetic nanoparticles, J. Biomed. Mater. Res. - Part B Appl. Biomater. 106 (2018) 2169–2176. https://doi.org/10.1002/jbm.b.34015

[29] L. de A.S. de Toledo, H.C. Rosseto, M.L. Bruschi, Iron oxide magnetic nanoparticles as antimicrobials for therapeutics, Pharm. Dev. Technol. 23 (2018) 316–323. https://doi.org/10.1080/10837450.2017.1337793

[30] E.N. Taylor, T.J. Webster, The use of superparamagnetic nanoparticles for prosthetic biofilm prevention., Int. J. Nanomedicine. 4 (2009) 145–152. https://doi.org/10.2147/ijn.s5976

[31] G. Subbiahdoss, S. Sharifi, D.W. Grijpma, S. Laurent, H.C. Van Der Mei, M. Mahmoudi, H.J. Busscher, Magnetic targeting of surface-modified superparamagnetic iron oxide nanoparticles yields antibacterial efficacy against biofilms of gentamicin-resistant staphylococci, Acta Biomater. 8 (2012) 2047–2055. https://doi.org/10.1016/j.actbio.2012.03.002

[32] C. Xu, O.U. Akakuru, J. Zheng, A. Wu, Applications of iron oxide-based magnetic nanoparticles in the diagnosis and treatment of bacterial infections, Front. Bioeng. Biotechnol. 7 (2019) 1–15. https://doi.org/10.3389/fbioe.2019.00141

[33] B.M. Geilich, I. Gelfat, S. Sridhar, A.L. van de Ven, T.J. Webster, Superparamagnetic iron oxide-encapsulating polymersome nanocarriers for biofilm eradication, Biomaterials. 119 (2017) 78–85. https://doi.org/10.1016/j.biomaterials.2016.12.011

[34] H. Maleki, A. Rai, S. Pinto, M. Evangelista, R.M.S. Cardoso, C. Paulo, T. Carvalheiro, A. Paiva, M. Imani, A. Simchi, L. Durães, A. Portugal, L. Ferreira, High Antimicrobial Activity and Low Human Cell Cytotoxicity of Core-Shell Magnetic Nanoparticles Functionalized with an Antimicrobial Peptide, ACS Appl. Mater. Interfaces. 8 (2016) 11366–11378. https://doi.org/10.1021/acsami.6b03355

[35] M. Chakravarty, A. Vora, Nanotechnology-based antiviral therapeutics, Drug Deliv. Transl. Res. 11 (2021) 748–787. https://doi.org/10.1007/s13346-020-00818-0

[36] K. Maduray, R. Parboosing, Metal Nanoparticles: a Promising Treatment for Viral and Arboviral Infections, Biol. Trace Elem. Res. 199 (2021) 3159–3176. https://doi.org/10.1007/s12011-020-02414-2

[37] S.J. Park, H.H. Park, S.Y. Kim, S.J. Kim, K. Woo, G.P. Ko, Antiviral properties of silver nanoparticles on a magnetic hybrid colloid, Appl. Environ. Microbiol. 80 (2014) 2343–2350. https://doi.org/10.1128/AEM.03427-13

[38] R.D. Jayant, Layer-by-Layer (LbL) assembly of anti HIV drug for sustained release to brain using magnetic nanoparticle, J. Neuroimmune Pharmacol. 9 (2014)

[39] R.D. Jayant, V.S.R. Atluri, M. Agudelo, V. Sagar, A. Kaushik, M. Nair, Sustained-release nanoART formulation for the treatment of neuroAIDS, Int. J. Nanomedicine. 10 (2015) 1077–1093. https://doi.org/10.2147/IJN.S76517

[40] R. Medhi, P. Srinoi, N. Ngo, H.V. Tran, T.R. Lee, Nanoparticle-Based Strategies to Combat COVID-19, ACS Appl. Nano Mater. 3 (2020) 8557–8580. https://doi.org/10.1021/acsanm.0c01978

[41] L. Bromberg, D.J. Bromberg, T.A. Hatton, I. Bandín, A. Concheiro, C. Alvarez-Lorenzo, Antiviral properties of polymeric aziridine- and biguanide-modified core-shell magnetic nanoparticles, Langmuir. 28 (2012) 4548–4558. https://doi.org/10.1021/la205127x

[42] K. Niemirowicz, B. Durnaś, G. Tokajuk, K. Głuszek, A.Z. Wilczewska, I. Misztalewska, J. Mystkowska, G. Michalak, A. Sodo, M. Wątek, B. Kiziewicz, S. Góźdź, S. Głuszek, R. Bucki, Magnetic nanoparticles as a drug delivery system that enhance fungicidal activity of polyene antibiotics, Nanomedicine Nanotechnology, Biol. Med. 12 (2016) 2395–2404. https://doi.org/10.1016/j.nano.2016.07.006

[43] M. Bushman, L. Morton, N. Duah, N. Quashie, B. Abuaku, K.A. Koram, P.R. Dimbu, M. Plucinski, J. Gutman, P. Lyaruu, S. Patrick Kachur, J.C. de Roode, V. Udhayakumar, Within-host competition and drug resistance in the human malaria parasite plasmodium falciparum, Proc. R. Soc. B Biol. Sci. 283 (2016). https://doi.org/10.1098/rspb.2015.3038

[44] X. Wang, Y. Xie, N. Jiang, J. Wang, H. Liang, D. Liu, N. Yang, X. Sang, Y. Feng, R. Chen, Q. Chen, Enhanced Antimalarial Efficacy Obtained by Targeted Delivery of Artemisinin in Heparin-Coated Magnetic Hollow Mesoporous Nanoparticles, ACS Appl. Mater. Interfaces. 13 (2021) 287–297. https://doi.org/10.1021/acsami.0c20070

[45] M.H. Kim, I. Yamayoshi, S. Mathew, H. Lin, J. Nayfach, S.I. Simon, Magnetic nanoparticle targeted hyperthermia of cutaneous staphylococcus aureus infection, Ann. Biomed. Eng. 41 (2013) 598–609. https://doi.org/10.1007/s10439-012-0698-x

[46] C. Chen, L. Chen, Y. Yi, C. Chen, L.F. Wu, T. Song, Killing of Staphylococcus aureus via magnetic hyperthermia mediated by magnetotactic bacteria, Appl. Environ. Microbiol. 82 (2016) 2219–2226. https://doi.org/10.1128/AEM.04103-15

[47] T.J. Yu, P.H. Li, T.W. Tseng, Y.C. Chen, Multifunctional Fe 3O 4/alumina core/shell MNPs as photothermal agents for targeted hyperthermia of nosocomial and antibiotic-resistant bacteria, Nanomedicine. 6 (2011) 1353–1363. https://doi.org/10.2217/nnm.11.34

[48] B. Chudzik, A. Miaskowski, Z. Surowiec, G. Czernel, T. Duluk, A. Marczuk, M. Gagoś, Effectiveness of magnetic fluid hyperthermia against Candida albicans cells, Int. J. Hyperth. 32 (2016) 842–857. https://doi.org/10.1080/02656736.2016.1212277

[49] S. Singh, K.C. Barick, D. Bahadur, Inactivation of bacterial pathogens under magnetic hyperthermia using Fe3O4-ZnO nanocomposite, Powder Technol. 269 (2015) 513–519. https://doi.org/10.1016/j.powtec.2014.09.032

[50] V. Grazú, A.M. Silber, M. Moros, L. Asín, T.E. Torres, C. Marquina, M.R. Ibarra, G.F. Goya, Application of magnetically induced hyperthermia in the model protozoan Crithidia fasciculata as a potential therapy against parasitic infections, Int. J. Nanomedicine. 7 (2012) 5351–5360. https://doi.org/10.2147/IJN.S35510

[51] S.L. Berry, K. Walker, C. Hoskins, N.D. Telling, H.P. Price, Nanoparticle-mediated magnetic hyperthermia is an effective method for killing the human-infective protozoan parasite Leishmania mexicana in vitro, Sci. Rep. 9 (2019) 1–9. https://doi.org/10.1038/s41598-018-37670-9

[52] A.F.R. Rodriguez, C.C. dos Santos, K. Lüdtke-Buzug, A.C. Bakenecker, Y.O. Chaves, L.A.M. Mariúba, J. V. Brandt, B.E. Amantea, R.C. de Santana, R.F.C. Marques, M. Jafelicci, M.A. Morales, Evaluation of antiplasmodial activity and cytotoxicity assays of amino acids functionalized magnetite nanoparticles: Hyperthermia and flow cytometry applications, Mater. Sci. Eng. C. 125 (2021). https://doi.org/10.1016/j.msec.2021.112097

[53] J. Mosayebi, M. Kiyasatfar, S. Laurent, Synthesis, functionalization, and design of magnetic nanoparticles for theranostic applications, 2017. https://doi.org/10.1002/adhm.201700306

[54] H.W. Cheng, H.Y. Tsao, C.S. Chiang, S.Y. Chen, Advances in Magnetic Nanoparticle-Mediated Cancer Immune-Theranostics, Adv. Healthc. Mater. 10 (2021) 1–20. https://doi.org/10.1002/adhm.202001451

[55] Y. Liu, F. Cao, B. Sun, J.A. Bellanti, S.G. Zheng, Magnetic nanoparticles: A new diagnostic and treatment platform for rheumatoid arthritis, J. Leukoc. Biol. 109 (2021) 415–424. https://doi.org/10.1002/JLB.5MR0420-008RR

[56] J. Gao, H. Gu, B. Xu, Multifunctional magnetic nanoparticles: Synthesis modification and biomedical applications, Acc. Chem. Res. 42 (2009) 1097–1107. https://doi.org/10.1021/ar9000026

[57] M.J. Gorbet, A. Singh, C. Mao, S. Fiering, A. Ranjan, Using nanoparticles for in situ vaccination against cancer: mechanisms and immunotherapy benefits, Int. J. Hyperth. 37 (2020) 18–33. https://doi.org/10.1080/02656736.2020.1802519

[58] L. Rao, S.K. Zhao, C. Wen, R. Tian, L. Lin, B. Cai, Y. Sun, F. Kang, Z. Yang, L. He, J. Mu, Q.F. Meng, G. Yao, N. Xie, X. Chen, Activating Macrophage-Mediated Cancer Immunotherapy by Genetically Edited Nanoparticles, Adv. Mater. 32 (2020) 1–9. https://doi.org/10.1002/adma.202004853

[59] Q. Jiang, K. Wang, X. Zhang, B. Ouyang, H. Liu, Z. Pang, W. Yang, Platelet Membrane-Camouflaged Magnetic Nanoparticles for Ferroptosis-Enhanced Cancer Immunotherapy, Small. 16 (2020) 1–17. https://doi.org/10.1002/smll.202001704

[60] A.I. Bocanegra Gondan, A. Ruiz-de-Angulo, A. Zabaleta, N. Gómez Blanco, B.M. Cobaleda-Siles, M.J. García-Granda, D. Padro, J. Llop, B. Arnaiz, M. Gato, D. Escors, J.C. Mareque-Rivas, Effective cancer immunotherapy in mice by polyIC-imiquimod complexes and engineered magnetic nanoparticles, Biomaterials. 170 (2018) 95–115. https://doi.org/10.1016/j.biomaterials.2018.04.003

[61] T.S.C. Ng, V. Gunda, R. Li, M. Prytyskach, Y. Iwamoto, R.H. Kohler, S. Parangi, R. Weissleder, M.A. Miller, Detecting immune response to therapies targeting PDL1 and BRAF by using ferumoxytol MRI and macrin in anaplastic thyroid cancer, Radiology. 298 (2020) 123–132. https://doi.org/10.1148/RADIOL.2020201791

[62] C.S. Chiang, Y.J. Lin, R. Lee, Y.H. Lai, H.W. Cheng, C.H. Hsieh, W.C. Shyu, S.Y. Chen, Combination of fucoidan-based magnetic nanoparticles and immunomodulators enhances tumour-localized immunotherapy, Nat. Nanotechnol. 13 (2018) 746–754. https://doi.org/10.1038/s41565-018-0146-7

[63] A.J. Grippin, B. Wummer, T. Wildes, K. Dyson, V. Trivedi, C. Yang, M. Sebastian, H.R. Mendez-Gomez, S. Padala, M. Grubb, M. Fillingim, A. Monsalve, E.J. Sayour, J. Dobson, D.A. Mitchell, Dendritic Cell-Activating Magnetic Nanoparticles Enable Early Prediction of Antitumor Response with Magnetic Resonance Imaging, ACS Nano. 13 (2019) 13884–13898. https://doi.org/10.1021/acsnano.9b05037

[64] B. Yu, B. Choi, W. Li, D.H. Kim, Magnetic field boosted ferroptosis-like cell death and responsive MRI using hybrid vesicles for cancer immunotherapy, Nat. Commun. 11 (2020). https://doi.org/10.1038/s41467-020-17380-5

[65] H. Lei, Y. Pan, R. Wu, Y. Lv, Innate Immune Regulation Under Magnetic Fields With Possible Mechanisms and Therapeutic Applications, Front. Immunol. 11 (2020) 1–10. https://doi.org/10.3389/fimmu.2020.582772

[66] K. Abdul, A.H.L. Abbas, P. Shiv, Cellular and Molecular Immunology, 8th ed., Saunders Elsevier, Philadelphia, PA, USA, 2015

[67] M. Ilves, J. Palomäki, M. Vippola, M. Lehto, K. Savolainen, T. Savinko, H. Alenius, Topically applied ZnO nanoparticles suppress allergen induced skin inflammation but induce vigorous IgE production in the atopic dermatitis mouse model, Part. Fibre Toxicol. 11 (2014) 1–12. https://doi.org/10.1186/s12989-014-0038-4

[68] H.C. Chuang, T.C. Hsiao, C.K. Wu, H.H. Chang, C.H. Lee, C.C. Chang, T.J. Cheng, Allergenicity and toxicology of inhaled silver nanoparticles in allergen-provocation mice models, Int. J. Nanomedicine. 8 (2013) 4495–4506. https://doi.org/10.2147/IJN.S52239

[69] T. Hirai, Y. Yoshioka, H. Takahashi, K. Ichihashi, T. Yoshida, S. Tochigi, K. Nagano, Y. Abe, H. Kamada, S. Tsunoda, H. Nabeshi, T. Yoshikawa, Y. Tsutsumi, Amorphous silica nanoparticles enhance cross-presentation in murine dendritic cells, Biochem. Biophys. Res. Commun. 427 (2012) 553–556. https://doi.org/https://doi.org/10.1016/j.bbrc.2012.09.095

[70] S. Hussain, J.A.J. Vanoirbeek, K. Luyts, V. De Vooght, E. Verbeken, L.C.J. Thomassen, J.A. Martens, D. Dinsdale, S. Boland, F. Marano, B. Nemery, P.H.M. Hoet, Lung exposure to nanoparticles modulates an asthmatic response in a mouse model, Eur. Respir. J. 37 (2011) 299–309. https://doi.org/10.1183/09031936.00168509

[71] M. Lu, M.H. Cohen, D. Rieves, R. Pazdur, FDA report: Ferumoxytol for intravenous iron therapy in adult patients with chronic kidney disease, Am. J. Hematol. 85 (2010) 315–319. https://doi.org/10.1002/ajh.21656

[72] Y.-X.J. Wang, Superparamagnetic iron oxide based MRI contrast agents: Current status of clinical application., Quant. Imaging Med. Surg. 1 (2011) 35–40. https://doi.org/10.3978/j.issn.2223-4292.2011.08.03

[73] T. Fülöp, R. Nemes, T. Mészáros, R. Urbanics, R.J. Kok, J.A. Jackman, N. Cho, G. Storm, J. Szebeni, Complement activation in vitro and reactogenicity of low-molecular weight dextran-coated SPIONs in the pig CARPA model: Correlation with physicochemical features and clinical information, J. Control. Release. 270 (2018) 268–274. https://doi.org/10.1016/j.jconrel.2017.11.043

[74] N.B. Alsaleh, J.M. Brown, Engineered Nanomaterials and Type I Allergic Hypersensitivity Reactions, Front. Immunol. 11 (2020) 1–14. https://doi.org/10.3389/fimmu.2020.00222

[75] F. Blank, P. Gerber, B. Rothen-Rutishauser, U. Sakulkhu, J. Salaklang, K. De Peyer, P. Gehr, L.P. Nicod, H. Hofmann, T. Geiser, A. Petri-Fink, C. Von Garnier, Biomedical nanoparticles modulate specific CD4 + T cell stimulation by inhibition of antigen processing in dendritic cells, Nanotoxicology. 5 (2011) 606–621. https://doi.org/10.3109/17435390.2010.541293

[76] L. Johnson, A. Duschl, M. Himly, Nanotechnology-based vaccines for allergen-specific immunotherapy: Potentials and challenges of conventional and novel adjuvants under research, Vaccines. 8 (2020). https://doi.org/10.3390/vaccines8020237

[77] B. Teste, F. Malloggi, J.M. Siaugue, A. Varenne, F. Kanoufi, S. Descroix, Microchip integrating magnetic nanoparticles for allergy diagnosis, Lab Chip. 11 (2011) 4207–4213. https://doi.org/10.1039/c1lc20809h

[78] X. Han, M. Cao, B. Zhou, C. Yu, Y. Liu, B. Peng, L. Meng, J.F. Wei, L. Li, W. Huang, Specifically immobilizing His-tagged allergens to magnetic nanoparticles for fast and quantitative detection of allergen-specific IgE in serum samples, Talanta. 219 (2020) 121301. https://doi.org/10.1016/j.talanta.2020.121301

[79] M. Cao, Y. Liu, C. Lu, M. Guo, L. Li, C. Yu, J.F. Wei, Ultrasensitive detection of specific IgE based on nanomagnetic capture and separation with a AuNP-anti-IgE nanobioprobe for signal amplification, Anal. Methods. 13 (2021) 2478–2484. https://doi.org/10.1039/d1ay00372k

[80] S. Ashraf, S. Qadri, B. Al-Ramadi, Y. Haik, Nanoparticles rapidly assess specific IgE in plasma, Nanotechnology. 23 (2012). https://doi.org/10.1088/0957-4484/23/30/305101

[81] Y.K. Hahn, Z. Jin, J.H. Kang, E. Oh, M.K. Han, H.S. Kim, J.T. Jang, J.H. Lee, J. Cheon, S.H. Kim, H.S. Park, J.K. Park, Magnetophoretic immunoassay of allergen-specific IgE in an enhanced magnetic field gradient, Anal. Chem. 79 (2007) 2214–2220. https://doi.org/10.1021/ac0615221

[82] L. Wu, S. Shen, What potential do magnetic iron oxide nanoparticles have for the treatment of rheumatoid arthritis?, Nanomedicine. 14 (2019) 927–930. https://doi.org/10.2217/nnm-2019-0071

[83] A.A. Yetisgin, S. Cetinel, M. Zuvin, A. Kosar, O. Kutlu, Therapeutic Nanoparticles and Their Targeted Delivery Applications, Molecules. 25 (2020) 1–31. https://doi.org/10.3390/molecules25092193

[84] X. Clemente-Casares, P. Santamaria, Nanomedicine in autoimmunity, Immunol. Lett. 158 (2014) 167–174. https://doi.org/10.1016/j.imlet.2013.12.018

[85] J.M. Thurman, N.J. Serkova, Nanosized Contrast Agents to Noninvasively Detect Kidney Inflammation by Magnetic Resonance Imaging, Adv. Chronic Kidney Dis. 20 (2013) 488–499. https://doi.org/10.1053/j.ackd.2013.06.001

Materials Research Forum LLC
https://doi.org/10.21741/9781644902332-9

[86] M.S. Dukhinova, A.Y. Prilepskii, V. V. Vinogradov, A.A. Shtil, Metal oxide nanoparticles in therapeutic regulation of macrophage functions, Nanomaterials. 9 (2019) 1–20. https://doi.org/10.3390/nano9111631

[87] Y.H. Luo, L.W. Chang, P. Lin, Metal-Based Nanoparticles and the Immune System: Activation, Inflammation, and Potential Applications, Biomed Res. Int. 2015 (2015). https://doi.org/10.1155/2015/143720

[88] K. Koushki, S.K. Shahbaz, M. Keshavarz, E.E. Bezsonov, T. Sathyapalan, A. Sahebkar, Gold nanoparticles: Multifaceted roles in the management of autoimmune disorders, 2021. https://doi.org/10.3390/biom11091289

[89] H. Ying, Y. Ruan, Z. Zeng, Y. Bai, J. Xu, S. Chen, Iron oxide nanoparticles size-dependently activate mouse primary macrophages via oxidative stress and endoplasmic reticulum stress, Int. Immunopharmacol. 105 (2022) 108533. https://doi.org/10.1016/j.intimp.2022.108533

[90] A.H. Lu, E.L. Salabas, F. Schüth, Magnetic nanoparticles: Synthesis, protection, functionalization, and application, Angew. Chemie - Int. Ed. 46 (2007) 1222–1244. https://doi.org/10.1002/anie.200602866

[91] D. Reichel, M. Tripathi, J.M. Perez, Biological effects of nanoparticles on macrophage polarization in the tumor microenvironment, Nanotheranostics. 3 (2019) 66–88. https://doi.org/10.7150/ntno.30052

[92] V. Mulens-Arias, J.M. Rojas, D.F. Barber, The Use of Iron Oxide Nanoparticles to Reprogram Macrophage Responses and the Immunological Tumor Microenvironment, Front. Immunol. 12 (2021). https://doi.org/10.3389/fimmu.2021.693709

[93] L. Liu, R. Sha, L. Yang, X. Zhao, Y. Zhu, J. Gao, Y. Zhang, L.P. Wen, Impact of Morphology on Iron Oxide Nanoparticles-Induced Inflammasome Activation in Macrophages, ACS Appl. Mater. Interfaces. 10 (2018) 41197–41206. https://doi.org/10.1021/acsami.8b17474

[94] Y. Zhang, Z. Zhang, The history and advances in cancer immunotherapy: understanding the characteristics of tumor-infiltrating immune cells and their therapeutic implications, Cell. Mol. Immunol. 17 (2020) 807–821. https://doi.org/10.1038/s41423-020-0488-6

[95] C.S. Nascimento, É.A.R. Alves, C.P. De Melo, R. Corrêa-Oliveira, C.E. Calzavara-Silva, Immunotherapy for cancer: Effects of iron oxide nanoparticles on polarization of tumor-associated macrophages, Nanomedicine. 16 (2021) 2633–2650. https://doi.org/10.2217/nnm-2021-0255

Materials Research Forum LLC
https://doi.org/10.21741/9781644902332-10

Chapter 10

Magnetic Nanoparticles for Nucleic Acid Delivery: Magnetofection, Gene Therapy and Vaccines

María V. Tuttolomondo[1], Sofia Municoy[1], María I. Alvarez Echazú[1][2], Lurdes M. López[1], Gisela S. Alvarez [1*]

[1] Universidad de Buenos Aires, Consejo Nacional de Investigaciones Científicas y Técnicas (CONICET). Instituto de Química y Metabolismo del Fármaco (IQUIMEFA), Facultad de Farmacia y Bioquímica, Junín 956, Piso 3°, (1113) Buenos Aires, Argentina

[2] Universidad de Buenos Aires, Facultad de Odontología, Cátedra de Anatomía Patológica, Marcelo T. de Alvear 2142 (1122), CABA, Argentina

* gialvarez@ffyb.uba.ar

Abstract

Gene therapy offers an alternative for the treatment of diseases such as cancer or neurodegenerative diseases by replacing, introducing or inactivating genes in the patient. In this aspect, the use of nanoparticles as nucleic acid delivery systems that prevents its degradation and facilitates its incorporation into target cells has been a subject of intense study in recent years. Among them, magnetic nanoparticles (MNP) offer advantages as new alternatives for non-viral transfection, such as guiding the nanoparticle and its content through magnetic fields towards target organs, increasing the efficiency and reducing transfection times. The use of MNP carrying genetic information to achieve transfection by the application of magnetic fields is known as magnetofection. In most cases, superparamagnetic iron oxide nanoparticles (SPIONs) are used as vehicles which also contain a cationic organic coating to increase their stability, gene incorporation and interaction with cell membranes. Another field in which the application of these new technologies is gaining attention is in the development of DNA and RNA- based vaccines for immunization and immunotherapy. In the following chapter, the use of magnetofection for *in vitro* experiments as well as the study of *in vivo* vaccine assays or gene therapy using MNP as nucleic acid carriers, will be discussed.

Keywords

Nanoparticles, Magnetofection, Vaccines, Gene Therapy, Immunotherapy

Contents

1.   Introduction

Magnetic nanoparticles (MNPs) are widely used for drug delivery, enzyme immobilization, and many other interesting biotechnological applications. They possess desirable properties like high surface area, size uniformity, biocompatibility, superparamagnetism, adsorption kinetics, and a magnetic moment that can be employed for several purposes. Among them, iron oxide nanoparticles are extensively studied since they do not retain any magnetization once the magnetic field is removed. This particularity has led to the application of iron oxide nanoparticles (Fe_2O_3 and Fe_3O_4) for biomedical applications. In addition, their easy functionalization with polymers represents another advantage [1]. Materials are classified according to their response to an external magnetic field in six basic forms of magnetism: diamagnetic, paramagnetic, superparamagnetic, ferromagnetic, ferrimagnetic and antiferromagnetic. However, superparamagnetic iron oxide nanoparticles (SPIONs) are the most extensively used for therapeutic applications due to the previously discussed advantage [2]. In the last decade, SPIONs have emerged in the medical and clinical field in a variety of formulations developed for both imaging and therapeutic purposes. They consist of cores made of iron oxides that can be targeted to a desired area by external magnets. They show attractive properties such as superparamagnetism, high saturation field, extra anisotropy contributions or shifted loops after field cooling. Because of this, the nanoparticles no longer show magnetic interaction once the external magnetic field is removed. Magnetite (Fe_3O_4), maghemite (γ-Fe_2O_3) and hematite (α-Fe_2O_3) are three main iron oxides that are classified under the category of SPIONs. Ferrites, which are mixed oxides of iron and other transition metal ions (i.e., Cu, Co, Mn and Ni), have also been reported as superparamagnetic [3]. Besides, iron oxides became the most popular materials due to their high chemical/colloidal stability and biocompatibility. On the other hand, pure metals like Iron, Nickel or Cobalt, which possess high magnetic moment and saturation magnetizations, are not chosen for *in vivo* applications because of their oxidative characteristics and high toxicity levels.

Consequently, iron oxides, despite their lower magnetization, offer a great potential to obtain more stable particles against oxidation [4].

The selection of the magnetic nanoparticles' synthesis method is also important since it determines the size, shape, surface properties and, in consequence, their applications. Iron oxides can exist in various polymorphs of which α-Fe_2O_3 (hematite), γ-Fe_2O_3, β-Fe_2O_3 (maghemite), and γ-Fe_2O_3, Fe_3O_4 (magnetite) are crystalline. The other forms are amorphous and generally exist at high pressure. To reduce toxicity and optimally accomplish their biomedical functions, these nanoparticles are usually functionalized with proteins, silica, polymers, surfactants, and organic materials. The iron oxide nanoparticles which are not functionalized generally degrade and leach. Because of this, they are not that biocompatible. Furthermore, they also easily agglomerate due to the magnetic dipole-dipole attraction, leading to a poor stability and dispersity. Hence, their use is rather limited [5].

As discussed, better colloidal stability, optimized suspension properties, or improved biocompatibility can be achieved through a wide range of strategies. Today, magnetic nanoparticles are usually protected with a layer composed by polymers, ceramics, or surfactants [6]. Furthermore, the coverage of magnetic nanoparticles with a shell presents other advantages, for example, achieving an optimized functionalization and conjugation to proteins, enzymes, or antibodies or even the possibility to obtain engineered anticancer drug delivery platforms. And, naturally, the accurate and precise monitoring over every step of the magnetic nanoparticles' synthesis process and their successful surface functionalization are mandatory, because these details will clearly define their chemical properties, their physical behavior, their colloidal stability, and, eventually, the biomedical performance [4,7]. For instance, it was reported that the surface of magnetic particles has been covalently functionalized with biodegradable and biocompatible polymers such as polysaccharides and linoleic acid [1], coated with citric acid [8], chitosan [9], SiO_2 [10] or biocompatible surfactants [11]. Polyethyleneimine (PEI) is commonly used to coat MNPs, especially when they are going to deliver genetic material for cell transfection, as it not only serves to stabilize the particles but also to increase the efficiency of gene delivery [12,13].

Other approaches to improve the clinical application of magnetic nanoparticles may involve a post-synthetic modification of the magnetic nanoparticles by their functionalization via the attachment of targeting ligands/imaging molecules, therapeutic modalities loading, and the obtention of nanocomposites with other biocompatible materials. The post-synthetic modification depends on the specific therapeutic or diagnostic application. And, not surprisingly, functionalization influences the pharmacokinetics and biodistribution of magnetic nanoparticles in biological fluids. For biomedical applications, various components are surface-anchored or assembled into the magnetic nanoparticles such as drugs, genes and viruses. Due to their inherent physicochemical properties, magnetic nanoparticles are usually stabilized by the coating with an inert and biocompatible material that provides core material protection and enhances aqueous solubility, crucial for biomedical applications. As previously mentioned, targeting ligands could also be anchored on the magnetic nanoparticle surface. Thus, a specific drug may be delivered to a target area by the recognition of specific receptors or biomarkers. To increase the specificity of the nanoparticles for targeting, their surfaces are conjugated by targeting species, such as low molecular weight ligands (i.e., sugars, thiamine, and folic acid), peptides, polyunsaturated fatty acids, hyaluronic acid, DNA, and proteins (lectins, transferrin, and antibodies) [14]. In some cases, fluorophores could also be added to the particles facilitating imaging and diagnosis using diagnostic techniques.

Regarding magnetic nanoparticles synthesis, different preparation methods were reported for the obtention of magnetic nanostructures in the form of nanorods, nanocubes, nanowires, including iron oxide magnetic nanoparticles by wet chemistry or "bottom-up" routes such a hydrothermal [15], solvothermal [16], sol-gel [17], co-precipitation [18], flow injection synthesis, electrochemical [19], microemulsion synthesis [20] and laser pyrolysis techniques [21].

As mentioned, magnetic nanoparticles are promising nanostructures, because of their simple synthesis and the ability to be manipulated thanks to an external magnetic field. For these reasons, magnetic nanoparticles are widely employed in the development of separation methods. For instance, several separation methods employing magnetic materials have been reported based on the selective molecular adsorption to various types of magnetic particles. For example, some advantageous properties of magnetic nanoparticles for protein purification from their natural and laboratory sources are the low remanence and costs without any harmful environmental effects. Also, due to the efficiency of magnetic nanoparticles in the separation and purification of proteins, the use of magnetic nanoparticles provides a great speed in the separation process and a better accuracy when compared to the traditional methods such as chromatography and electrophoresis. Additionally, there is no need to pretreat samples with centrifugation or filtration [22]. Among all these applications, magnetic separation of nucleic acids has several advantages compared to other traditional techniques such as column chromatography or extraction with organic solvents. The magnetic separation method provides the possibility of direct isolation from crude samples. In consequence, this provides the basis of several automated low-to high-throughput procedures which saves time and resources. Although various types of iron oxide nanoparticles-based kits are commercially available for DNA isolation, different functionalized iron oxide nanoparticles can offer new opportunities for improving separation efficiency. DNA extraction and purification is considered a critical step for many different biomedical applications such as gene therapy and clinical diagnosis. Many works in the field describe the synthesis and characterization of functionalized iron oxide nanoparticles. For instance, Sosa-Acosta *et al.* synthesized iron oxide nanoparticles with potential application in plasmid DNA isolation by a chemical coprecipitation method followed by a post-synthesis functionalization using silica and (3-aminopropyl) triethoxysilane. Afterwards, a second functionalization strategy was carried out by an *in situ* coprecipitation of Fe(II) and Fe(III) ions in presence of chitosan and tris(hydroxymethyl)aminomethane [23]. Regarding RNA, another study worth to mention for illustrating the use of magnetic nanoparticles, was published by Ullah Khan *et al.* They reported the synthesis of zinc ferrite nanoparticles surface-functionalized with silica and carboxyl modified with polyvinyl alcohol. The authors obtained the zinc ferrite nanoparticles by the synthesis through combustion, and their surface was functionalized with carboxyl modified polyvinyl alcohol and silica. Their morphology was nanocrystalline and spherical. Recently, this protocol was used for the COVID-19 diagnosis at a molecular level. Thanks to these nanoparticles, it was possible the isolation from a specimen of viral RNA with an automation process [24].

Hyperthermia for localized cancer treatment is another application of magnetic nanoparticles. In this case, the particles are activated by means of an external alternating magnetic field producing an increase in temperature in the tumor tissues of 40-45° C which leads to cell death or sensibilization to other therapeutic agents [25,26]. Diagnosis and drug delivery are two important areas, with abundant information, where MNPs are being applied which are exhaustively reviewed in the works by Farzin *et al.* [27], Avasthi and co-workers [28] or Gholami *et al.* [29] and will not be included in this chapter.

Among biomedical applications of magnetic nanoparticles, magnetofection is quite popular. It is a highly effective transfection technique in which DNA vectorized magnetic nanoparticles are introduced into the cells by an external magnetic field. Besides, it is a simple and fast process that offers transfection up to saturation level at a low dose. For cell uptake, the magnetic particles possibly transfect nonpermissive cells, because they do not need receptors or other cell membrane proteins. The surface of the magnetic nanoparticles used for this objective are usually coated with polycationic polymers for the vectorization of negatively charged DNA. Both nonviral and viral vector magnetofection of nucleic acids significantly improve the vector competence [14]. Some other commonly used techniques for cell transfection that will be further compared with magnetofection within this chapter include: lipofection, which is a chemical methodology where a positively charged chemical makes nucleic acid/chemical complexes with negatively charged nucleic acids which are then attracted to the negatively charged cell membrane; micro injection, which is a time consuming method that directly injects nucleic acid into the cytoplasm or nucleus with high efficiency or electroporation where a short electrical pulse disturbs cell membranes and makes holes in the membrane through which nucleic acids can pass [30].

Iron oxide nanoparticles are also used for gene therapy purposes. For instance, these particles could improve gene therapy efficacy, since they could carry specific nucleic acids which regulate altered gene expression, basically in the form of plasmid DNA or siRNA. Moreover, an iron oxide nanoparticle–gene complex may have a longer circulation time when compared to a free gene, avoiding the potential damage caused by nucleases and, in consequence, perform the nucleic acid delivery to the organ of interest or tumor. Furthermore, to enhance the success of a gene therapy, the iron oxide nanoparticle–gene complexes can be attracted near the tumor by a magnet, thus increasing gene transfer into cells, via the previously mentioned technique called magnetofection. However, it must be said that, nowadays, clinical trials which evaluate gene therapy for some diseases as cancer have been unsuccessful due to the existence of a vast number of different genes responsible for it, depending on the type of disease, its progression and type of patient, which can barely all be targeted with only one gene therapeutic agent. Nowadays, further studies are needed to improve the gene therapy efficacy. Some of the following aspects seem to be attractive approaches involving the use of iron magnetic nanoparticles: perform a pre-screening of genes for each patient taken individually and adapt gene therapy according to the detected genes, utilize a combination of different genes instead of only one gene with the aim of increasing the possibilities of targeting genes, or unite gene therapy with another modality of treatment [31].

Despite the different and varied applications of MNPs in the biomedical field, we will restrict the next sections to the latest advances in the use of MNPs for the optimization of cell culture transfection, the *in vivo* application of gene therapy and the development of genetically based vaccines as well as immunotherapy treatments for cancer that will be presented and discussed.

2. *In vitro* magnetofection

In 1978, Widder *et al.* introduced the concept of Magnetic Drug Targeting (MDT), where they described how magnetic therapeutic compounds could be concentrated in a target area by using an external magnetic field [32], laying the foundation for the development of the magnetofection technique. Magnetofection is a technique that enhances the introduction of gene vectors (viral and nonviral) by the application of magnetic fields into cells [33–35]. This is accomplished by associating these vectors to MNPs and has the major advantage of exploiting natural uptake pathways (endocytic mechanisms of cells during the transfection process without disrupting the cell membrane), resulting in high cell viability post transfection [36]. Nowadays, many approaches have been used to try and improve the transfection ratio and the efficiency of expression of the transferred genes. Even more, magnetofection can be used to transfect multiple genes at once [37]. Another example is the use of magnetofection to produce viable models to study neurodegenerative diseases which was provided by Jacquier *et al.* [38]. Besides plasmid DNA, also small interfering RNA (siRNA), short hairpin RNA (shRNA) and antisense oligonucleotides can be used for the magnetofection of cells [39]. MicroRNAs (small non-coding RNA molecules that are approximately 21-23 nucleotides in length) can bind to specific sequences in target genes to induce Dicer-mediated cleavage and silence target gene expression [40] (Table 1).

Even when magnetofection technique can be used for both viral and nonviral gene vectors, in this chapter we will focus on the latter. Concerning viral vectors, we can mention, as examples, the work of Pereyra *et al.* [41] and Scherer *et al.* [34]. Magnetofection in combination with viral vectors is usually used when target cells do not express the viral receptor needed for efficient infection, whether because the cell differentiated or matured and lost the receptors, or because certain cell types do not express the required receptors for viral transfection. Many cell types are known to be hard to transfect, such as NIH-3T3 fibroblasts, breast and mouse melanoma cancer cells, human peripheral blood lymphocytes and mature skeletal muscle cells. The entry of wild and recombinant Adenovirus into cells takes two steps: the coxsackievirus and adenovirus receptor (CAR), which mediates the attachment of the virus to the cell surface, and lower-affinity secondary receptors (integrins) that allow the internalization of the viral particles. In Pereyra´s work, in order to transfect C2C12 myotubes with an adenoviral vector, iron oxide nanoparticles covered with 25-kDA branched polyethylenimine (PEI-25) were added to recombinant adenoviral vectors. The average diameter of the magnetite core was 9 nm and the saturation magnetization was 62 emu/g iron. An applied magnetic field allowed the cells to be transfected with green fluorescent protein and insulin-like growth factor 1, that had their expression enhanced 400% after magnetofection when compared to the cells exposed to the non-magnetic vector. Scherer's work compares the rate of transfection using non-viral, adenoviral and retroviral magnetofection. They used commercial MNPs coated with PEI which, when combined with the vector, resulted in particle sizes of 400 to 1000 nm at the time of administration. Their results show that the use of paramagnetic nanoparticles in combination with an applied magnetic field increased the expression of the reporter gene up to 1000 times when compared to the non-magnetic vectors under the same conditions.

The association of magnetic nanoparticles with transfection agents such as poly(ethylenimine) (PEI) [42] or hyperbranched poly(ethylenimine)s (HPEI) solutions accelerates the transfection process and improves the transfection efficiency [37,43–45]. HPEI is a cationic polymer that can condense pDNA into compact nanoparticles that possess a unique endosomolytic activity for effective DNA release inside cells. Furthermore, HPEI has plenty of amine groups which can bind iron, making the preparation of iron oxide nanoparticles possible. Finally, the basicity of HPEI is high enough for the synthesis of iron oxide. Taking these advantages into consideration, Shi *et al.* [45,46] worked with *in situ* synthesized magnetite/HPEI (Mag-HPEI) nanocomposites by changing the weight ratio of HPEI to $FeSO_4 \cdot 7H_2O$ and the molecular weight of HPEI (HPEI60k and HPEI10k) and evaluated the composites performance by transfecting plasmid DNA into COS-7 cells. The cytotoxicity of Mag-HPEI gene vectors was compared to HPEI by MTT assay in the COS-7 cell line after 24 h incubation. Their results showed that the magnetic nanoparticles were superparamagnetic, observing a dose dependent and HPEI size dependent cytotoxicity and, additionally, that the transfection efficiency was enhanced when the size of the magnetic nanoparticle increased. The best results were obtained with Mag-HPEI10K vectors, due to the smaller size of the HPEI used and the reduction of positive charges due to the MNPs when compared to the HPEI alone. Expression of the luciferase gene showed that Mag-HPEI10K with the larger magnetic nanoparticles vectors had, in addition, the best magnetofection efficiency of the combinations analyzed by these authors, reaching 115-fold of that of standard HPEI transfection. PEI was used by Prosen and col. [39,47,48] to transfect murine B16F1 (low metastatic potential) and B16F10 (high metastatic potential) melanoma cells. Their experiments combined the use of superparamagnetic iron oxide nanoparticles (SPIONs), PEI, polyacrylic acid (PAA) to stabilize the SPIONs and, of course, plasmid DNA to transfect. They focused on the importance of the alkaline pH of the PEI water solution used needed to functionalize the SPION-PAA in order to obtain a significant gene expression. They also found that the differences in different storage times of iron salts, temperature of reaction and quality of water do not influence the reproducibility in the synthesis of the modified SPIONs. The size of the constructs was about 200–400 nm in diameter. To hinder the expression of MCAM (melanoma cell adhesion molecule) posttranscriptional specific gene silencing, employing RNA interference technology, was used. MCAM is involved in melanoma development and progression, thus making it a potential target for gene therapy of melanoma. Their results showed that the level of MCAM mRNA was reduced to 45% in B16F1 and 75% in B16F10 cells after magnetofection while the level of expression of MCAM protein was reduced by 69% in B16F1 and 27% in B16F10 cells after magnetofection. Magnetofection resulted in a statistically significant inhibition of cell proliferation for 77% in B16F1 and 65% in B16F10 cells at day 4. Moreover, SPIONs decorated with a PEI- chondroitin sulfate copolymer were studied by Lo *et al.* to target human cell lines (glioblastoma U87 cells; 293T embryonic kidney; CRL-5802 human lung cancer) for the transfection of microRNA-128 (sequence enriched in normal brain tissue and down-regulated in gliomas). The average vector size was around 136 nm and the exposition to the magnetic field was 20 minutes. The authors claim that the PEI-chondroitin sulfate copolymer surface significantly increases the gene expression when compared to the PEI gold standard surface in all the cell lines tested [49].

Despite PEI being considered the gold standard nonviral transfection agent, it presents disadvantages such as elevated toxicity and its transfection efficiency lacks in consistency and reproducibility. Several other agents are being studied. Dixon and col. achieved magnetofection by using a glycosaminoglycan (GAG)-binding enhanced transduction (GET) system based on combining the activities of peptide-cell membrane interaction with GAGs and cell penetrating peptides (CPPs) that were developed by their group. They used this development for *in vivo* bone repair and lung gene delivery. This system was not only used for DNA delivery, but also for RNA and oligopeptides, making it a versatile delivery system. Furthermore, they modified the GET system by adding the FGF2B-LK15-8R (FLR) peptide, which they claim has better transfection efficiency than PEI based vectors. This peptide consists of 3 domains: a fibroblast growth factor 2 (FGF2B) heparin-binding domain which acts as a membrane docking domain; LK15, a sequence able to complex DNA, and a cell penetrating peptide (CPP), which enhances endocytosis. These modifications, when under a magnetic field, allowed the NIH3T3 cell membrane binding in 5 seconds. Transfection was not hampered by the presence of the magnetic nanoparticles. In fact, it was enhanced. The rhodamine-labeled pDNA loaded magnetic vectors had a diameter of 244.7 ± 8.5 nm [50]. Liposomal magnetofection (LMF) can be used for the generation of induced pluripotent stem cells (iPS). In this method, ternary complexes of cationic lipids self-assembled with plasmid DNA were associated with superparamagnetic iron oxide nanoparticles. Mouse embryonic fibroblast cells were used, and virus free iPS cells were obtained. Results show that they achieved a 10-30-fold reproducible and effective transfection method when compared to viral vector systems. Signs of cell reprogramming appeared 2 days after transfection, in contrast to 3 to 5 weeks when using other non-viral vector alternatives [51]. Chitosan associated with a plasmid and MNPs was used as a vehicle for delivering TRAIL (tumor-necrosis factor-related apoptosis ligand) to mouse melanoma cell line B16F10 [52]. TRAIL is a molecule that selectively kills malignant cells and is expressed constitutively in many normal tissues. The vector size was between 200-250 nm and had a spherical and compact morphology. They measured the gene expression by the activity of caspase 3 and by observing the morphological changes in cells, which led the authors to conclude that the transfection of TRAIL was achieved by using this chitosan vector. Chitosan was also used in combination with PEI, SPIONs and PK11195 [53] to deliver genes into the mitochondria. PK11195 is a ligand of mitochondrial 18 kDa translocator protein (TSPO). By binding with TSPO, PK11195 could open mitochondrial permeability transition pores, resulting in mitochondrial dysfunction, leading to apoptosis, which makes it an ideal target site for cancer therapy. A549 (human lung carcinoma) and KB (human epidermal carcinoma), HeLa and HepG2 cells were used. The obtained vectors were 27.9 ± 5.4 nm in diameter and exhibited concentration-dependent cytotoxicity in all cell lines. The authors found that the association of chitosan to PEI showed less cytotoxicity than PEI alone, and the time of magnetofection was also shorter. Their results also showed that the PK11195 delivery system specifically targeted the mitochondria.

Therefore, it has been demonstrated by the aforementioned, that magnetofection technology is a useful tool for guided delivery of nucleic acid that allows highly efficient transfection rates which can be chosen for the study and treatment of multiple and varied diseases. In this respect, it is possible to find literature concerning tumor inhibition, neuronal or cardiovascular disease treatment or bone regeneration. Some examples are presented in this section.

Microglia are the resident immune cells of the brain. Since they have multiple functions in physiological and pathological conditions, including Alzheimer's disease (AD), their study is important. There are several microglia cell lines that present the advantage that they can be expanded several times but have the disadvantage of acquiring characteristics that are not present in primary microglia cells. iPS are an alternative, but they are not easy to obtain. A useful alternative is to work with primary microglia cells, which present the challenges of low yield when harvesting and limited survival period, coupled with low efficiency of transfection. Burguillos and col. describe a simple method based on magnetofection to modify the expression of different genes in primary microglia by using small interfering RNA (siRNA). These researchers used a commercial kit (Glial-Mag kit from OZ Biosciences) and specific siRNAs against TREM2. They found that the mRNA level was decreased to 40%, 48 h after transfection. The proposed system had no effect over the inflammatory response. This study allowed the researchers to study the effects of knocking down the TREM 2 gene, leading to the conclusion that partial knockdown of TREM2 expression caused a small (but statistically significant) increase in phagocytosis of neuronal debris, which is in agreement to those results obtained when using mice lacking one of the two TREM2 alleles [54,55]. In 2021, Umek et al. published the study of the effect of an anti-gene oligonucleotide (CAG19) approach to treat Huntington's disease, using magnetofection as transfection system. They delivered the oligonucleotide using a commercial magnetofection agent (NeuroMag Transfection reagent - OZ Biosciences) after the neural induction of iPS of both Huntington disease and non-Huntington disease patients. They studied the transfection on various stages of neural development and concluded that CAG19 oligonucleotide does not induce *in vitro* neurodevelopmental toxicity, and that magnetofection is an efficient delivery method for this oligonucleotide, achieving 40%– 45% mRNA downregulation [56].

Another important area of application is tumoral cell inhibition. In 2020, Hu *et al.* published results on their research using magnetofection with cationic lipopeptides attached to arginine and anionic magnetic nanoparticles to transfect microRNA 125b-1, with an average vector size of 258 nm. This group achieved a double effect when transfecting two different cells lines (mice macrophage RAW264.7 and mice breast tumor 4T1 cells) simultaneously; the effects observed were the promotion of the polarization by targeting interferon regulatory factor 4 (IRF4) in macrophages, and tumor cell inhibition, by targeting ETS proto-oncogene 1 and cyclin- J in tumor cells. The transfection time was 20 minutes, and the mechanism of action varied depending on the transfected cell. It was demonstrated that microRNA 125b-1 downregulates its target genes in both cell lines studied, resulting in M1-type macrophage polarization (increased expression of IL-1β, TNF-α, iNOS and IL-12) and the inhibition of tumor cell growth (downregulation of ETS1 and CCNJ and ROS generation). Furthermore, co-culture of both cell lines indicates that microRNA 125b-1 transfection can inhibit tumor growth by targeting both macrophage and tumor cells [57]. Furthermore, Mao and col. successfully altered proliferation and invasion of CD133$^+$ primary glioma stem cells. This was achieved by the transfection of a microRNA-374a using magnetofection. Even when they do not describe the vector used, based on their introduction it is safe to assume that a PEI-SPION based vector crosslinked with plasmid DNA for miR-374a was utilized. Their aim was to study if the overexpression of miR-374a regulates Neuritin protein-coding gene NRN1 and inhibits cell proliferation, invasion and tumorigenicity of CD133$^+$ human cells, which have low expression of miR-374a. Results show that once transfected, the over expression of miR-374a can inhibit the *in vitro* proliferation and invasion of CD133$^+$ human cells. Furthermore, the expression of neuritin and cell cycle regulatory proteins is inhibited, significantly reducing GSC tumorigenicity [40].

Some other examples of *in vitro* magnetofection include the work by Krötz *et al.* [58] who also applied magnetofection to the delivery of antisense oligonucleotides to human umbilical vein endothelial cells (HUVEC). As many other groups, they chose to use a primary culture to provide a more adequate model for the *in vivo* situation than that achievable with immortalized cell lines. Their aim was to enhance the delivery of antisense oligodesoxynucleotides (ODN) into primary cells, targeting the p22^{phox} subunit of endothelial NAD(P)H-oxidase, in order to decrease the basal levels and prevent superoxide release due to loss of NAD(P)H-oxidase activity. Superoxide (O_2^-) production by the vascular endothelium is assumed to interfere with endothelial function by scavenging nitric oxide, which forms peroxynitrite, and causes endothelial dysfunction, associated with atherosclerosis, intimal hyperplasia, hypertension, diabetes, and hypercholesterolemia. Magnetofection allowed these researchers to obtain high transfection rates at low toxicity and low transfection times, which is desirable when studying ROS-generating enzymes to obtain cleaner results. Results showed a positive outcome (15 minutes of magnetofection and 90% of ODN uptake) when compared to standard techniques. The survival of cells was also greater: viability of 96% versus only 78% for standard transfection under conditions that yielded 80% transfected cells. This approach resulted effective in knocking out proteins with functional consequences in order to characterize and confirm their physiological role.

It is also possible to mention the application of magnetofection for the treatment of bone disease. One of the major issues in bone regeneration therapy is that the hypoxic/ischemic wound environment compromises the viability of transplanted cells. However, overexpression of B-cell lymphoma 2 (Bcl-2) is known to inhibit apoptosis in implanted cells. Brett *et al.* used SPIONS-PEI and poly-β-amino ester vectors to transfect a plasmid encoding green fluorescent protein and Bcl-2 to human adipose-derived stromal cells (ASCs) in order to inhibit apoptosis, leading to better survival of the implanted cells for bone regeneration. They compared magnetofection to nucleofection transfection method which is an electroporation-based physical method that allows the transfer of nucleic acids by applying a specific voltage. Size of the obtained vectors was 159 nm, and magnetofection showed a 31% increase in transfection when compared to the nucleofection technique. Consequently, Bcl-2 overexpression led to better viability at 48 h in magnetofected vs nucleofected cells (73% vs 49% respectively). Even more, the osteogenic capacity of magnetofected cells was improved when compared to nucleofected cells [59].

Finally, magnetofection has also been used to study the restoration of functions such as loss of vision or hearing. Withing this group of applications, it is worth mentioning the work by Sen *et al.* for the magnetic delivery of siRNA to retina explant cultures that sustain the three-dimensional architecture and cellular connections withing the tissue [60]. The authors efficiently silenced the vasolin-containing protein (VCP) in RHO P23H rat retinal explants applying a magnetic force in a reverse magnetofection technique, from above the culture plate, concentrating the siRNA-MNP from the culture media to the cells. Moreover, they compared the results with another transfection technique, lipofection, which resulted in lower efficiency. They highlight the importance of this technique on reaching all the cell layers in the explant and propose it as an alternative for *in vivo* experiments in the absence of adverse effects on cell viability for the treatment of diseases such as hereditary retinitis pigmentosa which is characterized by gradual loss of photoreceptor neurons probably due to excessive activity of VCP.

Table 1. Combination of magnetic nanoparticles with viral or non-viral transfection vectors for in vitro magnetofection.

Vector	Advantages of combination with MNPs	Vector subtype	Type of delivered genetic material	Ref
Viral particle	Increases the transfection efficiency			[34,41,61]
Non-viral	Accelerates transfection process and improves transfection efficiency. Safer strategy	Synthetic polymers (PEI, HPEI)		[37,44–47,59,62]
			Interference RNA	[47–49]
			Antisense oligonucleotides	[56,58]
			siRNA	[54,55,63]
			Multiple gene delivery	[37]
		Lipids	Plasmid DNA for the generation of induced pluripotent stem cells	[51]
			siRNA delivery	[57]
		Carbohydrates	DNA	[50,52,53]
			siRNA	[64,65]

3. Gene therapy

Gene therapy has emerged as a new-generation technique to treat or prevent a wide range of hereditary diseases or congenital disorders by a direct transfection of genetic material to targeted cells or tissues [66,67]. This implies a replacement of mutated or altered genes by new healthy ones that encode proteins of interest [68]. These genetic advances together with progresses in bioengineering allow to treat many diseases that had inefficient treatments or were untreatable before [69].

The delivery of therapeutic genetic material requires safety, efficiency, precision and low cost. Different systems have been developed to transport genes to host cells [70]. The most commonly used technique consists of loading the healthy gene into a vector, which acts as a protector from enzymatic biodegradation and enhances its stability. Despite there are viral and non-viral vectors [71,72], the idyllic vector is one that transfers a precise amount of DNA into each host cell, thus guaranteeing an efficient introduction of the genetic material and allowing for encoding the specific product without producing a negative immune response and toxicity. Furthermore, the carrier should have a suitable diameter to enter the cells and the nucleus, go across the cytoplasm, escape the endosome/lysosome and allow the endocytosis process. As well as this, the vector should possess peptides on their surface to achieve a successful cell penetration.

Regardless of the type of gene therapy used, the first fundamental step is for the genetic material to reach the interior of the cell. For this, one of the major methods used is the *in vivo* delivery of gene therapy, *i.e.*, the genetic material is carried by the help of a vector to cells that remain inside the body. However, there are many intracellular and extracellular barriers to *in vivo* gene delivery [73], such as intracellular and nuclear uptake and targeting to specific tissues or cells, that must be overcome to increase the efficiency of therapies. In this sense, to promote cellular uptake of gene vectors, MNPs are already in use through their incorporation with both viral and non-viral gene vectors to improve the *in vivo* delivery to target locations within the body by the application of an external magnetic field (EMF) [74]. Normally, vector complexes containing MNPs and the therapeutic genes are administered by an intravenous injection, and then with the application of EMF they are oriented through the bloodstream towards the site of interest. After being captured by the field, the magnetic vectors stop at the target, where they are absorbed by the tissue [75]. Therefore, under the action of a magnetic field, MNPs could reduce the delivery time and doses for gene transfection and promote DNA shuffling [76] into the cells, leading at the same time to less side effects caused by an excessive exposure to gene carrier vectors. For example, the previously mentioned superparamagnetic iron oxide (SPIONs) nanoparticles [77] are ideal to prepare gene delivery systems with magnetic response, due to their high biocompatibility, superparamagnetic properties, accelerated transfection process and magnetic directing [78].

3.1   Non-viral vectors for *in vivo* magnetofection

As mentioned above, MNPs can be combined with viral and non-viral vectors. Since the last ones are safer, this section will focus on the method of non-viral gene delivery— *in vivo* magnetofection. The non-viral magnetofection for the delivery of genetic material to cells or tissues *in vivo* is a relatively young technique and some of the few reported works have been accurately analyzed by Sizikov *et al.* [79].

Among the most commonly used non-viral vectors for *in vivo* magnetofection, we can mentioned magnetopolyplexes formed by biodegradable polymers [80], magnetolipoplexes based on cationic liposomes and niosomes [81], and other magnetic carriers that use an unusual magnetic media. Magnetopolyplexes consist of MNPs coated by cationic polymers. They have been recently used, for example, to develop a novel therapy for rectoanal motility disorders [82]. In this work, Singh *et al.* mixed PolyMag, a commercial formulation composed of magnetic nanoparticles coated with cationic polymers[83], with microRNA-139-5p, a miRNA that regulates RhoA-associated kinase (RhoA/ROCK) normalizing the internal anal sphincter (IAS). This magnetic polyplex containing the genetic information was then injected in the perianal region of rats and exposed to a magnetic field for 20 minutes. Magnetofection effectiveness was evaluated and confirmed by confocal microscopy. *In vivo* data revealed that by using this technique miR-139-5p is concentrated in the IAS, thus reducing IAS-associated constipation and other rectoanal motility disorders. Hence, they demonstrated that this magnetofection-mediated nucleic acids delivery is a successful *in vivo* method to enhance site-directed transfection and concentration of miR139-5p in the IAS. Their results were promising as a new therapy for rectoanal continence disorders.

Another obvious target disease for magnetic gene transfection is cancer [34]. Prosen and coworkers, developed novel SPIONs coated with poly-acrylic acid (PAA) and functionalized with polyethylenimine (PEI) for further binding therapeutic plasmids against the melanoma cell adhesion molecule (MCAM). After proving the safety and efficacy of the genetic polyplexes (SPIONs-PAA-PEI- pDNAanti-MCAM) as carriers for the reporter and therapeutic plasmids *in vitro* [42], they demonstrated *in vivo* their antitumor effect in murine tumor models after silencing MCAM by magnetofection [39]. In a more recent work, they studied the distribution, accumulation, uptake and the consequent therapeutic effect of intratumorally injected SPIONs-PAA-PEI-pDNAanti-MCAM by the application (magnetofection) or not (nanofection) of an EMF in B16F10 murine melanoma tumor *in vivo* [84]. For magnetofection test, they applied Neodymium-Iron-Boron (NdFeB) permanent magnets with a surface magnetic flux density of 403 mT and magnetic gradient of 38 T/m above the tumors, immediately after the intratumorally injection of the magnetopolyplexes. The results indicated that the intratumorally injected magnetofection complexes remain locally at the injection site and the exposure of tumors to an external magnetic field enhanced the uptake of genetic polyplexes from the extracellular matrix into the cells. Furthermore, they proved that magnetofection complexes were internalized only by melanoma and immune cells, indicating the selectivity of *in vivo* magnetofection. Finally, they demonstrated that three consecutive magnetofections of murine melanoma with SPIONs-PAA-PEI-pDNAanti-MCAM led to a significant antitumor effect.

Magnetopolyplexes have also been applied for the magnetofection of plasmids for tissue-engineering technology. For example, in 2018 it was reported a novel strategy for *in vivo* angiogenesis and osteogenesis by the use of a new artificial bone scaffold containing superparamagnetic plasmid gene microspheres [85]. In this work, the human vascular endothelial growth factor (VEGF)-green fluorescent fusion protein eukaryotic expression plasmid (pReceiver-VEGF165/DH5a) was first combined with superparamagnetic chitosan nanoparticles, then incorporated in gelatin microspheres (SPCPGM) and finally introduced into a bone scaffold made of nano-hydroxyapatite/polyamide bone cement (Fig. 1.A,B). With the main objective of repairing large bone defects, a model of a large radius defect was generated in New Zealand rabbits and the bone scaffold containing SPCPGM was then implanted (Fig. 1.C). To study *in vivo* angiogenesis and osteogenesis, a NdFeB permanent magnet with a static EMF of 200 mT was located outside the bone defect of the rabbits for eight weeks. After 12 weeks of the operation, X-ray images (Fig. 1.E) revealed that the radial bone defect was completely rectified and most of the bone marrow cavity had recanalized in those rabbits exposed to the magnetic field. However, in the absence of an EMF, the connection of the bone fragments of the radial bone defect was incomplete. Based on these results, they demonstrated that the application of an external magnetic field enhanced the transfection efficiency and thus, the local concentration of plasmid was increased, improving angiogenesis and osteogenesis in the bone defect. This study provides a new type of angiogenesis approach for bone tissue engineering research.

Magnetolipoplexes are also widely used for *in vivo* magnetofection as they have the capability of saving the genetic material from adverse degradation during the transfection [86]. Cationic lipids, which behave as amphiphilic lipids, are preferable over viral gene vectors since they are non-immunogenic, can be easily prepared, form spontaneous complexes with the negatively charged DNA by electrostatic interactions and interact with the cell membrane inducing the endocytosis of the lipoplex and the following release of the genetic material into the cytoplasm. Some of the most extensively used cationic lipids for lipofection purposes are DODAC (N,N-dioleyl-N,N-dimethylammonium chloride), DOTMA (2,3-bis(oleoyl)oxipropyltrimethylammonium chloride) and DOTAP (N-[1-(2,3-Dioleoyloxy)propyl]-N,N,N-trimethylammoniummethyl sulfate) [87–89]. For example, DOTAP-based magnetic cationic liposomes containing SPIONs were constructed to deliver the pGL4.50 [luc2/CMV/Hygro] vector encoding the luciferase reporter gene *luc2* (*Photinus pyralis*) [90], to develop an effective macrophage-based cell therapy for colonic inflammatory disorders [91]. In this work, researchers first evaluated the uptake and gene transfection efficiency in RAW264 murine macrophage-like cells by the application of a magnetic field intensity of 300 mT. The obtained transfected macrophages were then intrarectally injected in ICR mice and immediately after the cell injection, an EMF was applied to the abdominal area for 1 h. After 6 or 24 h, the mice were sacrificed, and the luciferase activity of their colons was measured. They observed that the number of cells associated to the magnetic cationic liposomes was significantly increased by the application of an EMF, without cytotoxicity and without affecting the phenotypic characteristics of the macrophages. Furthermore, they observed a remarkable adhesion of the magnetized RAW264 cells in the murine colon in the presence of EMF. In fact, approximately 40 % of the injected macrophages containing the magnetolipoplexes remained in the colon until 24 h after the injection, despite the exposure to the magnetic field was for just 1 h. These results revealed a strong adhesion of the magnetized cells to the tissue thanks to the application of an EMF, representing a valuable approach to improve the delivery of macrophages using magnetic guidance and its therapeutic value against colonic inflammatory diseases.

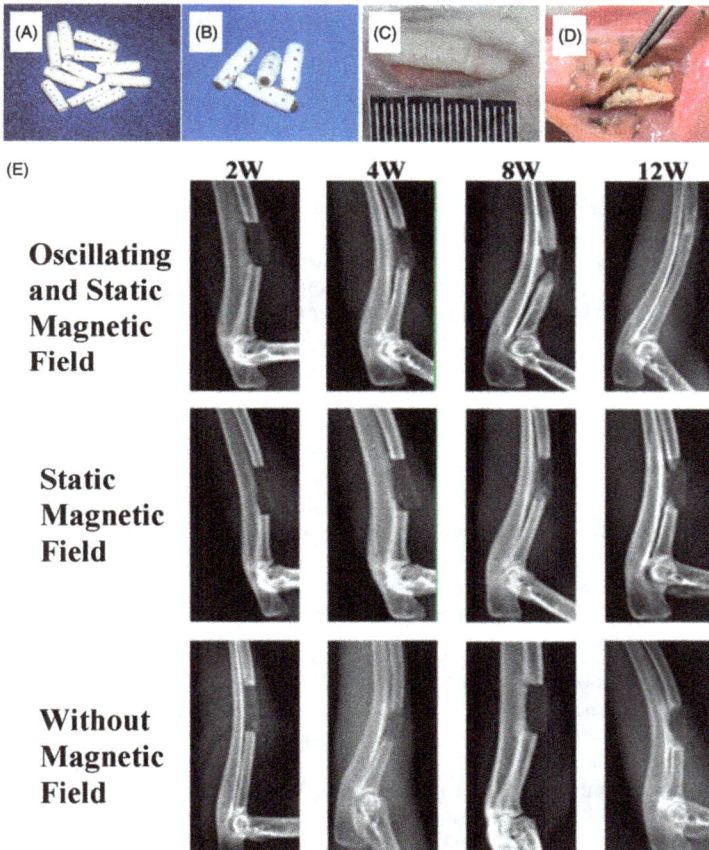

Fig. 1. Magnetofection repair of a radius defect in a New Zealand white rabbit with an artificial bone scaffold loaded with superparamagnetic chitosan pDsVEGF165-Red1-N1 gelatin microspheres. (A) Artificial bone scaffold (B) Artificial bone scaffold loaded with superparamagnetic chitosan pDsVEGF165-Red1-N1 gelatin microspheres. (C) The artificial bone scaffold loaded with superparamagnetic chitosan pDsVEGF165-Red1-N1 gelatin microspheres was implanted in the radial bone defect of a New Zealand rabbit. (D) General observation at 6 postoperative weeks. (E) X-ray images [85]

In another work, Wang *et al.* compared transfection efficiency between lipofection technique and liposomal magnetofection of a specific plasmid that can knockdown the type1 insulin-like growth factor receptor (IGF-1R), an oncogene overexpressed in lung cancer [92]. In both cases, they used the commercial formulation Lipofectamine™ 2000 (Lp2000, Invitrogen) to deliver the plasmid expressing GFP and shRNAs targeting IGF-1R (pGFPshIGF-1Rs) and transfer *in vitro* A549 cells, and then tumor-bearing mice *in vivo*. However, while for lipofection the transfection mixture contained Lp2000 and the plasmid and no external magnetic field was applied, for liposomal magnetofection, they also included SPIONs to the formulation and applied NdFeB permanent magnets to promote both *in vitro* and *in vivo* transfection. Particularly, the *in vivo* assays consisted in injecting the complexes containing or not he SPIONs into the tail vein of the mice, and, in case of magnetolipofection, a NdFeB magnet of 400 mT was placed onto the subcutaneous tumor surface for 14 min. After 24 h, the mice treated with lipofection and liposomal magnetofection were sacrificed and the tumors were removed to analyze the GFP-expressing cells. As a result, they found that the lipofection efficiency of pGFPshIGF-1R *in vitro* was only 23.3 ± 3.5 % and silenced IGF-1R gene expression by 56.1 ± 6 %, while the liposomal magnetofection efficiency of pGFPshIGF-1R was 59.7 ± 4.4 % and showed to be effective in downregulating the IGF-1R gene by 85.1 ± 3 % after 15 min exposure of cells to the EMF. *In vivo* assays revealed similar results: when using liposomal magnetofection a higher GFP expression and a more silenced IGF-1R expression were observed in the subcutaneous tumors, whereas only weak GFP expression and lower silencing of IGF-1R were observed in tumors with lipofection. These results suggest that liposomal magnetofection induced specific targeting and uptake of genetic material into the site of interest, which provides a novel method for gene therapy of lung cancer.

Apart from magnetopoliplexes and magnetolipolexes, there are other magnetic carriers that use uncommon magnetic media or other active targeting (in addition to the EMF). For example, the group of Shubhra *et al.* designed and developed a one-pot fabrication of magnetic calcium phosphate (CaP) nanoparticles (NPs) co-immobilizing DNA and iron oxide (IO) nanocrystals (DNA-IO-CaP NPs) [93]. They obtained preliminary results of the *in vivo* transfection of the plasmid including cDNA of luciferase (pGL3 control vector) to mice ischemic brain tissue [94]. After inducing an infarction on the right side of the mouse brain, the DNA-IO-CaP NPs were injected and a magnetic field of 240 mT was applied. Successfully, they found significant luciferase activity in the brain tissue as a result of targeted NP delivery. Thus, they demonstrated that this new coprecipitation process for developing novel multifunctional CaP-based NPs is really attractive to achieve magnetofection and improve targeted DNA delivery. Another example of unusual magnetic carriers used for magnetofection includes the self-assembled complexes based on polyethylenimine-coated cationic magnetic iron beads (PolyMag), plasmid DNA, and a bis(cysteinyl) histidine-rich Tat peptide [95]. In this case, the endosomolytic Tat peptide, a cationic cell-penetrating peptide, was used as an additional active targeting, as it enhances the cytoplasmic delivery of different types of cargos. The researchers hypothesized that the combination of a magnetic field with the Tat peptide would improve the gene transfer efficiency as a result of the combination of both effects: enhanced accumulation of magnetofection vectors and the effective transmembrane delivery. To verify this, they examined the *in vivo* gene transfection efficiency of the plasmid DNA encoding a firefly luciferase gene under the control of the CAG promoter (pCAG-luc) by binary (without Tat) and ternary complexes. For this, they injected the pCAG-luc-containing PolyMag/DNA or PolyMag/DNA/Tat complexes in the spinal cord of adult male Wistar rats and after injection, an externally NdFeB permanent magnet of 1.21 T was applied on the top of the injection site for 15 min. They found that the level of transgene expression obtained using ternary complexes was approximately 2 times higher than that of binary complexes. Besides, the improved magnetofection resulting from the inclusion of the Tat peptide was confirmed by immunostaining using antibodies against luciferase. Again, the ternary complexes gave better results, as a stronger staining in spinal meninges was observed after injection, compared to binary complexes. Thus, they concluded that the mechanisms by which the Tat peptide improves cargos uptake across the plasma membrane of cells are complementary to magnetofection that enhances accumulation of genetic material at specific sites. In conclusion, the combination of the two methods can significantly improve the efficiency of targeted transgene expression. In another work, instead of using iron-based magnetic nanoparticles, a novel non-viral gene carrier was created by the combination of magnetic gold nanoparticles with the plasmid pGPH1/GFP/ Neo-Bag-1-homo-825 silencing Bag-1 gene to evaluate the knockdown of Bag-1 on colorectal cancer therapy [96]. With this novel formulation, a significant silencing and downregulation of Bag-1 was achieved *in vivo*, thus successfully inhibiting the growth of tumors induced in male Balb c/nude mice. This is an interesting example of an alternative non-viral carrier based on magnetic gold nanoparticles to be applied in gene therapy for colorectal cancer.

The efficacy of MNP-based genetic material delivery has been demonstrated most clearly *in vitro*. However, here we presented some of the developments successfully tested *in vivo*. There is great potential for non-viral *in vivo* magnetofection to be applied as an alternative method to treat a wide variety of disorders that nowadays do not have an efficient treatment. The designing and creation of new combinations of magnetic particles with different carriers and genetic materials, together with the optimization of magnetic field parameters is already beginning to evidence important advances of this promising technique. While scale-up to clinical application is likely to be a very ambitious goal, the potential clinical use of *in vivo* magnetofection to facilitate delivery of therapeutic genes is surely enticing.

4.   MNPs for vaccines and immunotherapy

The main applications of magnetic nanoparticles in the biomedical area are based on 1) cell labeling and cell separation; 2) drug delivery; 3) hyperthermia and 4) contrast enhancement of magnetic resonance imaging [97]. On the other hand, magnetofection for gene therapy has already been extensively discussed in the previous sections. Herein, the use of MNPs for immunotherapy and development of vaccines will be presented.

Regarding their use in vaccines, MNPs are implemented to a greater extent in third-generation vaccines which consist of the incorporation of nucleic acids in vectors which code for an antigen capable of producing a humoral and cellular immune response. They present some advantages over traditional vaccines, consisting of proteins or inactivated/attenuated pathogens, such as their easy production, stability, lower cost, and greater safety with respect to second-generation vaccines [98]. In the same way that occurs with gene therapy, the entry of naked DNA does not often overcome certain biological barriers, giving a weak immune response, so, to improve its delivery, many viral and non-viral vectors that direct and protect nucleic acids have been proposed. One option consists of magnetic nanoparticles, which have been shown to increase the efficiency and speed of gene delivery in different tissues by the method called magnetofection as it was previously described [99,100]. This method was reviewed by Plank *et al.* in 2011 and they define it as the delivery of nucleic acid under the influence of a magnetic field which acts on nucleic acid vectors made of magnetic nanoparticles directing the content to target cells [101].

The application of magnetic nanoparticles for the delivery of nucleic acids in the area of vaccine development or immunotherapy against cancer is still limited and only a reduced number of *in vivo* applications can be found in literature. However, its future use is promising, opening the doors to new forms of therapies.

Malaria DNA vaccines are the most studied examples in this field. Malaria is a tropical, life-threatening disease caused by the *Plasmodium* parasite and transmitted by the bite of infected mosquitos. Al-Deen *et al.* in 2011 and 2014, studied *in vitro* vectors made up of superparamagnetic iron oxide nanoparticles of approximately 100-300 nm coated by the polymer polyethyleneamine, conjugated to plasmid DNA containing a 19 kDa terminal carboxylic fragment associated with the membrane of merozoite surface protein 1 (MSP1$_{19}$). In both cases, after evaluating the optimal conditions for its development, they concluded that this system could be used in *in vivo* studies due to the high transfection potential demonstrated by the expression of the protein *in vitro* after magnetofection of eukaryotic cells [102,103]. Years later, Al-Deen and collaborators in 2017, continued with the study of these vectors, and to improve their stability and their intrinsic transfection efficiency, they incorporated a new component, hyaluronic acid, which is known to interact with multiple cell receptors on key antigen-presenting cells like dendritic cells and macrophages. The authors evaluated their *in vivo* immunogenicity through two different routes of administration, intraperitoneal and intramuscular in BALB/c mice observing that hyaluronic acid coated particles produced a greater immune response which was even higher for intraperitoneal administration. This broad-spectrum cellular response was mainly Th1 although, to a lesser extent, it also produced Th2 and Th17 responses. They also observed an important humoral response with high levels of IgG2a, IgG1 and IgG2b cytophilic antibodies when the IgG subclass distribution of anti-PyMSP1$_{19}$ antibodies was studied. Surprisingly, it was also noted that the impact of magnetic field for the antibody production and T cell responses were different with an enhanced antibody production but no changes in the T cell response. These *in vitro* and *in vivo* results indicate that magnetic nanoparticles can be used as DNA carriers in malaria vaccines due to their broad immune response that improve the potency of traditional DNA vaccines [99].

DNA vaccination has also been studied in the last years for HIV vaccine development and magnetic nanoparticles have been proposed as alternative non-viral vectors. Zhou *et al.*, studied the distribution of DNA (pCMV cytomegalovirus vector having the Luciferase gene) associated with magnetic particles after injection into rabbit muscle with or without the application of a magnetic field. They found that after exposure to the magnetic field, a broad distribution of the magnetic particles was acquired. However, these particles were retained in the regions close to the injection site when no magnetic field was applied. After collecting this evidence, they continued evaluating the immunogenicity of an HIV-1 gag DNA vaccine for human immunodeficiency virus. In this case, they observed that magnetofection significantly enhanced and accelerated antibody responses to HIV-1 gag protein compared to naked DNA as well as a greater cellular response [100]. Similarly, Vahid *et al.*, used SPIONs coated with trimethyl chitosan (TMC) and polyethylenimine (PEI) to prevent the agglomeration, to enhance the solubility of these carrier particles and to promote cell uptake. These particles were employed to deliver a potent anti-tat siRNA (a critical viral regulatory gene) as a safe anti-HIV therapeutic approach. After evaluating cell transfection and gene silencing *in vitro*, the authors concluded that the siRNA-loaded SPION-TMC-PEI at a concentration of 50 µg/mL of nanoparticle with a dose of 100 pmol/mL of siRNA, increased the siRNA quantity into the cells and dramatically decreased the expression of HIV-1 tat gene in comparison with the control groups [104]. For their part, Kamalzare and collaborators, also used siRNA technology to study new therapeutic systems against HIV-1 infection [105]. In this case, the authors made use of a superparamagnetic iron oxide nanoparticle for delivering siRNA against HIV-1 nef coated with trimethyl chitosan (TMC), and carboxymethyl dextran (CMD) at different ratios. The particles had a size of 112 ± 3.0 nm and a zeta potential of 18.7 ± 2.6 mV as the particle size under 150 nm and a moderate positive zeta potential are efficient factors for a successful cell internalization. The results showed that the proposed formulation CMD–TMC–SPIONs had low cytotoxicity whereas the transfection efficiency, as well as the gene expression inhibition, were significantly increased in comparison to control groups.

Another example of magnetic nanoparticles in third-generation vaccines is currently being used to combat the coronavirus disease 2019 (COVID-19) caused by the severe acute respiratory syndrome coronavirus-2 (SARS-Cov-2) infection. Erasmus and collaborators in 2020, developed an Alphavirus-derived replicon RNA vaccine candidate which encodes the S protein of SARS-CoV-2 and demonstrated the expression of S protein in BHK cells. After that, the authors formulated LION lipid inorganic nanoparticles in order to optimize the stability, immunogenicity and administration of the vaccine for *in vivo* assays. This LION formulation is a stable cationic squalene emulsion (which also acts as adjuvant) containing superparamagnetic iron oxide nanoparticles in the oil phase, and a key component cationic lipid called 1,2-dioleoyl-3-trimethylammonium propane (DOTAP), which allows electrostatic association with RNA molecules when combined by a simple step of 1:1 mix. The authors performed an *in vivo* immunization test in female C57BL/6 or

BALB/c mice between 6 and 12 weeks of age and in adult male pigtail macaques between 3 and 6 years of age. The vaccine induced TH1-biased antibody and T cell responses in both mice and pig-tailed macaques. However, the authors caution that more studies are required to evaluate the safety, kinetics, and durability of immune responses and protection against the disease. They concluded that their vaccine is a good candidate for protection against COVID-19 and proposed its production in a two-vial approach with one vial containing the LION formulation and the other one having the repRNA vaccine and they also claim it will soon enter clinical trials under the name HDT-301 [106]. Nevertheless, it is not clear why the authors choose to include SPIONs in their formulation as they are not used for magnetofection. It can be assumed, according to information gathered in literature, that SPIONs may be acting as adjuvants to enhanced and prolong the immune response as subunit vaccines are less immunogenic than virus attenuated vaccines. In this sense, it has been recently reviewed the use of iron nanoparticles in the formulation of vaccines as novel adjuvants. The authors describe their immunomodulatory effect as well as the influence of physico-chemical properties in their performance [107].

Immunotherapy is a type of cancer treatment that stimulates the body's natural defenses to fight cancer. Another example of third-generation vaccines under development, where magnetic nanoparticles are implemented, is in the production of immunotherapies against cancer. Grippin *et al.* in 2019 created a library of lipid nanoparticle formulations and evaluated their capacity to transfect and activate bone marrow-derived dendritic cells. After carrying out *in vitro* tests they concluded that the lipid system composed of 1,2-dioleoyl-3-trimethylammonium propane liposomes with cholesterol and loaded with mRNA encoding tumor antigens was the best option for the activation of dendritic cells (CD). Then, they incorporated iron oxide nanoparticles inside the liposomes (IO-RNA-NPs) which, due to their properties, improved RNA transfection to DC after application of a magnetic field. The formulation also improved dendritic cell activation (in comparison to electroporation) measured by the presence of costimulatory markers and production of inflammatory cytokines. When a tumor model was used for the application of a single vaccine with DCs loaded with IO-RNA-NPs, the authors found an enhanced inhibition of tumor growth and cell migration to lymph nodes. Iron oxide particles allowed cell monitoring by magnetic resonance imaging (MRI) as DC migration was found to be a biomarker of antitumor response to vaccination. They concluded that their system holds great promise to be used in the prediction of tumor regression as they claim that the reduction in T2*-weighted MRI intensity observed in treated lymph nodes 2 days after vaccination correlates strongly with reduced tumor size 2-5 weeks after vaccination and it is also associated with a 73% increase in median survival compared to treated mice that do not exhibit this change in MRI images (Fig. 2) [108] . In this work, the incorporation of iron oxide nanoparticles played a double role allowing *in vitro* transfection by magnetofection and tracking cell migration to lymph nodes.

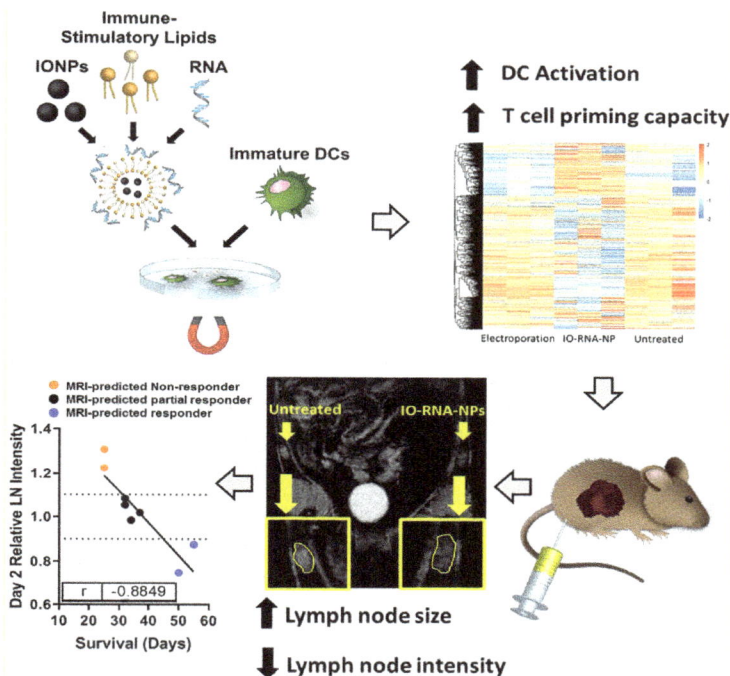

Fig. 2. Schematic Illustration of IO-RNA-NPs generated by combining commercially available IONPs and mRNA encoding tumor antigens with a combination of previously translated lipids with exceptional capacity for mRNA delivery and DC activation. Incubation of these particles with DCs in the presence or absence of a magnetic field led to profound DC activation characterized by dramatic changes in RNA expression and enhanced capacity to stimulate antigen specific T cells. Heat map comparing mRNA expression in BMDCs 24 hours after treatment with GFP mRNA via IO-RNA-NPs or Electroporation. IO-RNA-NPs enabled MRI-based detection of DC migration to lymph nodes that correlated directly with survival in murine tumor models. T2*-weighted MRI image 48 hours after vaccination with IO-RNA-NP-loaded DsRed+ DCs in the left inguinal area. Yellow borders indicate lymph nodes on treated (right) and untreated (left) sides. Correlation of the relative change in MRI-detected lymph node intensity in treated compared to untreated lymph nodes (Relative LN Intensity) on Day 2 with survival. "Reprinted with permission from Grippin, A J et al. "Dendritic Cell-Activating Magnetic Nanoparticles Enable Early Prediction of Antitumor Response with Magnetic Resonance Imaging." ACS nano 13,12 (2019). 13884-13898. Copyright 2019 American Chemical Society."

That same year, Yoo *et al.* performed *in vivo* tests on C57Bl/6 mice implanted with a pancreatic cancer cell line. The authors combined gemcitabine (antitumoral) with small interfering RNA (siRNA) antagonist of programmed death ligand 1 (PD-L1), which is known as an immune checkpoint inhibitor of cancer, conjugated to a dextran-coated magnetic nanocarrier. To ensure long circulation times and efficient diffusion across the vascular endothelium and throughout the tumor the nanoparticles had a final size of around 20 nm. When treating the mice with high and low doses of this combination, they observed a significant inhibition of tumor growth with respect to the control, which led to a longer survival. In the case of mice treated with the lower dose, they survived up to 8 weeks, while the survival for those treated with the higher dose was 12 weeks. After treatment, the authors observed an increase in the recruitment of CD8 + tumor infiltrating lymphocytes, a decrease in the expression of PD-L1 and an increase in cell-mediated cytotoxicity evidenced by higher levels of Granzyme B and a decrease in the infiltration by immunosuppressive Foxp3 + regulatory T cells, finally leading to a good antitumor immune response. The authors chose to use MNPs as they allow highly efficient delivery of RNA to tumor cells and, at the same time, can be useful for MRI imaging of the tumor [109].

Prijic *et al.* modified the surface of SPIONs with a double layer of polyacrylic acid (PAA) and PEI endosomolytic polymers (SPIONs-PAA-PEI) and included plasmid DNA encoding either reporter gene for enhanced green fluorescent protein (GFP) or therapeutic gene for murine interleukin 12 (IL-12). They evaluated the particles *in vitro* toxicity and cellular internalization after magnetofection as well as *in vivo* toxicity and biodistribution before employing them for the treatment of mice tumors [42]. They found that SPIONs-PAA-PEI-pDNA decreased survival of cells to approximately 70%, which indicated that they were not vastly cytotoxic, in comparison to PEI-pDNA that decreased 50%, increased the transfection efficacy with respect to other methods (lipofection and electroporation) and they were nontoxic *in vivo*. Afterwards, the authors evaluated the antitumor effectiveness of pDNA^{IL-12} after magnetofection with the synthetized SPIONs-PAA-PEI in weakly immunogenic TS/A mammary adenocarcinoma. They observed that the therapeutic effect of SPIONs-PAA-PEI-pDNA^{IL-12} after magnetofection of TS/A tumors was significantly better than in the absence of an external magnetic field, measured as reduction of tumor volume.

Finally, bacterial magnetic nanoparticles (BMPs), found in magnetotactic bacteria *Magnetospirillum gryphiswaldense*, have been used as alternative magnetic particles which offer good biocompatibility as they are covered with a stable cytoplasmic membrane. In this case, BMPs of 45-55 nm were combined with PEI to obtain a DNA vaccine system against the human papillomavirus type-16 E7 (HPV-16E7). The plasmid construct (pSLC-E7-Fc) contained a fused gene, in which the E7 gene was sandwiched by the secondary lymphoid tissue chemokine (SLC) and the IgG Fc fragment genes which can synergistically enhance the immunogenicity of an E7-expressing plasmid DNA vaccine. They found that under a magnetic magnetic force (600 mT, 10 min), BMP-DNA complexes reach the surface of the cells more quickly and promote the entry of plasmid DNA into the cells more efficiently. As mentioned in other studies, the authors also observed that subcutaneous administration of the vaccine with magnet application resulted in an increased tumor protection when compared with intramuscular injection and that the SLC-E7-Fc fusion protein was expressed and secreted recruiting a large number of dendritic cells at the vaccination site, which facilitate antigen presentation [110].

Although this chapter is mainly focused on the applications of magnetic nanoparticles in the delivery of genetic material, in cancer treatment, magnetic nanoparticles are also used to produce other type of vaccines that do not involve DNA / RNA delivery. For example, Schreiber *et al.* in 2010 developed non-toxic fluorescent carbon magnetic nanoparticles capable of interacting both *in vitro* and *in vivo* with dendritic cells, as well as with an abundant variety of proteins and adjuvants due to their surface chemistry. This system was able to function as an antigen carrier towards dendritic cells and to generate a Th1 response by the CpG codelivery [111]. Ho *et al.*, in their 2011 work, also concluded that iron oxide nanoparticles show promise for their biocompatibility, their low toxicity, their stability and their ability to conjugate with various antigens for the production of vaccines [112]. In 2015, Sungsuwan and collaborators also proposed the use of magnetic nanoparticles coated with mucin 1, a glycopeptide overexpressed in many types of cancer, as carriers of tumor-associated carbohydrate antigens (TACA) for the development of cancer vaccines. The authors observed that magnetic nanoparticles could enhance the humoral response and that they could be used as a contrast agent for *in vivo* monitoring to elucidate the mechanism of immune activation as well [113].

Genetic vaccines using magnetic particle carriers for veterinary use are also under study. In this group, it is possible to find a vaccine development for bovine herpesvirus type 1 (BoHV-1) which is one of the causative pathogens for the bovine respiratory syndrome [114]. The authors propose an alternative route of immunization, intranasal, due to low mucosal immunity of inactivated and gene-deleted vaccines after muscle injection for commercial formulations. In their study, plasmids carrying gB, gC and gD genes of BoHV-1, which are immunodominant antigens, were mixed with PEI magnetic beads of 50 nm protected with a layer of PEG. The immunohistochemistry showed a high expression of viral proteins in the lungs of vaccinated mice with the PEI-DNA-PEG magnetic beads but not viral proteins were detected in the other groups (PBS, DNA, PEI and PEI-DNA). On the contrary, viral proteins were found in the spleen for DNA and PEI-DNA groups but their expression was much weaker in the PEI-DNA-PEG formulation. Furthermore, the antibody level of anti-BoHV1 was higher in the PEI-DNA-PEG group than the DNA group as well as the level of IgA, indicating mucosal immunity stimulation. The vaccine also showed good cellular response and safety after *in vivo* experiments. The authors did not use an external magnetic force to direct the destiny of the prepared magnetic beads or perform magnetofection. They chose to use magnetic nanoparticles in their formulation because SPION-PEI particles have lower toxicity and a higher efficiency for gene delivery compared to PEI.

Another veterinary application includes the phase I trial by Hüttinger *et al.* for the treatment of feline fibrosarcomas [115]. They proposed to administer cytokine genes for *in situ* vaccination into tumor cells to enhance the host's immune response to primary tumors and distant metastases. For that purpose, feline granulocyte-macrophage colony-stimulating factor (GM-CSF) gene was chosen as it stimulates anti-tumor immunity by augmenting the antigen-presenting activity of macrophages. Solutions of the plasmid carrying the feGM-CSF gene were mixed with PEI-coated iron oxide magnetic nanoparticles and injected in the tumor of cats followed by the application of a magnetic gradient field for 1 h. Magnetofection was chosen for local gene transfer to avoid systemic transfection which can lead to an immunosuppressive effect. The study aimed to determine to possible toxicity and feasibility of this therapy and the authors only found mild changes in some blood parameters which were not related to immunization, no systemic levels of feGM-CSF and the expression of the transfected gene in tumors (as it was not possible to distinguish from exogenous or endogenous expression of feGM-CSF the authors obtained supportive evidence by transfecting more animals with human GM-CSF). A subsequent phase II trial will be tested with the highest dose of 1250 μg feGM-CSF found to be safe and effective in the first trial.

Conclusions

Magnetofection has been adapted to all types of nucleic acids (DNA, mRNA, siRNA, shRNA), non-viral and viral transfection systems and has been successfully tested on a wide range of cell lines, including cells which are difficult to transfect or primary cultures. The success found in *in vitro* assays has encouraged its *in vivo* application in biomedical developments such as gene therapy, cancer immunotherapy or third generation vaccines.

Although the increase in the use of this technology for the treatment of diseases has been showing promising results, very few cases have entered the clinical trial stage and factors like its safety and possible long-term side effects remain to be evaluated.

References

[1] E.M. Materón, C.M. Miyazaki, O. Carr, N. Joshi, P.H.S. Picciani, C.J. Dalmaschio, F. Davis, F.M. Shimizu, Magnetic nanoparticles in biomedical applications: A review, Appl. Surf. Sci. Adv. 6 (2021) 100163. https://doi.org/10.1016/j.apsadv.2021.100163

[2] P. Kush, P. Kumar, R. Singh, A. Kaushik, Aspects of high-performance and bio-acceptable magnetic nanoparticles for biomedical application, Asian J. Pharm. Sci. (2021). https://doi.org/https://doi.org/10.1016/j.ajps.2021.05.005. https://doi.org/10.1016/j.ajps.2021.05.005

[3] M. Mahmoudi, S. Sant, B. Wang, S. Laurent, T. Sen, Superparamagnetic iron oxide nanoparticles (SPIONs): Development, surface modification and applications in chemotherapy, Adv. Drug Deliv. Rev. 63 (2011) 24-46. https://doi.org/10.1016/j.addr.2010.05.006

[4] P.M. Martins, A.C. Lima, S. Ribeiro, S. Lanceros-Mendez, P. Martins, Magnetic Nanoparticles for Biomedical Applications: From the Soul of the Earth to the Deep

Materials Research Forum LLC
https://doi.org/10.21741/9781644902332-10

History of Ourselves, ACS Appl. Bio Mater. 4 (2021) 5839-5870.
https://doi.org/10.1021/acsabm.1c00440

[5] U.S. Ezealigo, B.N. Ezealigo, S.O. Aisida, F.I. Ezema, Iron oxide nanoparticles in biological systems: Antibacterial and toxicology perspective, JCIS Open. (2021) 100027. https://doi.org/10.1016/j.jciso.2021.100027

[6] A. Ditsch, P.E. Laibinis, D.I.C. Wang, T.A. Hatton, Controlled Clustering and Enhanced Stability of Polymer-Coated Magnetic Nanoparticles, Langmuir. 21 (2005) 6006-6018. https://doi.org/10.1021/la047057+

[7] L. Mohammed, H.G. Gomaa, D. Ragab, J. Zhu, Magnetic nanoparticles for environmental and biomedical applications: A review, Particuology. 30 (2017) 1-14. https://doi.org/10.1016/j.partic.2016.06.001

[8] M.A. Dheyab, A.A. Aziz, M.S. Jameel, O.A. Noqta, P.M. Khaniabadi, B. Mehrdel, Simple rapid stabilization method through citric acid modification for magnetite nanoparticles, Sci. Rep. 10 (2020) 10793. https://doi.org/10.1038/s41598-020-67869-8

[9] I. Khmara, O. Strbak, V. Zavisova, M. Koneracka, M. Kubovcikova, I. Antal, V. Kavecansky, D. Lucanska, D. Dobrota, P. Kopcansky, Chitosan-stabilized iron oxide nanoparticles for magnetic resonance imaging, J. Magn. Magn. Mater. 474 (2019) 319-325. https://doi.org/10.1016/j.jmmm.2018.11.026

[10] J. Ning, M. Wang, X. Luo, Q. Hu, R. Hou, W. Chen, D. Chen, J. Wang, J. Liu, SiO_2 Stabilized Magnetic Nanoparticles as a Highly Effective Catalyst for the Degradation of Basic Fuchsin in Industrial Dye Wastewaters, Molecules. 23 (2018) 2573. https://doi.org/10.3390/molecules23102573

[11] A. Rajan, M. Sharma, N.K. Sahu, Assessing magnetic and inductive thermal properties of various surfactants functionalised Fe_3O_4 nanoparticles for hyperthermia, Sci. Rep. 10 (2020) 15045. https://doi.org/10.1038/s41598-020-71703-6

[12] L.L. Félix, M.A. Rodriguez Martínez, D.G. Pacheco Salazar, J.A. Huamani Coaquira, One-step synthesis of polyethyleneimine-coated magnetite nanoparticles and their structural, magnetic and power absorption study, RSC Adv. 10 (2020) 41807-41815. https://doi.org/10.1039/D0RA08872B

[13] S.S. Rohiwal, N. Dvorakova, J. Klima, M. Vaskovicova, F. Senigl, M. Slouf, E. Pavlova, P. Stepanek, D. Babuka, H. Benes, Z. Ellederova, K. Stieger, Polyethylenimine based magnetic nanoparticles mediated non-viral CRISPR/Cas9 system for genome editing, Sci. Rep. 10 (2020) 4619. https://doi.org/10.1038/s41598-020-61465-6

[14] S. Rahim, F. Jan Iftikhar, M.I. Malik, Chapter 16 - Biomedical applications of magnetic nanoparticles, in: M.R. Shah, M. Imran, S.B.T.-M.N. for D.D. and D.A. Ullah (Eds.), Micro Nano Technol., Elsevier, 2020: pp. 301-328. https://doi.org/10.1016/B978-0-12-816960-5.00016-1

[15] T.J. Daou, G. Pourroy, S. Bégin-Colin, J.M. Grenèche, C. Ulhaq-Bouillet, P. Legaré, P. Bernhardt, C. Leuvrey, G. Rogez, Hydrothermal Synthesis of Monodisperse

Magnetite Nanoparticles, Chem. Mater. 18 (2006) 4399-4404.
https://doi.org/10.1021/cm060805r

[16] W. Zhang, F. Shen, R. Hong, Solvothermal synthesis of magnetic Fe_3O_4 microparticles via self-assembly of Fe_3O_4 nanoparticles, Particuology. 9 (2011) 179-186. https://www.sciencedirect.com/science/article/pii/S1674200111000034. https://doi.org/10.1016/j.partic.2010.07.025

[17] J. Xu, H. Yang, W. Fu, K. Du, Y. Sui, J. Chen, Y. Zeng, M. Li, G. Zou, Preparation and magnetic properties of magnetite nanoparticles by sol-gel method, J. Magn. Magn. Mater. 309 (2007) 307-311. https://www.sciencedirect.com/science/article/pii/S0304885306009486. https://doi.org/10.1016/j.jmmm.2006.07.037

[18] I.L. Ardelean, L.B.N. Stoencea, D. Ficai, A. Ficai, R. Trusca, B.S. Vasile, G. Nechifor, E. Andronescu, Development of Stabilized Magnetite Nanoparticles for Medical Applications, J. Nanomater. 2017 (2017) 6514659. https://doi.org/10.1155/2017/6514659. https://doi.org/10.1155/2017/6514659

[19] M. Starowicz, P. Starowicz, J. Zukrowski, J. Przewoźnik, A. Lemański, C. Kapusta, J. Banaś, Electrochemical synthesis of magnetic iron oxide nanoparticles with controlled size, J. Nanopart. Res. 13 (2011) 7167-7176. https://pubmed.ncbi.nlm.nih.gov/22207821. https://doi.org/10.1007/s11051-011-0631-5

[20] M.V. Tuttolomondo, M.E. Villanueva, G.S. Alvarez, M.F. Desimone, L.E. Díaz, Preparation of submicrometer monodispersed magnetic silica particles using a novel water in oil microemulsion: Properties and application for enzyme immobilization, Biotechnol. Lett. 35 (2013) 1571-1577. https://doi.org/10.1007/s10529-013-1259-6

[21] O. Bomati, M.P. Morales, C.J. Serna, S. Veintemillas, Magnetic nanoparticles prepared by laser-induced pyrolysis, INTERMAG Eur. 2002 - IEEE Int. Magn. Conf. (2002). https://doi.org/10.1109/INTMAG.2002.1001479. https://doi.org/10.1109/INTMAG.2002.1001479

[22] R. Eivazzadeh-Keihan, H. Bahreinizad, Z. Amiri, H.A.M. Aliabadi, M. Salimi-Bani, A. Nakisa, F. Davoodi, B. Tahmasebi, F. Ahmadpour, F. Radinekiyan, A. Maleki, M.R. Hamblin, M. Mahdavi, H. Madanchi, Functionalized magnetic nanoparticles for the separation and purification of proteins and peptides, TrAC Trends Anal. Chem. 141 (2021) 116291. https://doi.org/10.1016/j.trac.2021.116291

[23] J.R. Sosa-Acosta, J.A. Silva, L. Fernández-Izquierdo, S. Díaz-Castañón, M. Ortiz, J.C. Zuaznabar-Gardona, A.M. Díaz-García, Iron Oxide Nanoparticles (IONPs) with potential applications in plasmid DNA isolation, Colloids Surfaces A Physicochem. Eng. Asp. 545 (2018) 167-178. https://doi.org/10.1016/j.colsurfa.2018.02.062

[24] A. Ullah Khan, L. Chen, G. Ge, Recent development for biomedical applications of magnetic nanoparticles, Inorg. Chem. Commun. 134 (2021) 108995. https://doi.org/10.1016/j.inoche.2021.108995

[25] T. Kobayashi, Cancer hyperthermia using magnetic nanoparticles., Biotechnol. J. 6 (2011) 1342-1347. https://doi.org/10.1002/biot.201100045

[26] D. Chang, M. Lim, J.A.C.M. Goos, R. Qiao, Y.Y. Ng, F.M. Mansfeld, M. Jackson, T.P. Davis, M. Kavallaris, Biologically Targeted Magnetic Hyperthermia: Potential and Limitations , Front. Pharmacol. . 9 (2018) 831. https://www.frontiersin.org/article/10.3389/fphar.2018.00831. https://doi.org/10.3389/fphar.2018.00831

[27] A. Farzin, S.A. Etesami, J. Quint, A. Memic, A. Tamayol, Magnetic Nanoparticles in Cancer Therapy and Diagnosis, Adv. Healthc. Mater. 9 (2020) 1901058. https://doi.org/10.1002/adhm.201901058. https://doi.org/10.1002/adhm.201901058

[28] A. Avasthi, C. Caro, E. Pozo-Torres, M.P. Leal, M.L. García-Martín, Magnetic Nanoparticles as MRI Contrast Agents, Top. Curr. Chem. 378 (2020) 40. https://doi.org/10.1007/s41061-020-00302-w. https://doi.org/10.1007/s41061-020-00302-w

[29] A. Gholami, S.M. Mousavi, S.A. Hashemi, Y. Ghasemi, W.-H. Chiang, N. Parvin, Current trends in chemical modifications of magnetic nanoparticles for targeted drug delivery in cancer chemotherapy., Drug Metab. Rev. 52 (2020) 205-224.

[30] T.K. Kim, J.H. Eberwine, Mammalian cell transfection: the present and the future, Anal. Bioanal. Chem. 397 (2010) 3173-3178. https://doi.org/10.1007/s00216-010-3821-6. https://doi.org/10.1007/s00216-010-3821-6

[31] E. Alphandéry, Iron oxide nanoparticles for therapeutic applications, Drug Discov. Today. 25 (2020) 141-149. https://doi.org/10.1016/j.drudis.2019.09.020

[32] K.J. Widder, A.E. Senyei, D.G. Scarpelli, Magnetic Microspheres: A Model System for Site Specific Drug Delivery in Vivo, Proc. Soc. Exp. Biol. Med. 158 (1978) 141-146. https://journals.sagepub.com/doi/abs/10.3181/00379727-158-40158. https://doi.org/10.3181/00379727-158-40158

[33] C.H. Lee, E.Y. Kim, K. Jeon, J.C. Tae, K.S. Lee, Y.O. Kim, M.-Y. Jeong, C.-W. Yun, D.K. Jeong, S.K. Cho, J.H. Kim, H.Y. Lee, K.Z. Riu, S.G. Cho, S.P. Park, Simple, Efficient, and Reproducible Gene Transfection of Mouse Embryonic Stem Cells by Magnetofection, Stem Cells Dev. 17 (2008) 133-142. https://doi.org/10.1089/scd.2007.0064. https://doi.org/10.1089/scd.2007.0064

[34] F. Scherer, M. Anton, U. Schillinger, J. Henke, C. Bergemann, A. Krüger, B. Gänsbacher, C. Plank, Magnetofection: Enhancing and targeting gene delivery by magnetic force in vitro and in vivo, Gene Ther. 9 (2002) 102-109. https://doi.org/10.1038/sj.gt.3301624

[35] N. Laurent, C. Sapet, L. Le Gourrierec, E. Bertosio, O. Zelphati, Nucleic acid delivery using magnetic nanoparticles: the MagnetofectionTM technology, Ther. Deliv. 2 (2011) 471-482. https://doi.org/10.4155/tde.11.12. https://doi.org/10.4155/tde.11.12

[36] J. Estelrich, E. Escribano, J. Queralt, M.A. Busquets, Iron Oxide Nanoparticles for Magnetically-Guided and Magnetically-Responsive Drug Delivery, Int. J. Mol. Sci. 16 (2015). https://doi.org/10.3390/ijms16048070

[37] Y. Wang, H. Cui, K. Li, C. Sun, W. Du, J. Cui, X. Zhao, W. Chen, A magnetic nanoparticle-based multiple-gene delivery system for transfection of porcine kidney cells, PLoS One. 9 (2014) 1-9. https://doi.org/10.1371/journal.pone.0102886

[38] A. AU - Jacquier, V. AU - Risson, L. AU - Schaeffer, Modeling Charcot-Marie-Tooth Disease In Vitro by Transfecting Mouse Primary Motoneurons, JoVE. (2019) e57988. https://www.jove.com/t/57988. https://doi.org/10.3791/57988

[39] L. Prosen, B. Markelc, T. Dolinsek, B. Music, M. Cemazar, G. Sersa, Mcam Silencing With RNA Interference Using Magnetofection has Antitumor Effect in Murine Melanoma, Mol. Ther. Nucleic Acids. 3 (2014) e205-e205. https://doi.org/10.1038/mtna.2014.56

[40] Z. Pan, Z. Shi, H. Wei, F. Sun, J. Song, Y. Huang, T. Liu, Y. Mao, Magnetofection Based on Superparamagnetic Iron Oxide Nanoparticles Weakens Glioma Stem Cell Proliferation and Invasion by Mediating High Expression of MicroRNA-374a, J. Cancer. 7 (2016) 1487-1496. https://www.jcancer.org/v07p1487.htm. https://doi.org/10.7150/jca.15515

[41] A. Soledad Pereyra, O. Mykhaylyk, Magnetofection Enhances Adenoviral Vector-based Gene Delivery in Skeletal Muscle Cells, J. Nanomed. Nanotechnol. 07 (2016). https://doi.org/10.4172/2157-7439.1000364

[42] S. Prijic, L. Prosen, M. Cemazar, J. Scancar, R. Romih, J. Lavrencak, V.B. Bregar, A. Coer, M. Krzan, A. Znidarsic, G. Sersa, Surface modified magnetic nanoparticles for immuno-gene therapy of murine mammary adenocarcinoma., Biomaterials. 33 (2012) 4379-4391. https://doi.org/10.1016/j.biomaterials.2012.02.061

[43] A. Villanueva, M. Cañete, A.G. Roca, M. Calero, S. Veintemillas-Verdaguer, C.J. Serna, M. del Puerto Morales, R. Miranda, The influence of surface functionalization on the enhanced internalization of magnetic nanoparticles in cancer cells, Nanotechnology. 20 (2009) 115103. http://dx.doi.org/10.1088/0957-4484/20/11/115103. https://doi.org/10.1088/0957-4484/20/11/115103

[44] M. Lee, K. Chea, R. Pyda, M. Chua, I. Dominguez, Comparative Analysis of Non-viral Transfection Methods in Mouse Embryonic Fibroblast Cells., J. Biomol. Tech. 28 (2017) 67-74. https://doi.org/10.7171/jbt.17-2802-003

[45] Y. Shi, J. Du, L. Zhou, X. Li, Y. Zhou, L. Li, X. Zang, X. Zhang, F. Pan, H. Zhang, Z. Wang, X. Zhu, Size-controlled preparation of magnetic iron oxide nanocrystals within hyperbranched polymers and their magnetofection in vitro, J. Mater. Chem. 22 (2012) 355-360. http://dx.doi.org/10.1039/C1JM14079E. https://doi.org/10.1039/C1JM14079E

[46] Y. Shi, L. Zhou, R. Wang, Y. Pang, W. Xiao, H. Li, Y. Su, X. Wang, B. Zhu, X. Zhu, D. Yan, H. Gu, In situ preparation of magnetic nonviral gene vectors and

magnetofection in vitro., Nanotechnology. 21 (2010) 115103.
https://doi.org/10.1088/0957-4484/21/11/115103

[47] L. Prosen, M. Čemažar, G. Sersa, Magnetofection: An Effective, Selective and Feasible Non-viral Gene Delivery Method BT - 1st World Congress on Electroporation and Pulsed Electric Fields in Biology, Medicine and Food & Environmental Technologies, in: T. Jarm, P. Kramar (Eds.), Springer Singapore, Singapore, 2016: pp. 335-338. https://doi.org/10.1007/978-981-287-817-5_74

[48] L. Prosen, S. Prijic, B. Music, J. Lavrencak, M. Cemazar, G. Sersa, Magnetofection: A Reproducible Method for Gene Delivery to Melanoma Cells, Biomed Res. Int. 2013 (2013) 209452. https://doi.org/10.1155/2013/209452. https://doi.org/10.1155/2013/209452

[49] Y.-L. Lo, H.-L. Chou, Z.-X. Liao, S.-J. Huang, J.-H. Ke, Y.-S. Liu, C.-C. Chiu, L.-F. Wang, Chondroitin sulfate-polyethylenimine copolymer-coated superparamagnetic iron oxide nanoparticles as an efficient magneto-gene carrier for microRNA-encoding plasmid DNA delivery, Nanoscale. 7 (2015) 8554-8565. http://dx.doi.org/10.1039/C5NR01404B. https://doi.org/10.1039/C5NR01404B

[50] L.A. Blokpoel Ferreras, S.Y. Chan, S. Vazquez Reina, J.E. Dixon, Rapidly Transducing and Spatially Localized Magnetofection Using Peptide-Mediated Non-Viral Gene Delivery Based on Iron Oxide Nanoparticles, ACS Appl. Nano Mater. 4 (2021) 167-181. https://doi.org/10.1021/acsanm.0c02465. https://doi.org/10.1021/acsanm.0c02465

[51] H.Y. Park, E.H. Noh, H.M. Chung, M.J. Kang, E.Y. Kim, S.P. Park, Efficient Generation of Virus-Free iPS Cells Using Liposomal Magnetofection, PLoS One. 7 (2012). https://doi.org/10.1371/journal.pone.0045812

[52] C.A. Alvizo-Baez, I.E. Luna-Cruz, N. Vilches-Cisneros, C. Rodríguez-Padilla, J.M. Alcocer-González, Systemic delivery and activation of the TRAIL gene in lungs, with magnetic nanoparticles of chitosan controlled by an external magnetic field, Int. J. Nanomedicine. 11 (2016) 6449-6458. https://doi.org/10.2147/IJN.S118343

[53] Y.-K. Kim, M. Zhang, J.-J. Lu, F. Xu, B.-A. Chen, L. Xing, H.-L. Jiang, PK11195-chitosan-graft-polyethylenimine-modified SPION as a mitochondria-targeting gene carrier, J. Drug Target. 24 (2016) 457-467. https://doi.org/10.3109/1061186X.2015.1087527. https://doi.org/10.3109/1061186X.2015.1087527

[54] J. Venero, M. Burguillos, Magnetofection as a new tool to study microglia biology, Neural Regen. Res. 14 (2019) 767-768. http://www.nrronline.org/article.asp?issn=1673-5374. https://doi.org/10.4103/1673-5374.249221

[55] A. Carrillo-Jimenez, M. Puigdellívol, A. Vilalta, J.L. Venero, G.C. Brown, P. StGeorge-Hyslop, M.A. Burguillos, Effective Knockdown of Gene Expression in Primary Microglia With siRNA and Magnetic Nanoparticles Without Cell Death or

Inflammation , Front. Cell. Neurosci. . 12 (2018) 313.
https://www.frontiersin.org/article/10.3389/fncel.2018.00313.
https://doi.org/10.3389/fncel.2018.00313

[56] T. Umek, T. Olsson, O. Gissberg, O. Saher, E.M. Zaghloul, K.E. Lundin, J. Wengel, E. Hanse, H. Zetterberg, D. Vizlin-Hodzic, C.I.E. Smith, R. Zain, Oligonucleotides Targeting DNA Repeats Downregulate Huntingtin Gene Expression in Huntington's Patient-Derived Neural Model System., Nucleic Acid Ther. (2021). https://doi.org/10.1089/nat.2021.0021

[57] A. Hu, X. Chen, Q. Bi, Y. Xiang, R. Jin, H. Ai, Y. Nie, A parallel and cascade control system: magnetofection of miR125b for synergistic tumor-association macrophage polarization regulation and tumor cell suppression in breast cancer treatment, Nanoscale. 12 (2020) 22615-22627. http://dx.doi.org/10.1039/D0NR06060G. https://doi.org/10.1039/D0NR06060G

[58] F. Krötz, C. de Wit, H.Y. Sohn, S. Zahler, T. Gloe, U. Pohl, C. Plank, Magnetofection - A highly efficient tool for antisense oligonucleotide delivery in vitro and in vivo, Mol. Ther. 7 (2003) 700-710. https://doi.org/10.1016/S1525-0016(03)00065-0

[59] E. Brett, E.R. Zielins, A. Luan, C.C. Ooi, S. Shailendra, D. Atashroo, S. Menon, C. Blackshear, J. Flacco, N. Quarto, S.X. Wang, M.T. Longaker, D.C. Wan, Magnetic Nanoparticle-Based Upregulation of B-Cell Lymphoma 2 Enhances Bone Regeneration, Stem Cells Transl. Med. 6 (2017) 151-160. https://doi.org/10.5966/sctm.2016-0051. https://doi.org/10.5966/sctm.2016-0051

[60] M. Sen, M. Bassetto, F. Poulhes, O. Zelphati, M. Ueffing, B. Arango-Gonzalez, Efficient Ocular Delivery of VCP siRNA via Reverse Magnetofection in RHO P23H Rodent Retina Explants., Pharmaceutics. 13 (2021). https://doi.org/10.3390/pharmaceutics13020225

[61] S.M. Shalaby, M.K. Khater, A.M. Perucho, S.A. Mohamed, I. Helwa, A. Laknaur, I. Lebedyeva, Y. Liu, M.P. Diamond, A.A. Al-Hendy, Magnetic nanoparticles as a new approach to improve the efficacy of gene therapy against differentiated human uterine fibroid cells and tumor-initiating stem cells., Fertil. Steril. 105 (2016) 1638-1648.e8. https://doi.org/10.1016/j.fertnstert.2016.03.001

[62] M. Takafuji, K. Kitaura, T. Nishiyama, S. Govindarajan, V. Gopal, T. Imamura, H. Ihara, Chemically tunable cationic polymer-bonded magnetic nanoparticles for gene magnetofection, J. Mater. Chem. B. 2 (2014) 644-650. http://dx.doi.org/10.1039/C3TB21290D. https://doi.org/10.1039/C3TB21290D

[63] R. Panday, A.M.E. Abdalla, M. Yu, X. Li, C. Ouyang, G. Yang, Functionally modified magnetic nanoparticles for effective siRNA delivery to prostate cancer cells in vitro, J. Biomater. Appl. 34 (2019) 952-964. https://doi.org/10.1177/0885328219886953. https://doi.org/10.1177/0885328219886953

[64] C. Sardo, E.F. Craparo, B. Porsio, G. Giammona, G. Cavallaro, Combining Inulin Multifunctional Polycation and Magnetic Nanoparticles: Redox-Responsive siRNA-Loaded Systems for Magnetofection, Polymers (Basel). 11 (2019). https://doi.org/10.3390/polym11050889

[65] M. Dowaidar, H. Nasser Abdelhamid, M. Hällbrink, Ü. Langel, X. Zou, Chitosan enhances gene delivery of oligonucleotide complexes with magnetic nanoparticles-cell-penetrating peptide, J. Biomater. Appl. 33 (2018) 392-401. https://doi.org/10.1177/0885328218796623. https://doi.org/10.1177/0885328218796623

[66] K. Bulaklak, C.A. Gersbach, The once and future gene therapy, Nat. Commun. 11 (2020) 11-14. https://doi.org/10.1038/s41467-020-19505-2

[67] E. Papanikolaou, A. Bosio, The Promise and the Hope of Gene Therapy, Front. Genome Ed. 3 (2021) 1-14. https://doi.org/10.3389/fgeed.2021.618346

[68] Thomas Blankenstein, Gene Therapy: Principles and Applications, Birkhäuser Basel, Berlin, 2012.

[69] M.R. Cring, V.C. Sheffield, Gene therapy and gene correction: targets, progress, and challenges for treating human diseases, Gene Ther. (2020). https://doi.org/10.1038/s41434-020-00197-8

[70] S. Mali, Delivery systems for gene therapy, Indian J. Hum. Genet. 19 (2013) 3-8. https://doi.org/10.4103/0971-6866.112870

[71] J.T. Bulcha, Y. Wang, H. Ma, P.W.L. Tai, G. Gao, Viral vector platforms within the gene therapy landscape, Signal Transduct. Target. Ther. 6 (2021). https://doi.org/10.1038/s41392-021-00487-6

[72] H. Yin, R.L. Kanasty, A.A. Eltoukhy, A.J. Vegas, J.R. Dorkin, D.G. Anderson, Non-viral vectors for gene-based therapy, Nat. Rev. Genet. 15 (2014) 541-555. https://doi.org/10.1038/nrg3763

[73] M. Ruponen, P. Honkakoski, S. Rönkkö, J. Pelkonen, M. Tammi, A. Urtti, Extracellular and intracellular barriers in non-viral gene delivery, J. Control. Release. 93 (2003) 213-217. https://doi.org/10.1016/j.jconrel.2003.08.004

[74] S. Majidi, F. Zeinali Sehrig, M. Samiei, M. Milani, E. Abbasi, K. Dadashzadeh, A. Akbarzadeh, Magnetic nanoparticles: Applications in gene delivery and gene therapy, Artif. Cells, Nanomedicine Biotechnol. 44 (2016) 1186-1193. https://doi.org/10.3109/21691401.2015.1014093

[75] J. Dobson, Gene therapy progress and prospects: Magnetic nanoparticle-based gene delivery, Gene Ther. 13 (2006) 283-287. https://doi.org/10.1038/sj.gt.3302720

[76] D.P. Clark, N.J. Pazdernik, Protein Engineering, in: Biotechnology, second, Academic Cell, 2016: pp. 365-392. https://doi.org/10.1016/B978-0-12-385015-7.00011-9

[77] J. Dulińska-Litewka, A. Łazarczyk, P. Hałubiec, O. Szafrański, K. Karnas, A. Karewicz, Superparamagnetic iron oxide nanoparticles-current and prospective medical applications, Materials (Basel). 12 (2019) 1-26. https://doi.org/10.3390/ma12040617

[78] M. Suciu, C.M. Ionescu, A. Ciorita, S.C. Tripon, D. Nica, H. Al-Salami, L. Barbu-Tudoran, Applications of superparamagnetic iron oxide nanoparticles in drug and therapeutic delivery, and biotechnological advancements, Beilstein J. Nanotechnol. 11 (2020) 1092-1109. https://doi.org/10.3762/bjnano.11.94

[79] A.A. Sizikov, M. V. Kharlamova, M.P. Nikitin, P.I. Nikitin, E.L. Kolychev, Nonviral locally injected magnetic vectors for in vivo gene delivery: A review of studies on magnetofection, Nanomaterials. 11 (2021) 1-17. https://doi.org/10.3390/nano11051078

[80] C. Vasile, Polymeric Nanomaterials: Recent Developments, Properties and Medical Applications, in: Polym. Nanomater. Nanotherapeutics, Elsevier, 2019: pp. 1-66. https://doi.org/10.1016/B978-0-12-813932-5.00001-7

[81] C. Tros de Ilarduya, Y. Sun, N. Düzgüneş, Gene delivery by lipoplexes and polyplexes, Eur. J. Pharm. Sci. 40 (2010) 159-170. https://doi.org/10.1016/j.ejps.2010.03.019

[82] J. Singh, I. Mohanty, S. Rattan, In vivo magnetofection: A novel approach for targeted topical delivery of nucleic acids for rectoanal motility disorders, Am. J. Physiol. - Gastrointest. Liver Physiol. 314 (2018) G109-G118. https://doi.org/10.1152/ajpgi.00233.2017

[83] OZBIOSCIENCES, PolyMag, (n.d.).

[84] L. Prosen, S. Hudoklin, M. Cemazar, M. Stimac, U. Lampreht Tratar, M. Ota, J. Scancar, R. Romih, G. Sersa, Magnetic field contributes to the cellular uptake for effective therapy with magnetofection using plasmid DNA encoding against Mcam in B16F10 melanoma in vivo, Nanomedicine. 11 (2016) 627-641. https://doi.org/10.2217/nnm.16.4

[85] C. Luo, X. Yang, M. Li, H. Huang, Q. Kang, X. Zhang, H. Hui, X. Zhang, C. Cen, Y. Luo, L. Xie, C. Wang, T. He, D. Jiang, T. Li, H. An, A novel strategy for in vivo angiogenesis and osteogenesis: magnetic micro-movement in a bone scaffold, Artif. Cells, Nanomedicine Biotechnol. 46 (2018) 636-645. https://doi.org/10.1080/21691401.2018.1465947

[86] V. Singh, P.K. Sharma, M.A. Alam, Role of cationic lipids for the formulation of lipoplexes, Int. J. Res. Pharm. Sci. Technol. 1 (2018) 1-8. https://doi.org/10.33974/ijrpst.v1i1.1

[87] S. Putzke, E. Feldhues, I. Heep, T. Ilg, A. Lamprecht, Cationic lipid/pDNA complex formation as potential generic method to generate specific IRF pathway stimulators, Eur. J. Pharm. Biopharm. 155 (2020) 112-121. https://doi.org/10.1016/j.ejpb.2020.08.010

[88] J.L. Bramson, C.A. Bodner, R.W. Graham, Activation of host antitumoral responses by cationic lipid/DNA complexes, Cancer Gene Ther. 7 (2000) 353-359. https://doi.org/10.1038/sj.cgt.7700143

[89] M. Buñuales, N. Düzgüne, S. Zalba, M.J. Garrido, C. Tros De Ilarduya, Efficient gene delivery by EGF-lipoplexes in vitro and in vivo, Nanomedicine. 6 (2011) 89-98. https://doi.org/10.2217/nnm.10.100

[90] Promega, pGL4.50[luc2/CMV/Hygro] Vector, (n.d.).

[91] Y. Kono, S. Gogatsubo, T. Ohba, T. Fujita, Enhanced macrophage delivery to the colon using magnetic lipoplexes with a magnetic field, Drug Deliv. 26 (2019) 935-943. https://doi.org/10.1080/10717544.2019.1662515

[92] C. Wang, C. Ding, M. Kong, A. Dong, J. Qian, D. Jiang, Z. Shen, Tumor-targeting magnetic lipoplex delivery of short hairpin RNA suppresses IGF-1R overexpression of lung adenocarcinoma A549 cells in vitro and in vivo, Biochem. Biophys. Res. Commun. 410 (2011) 537-542. https://doi.org/10.1016/j.bbrc.2011.06.019

[93] Q.T.H. Shubhra, A. Oyane, M. Nakamura, S. Puentes, A. Marushima, H. Tsurushima, Rapid one-pot fabrication of magnetic calcium phosphate nanoparticles immobilizing DNA and iron oxide nanocrystals using injection solutions for magnetofection and magnetic targeting, Mater. Today Chem. 6 (2017) 51-61. https://doi.org/10.1016/j.mtchem.2017.10.001

[94] Q.T.H. Shubhra, A. Oyane, M. Nakamura, S. Puentes, A. Marushima, H. Tsurushima, Preliminary in vivo magnetofection data using magnetic calcium phosphate nanoparticles immobilizing DNA and iron oxide nanocrystals, Data Br. 18 (2018) 1696-1701. https://doi.org/10.1016/j.dib.2018.04.058

[95] H.P. Song, J.Y. Yang, S.L. Lo, Y. Wang, W.M. Fan, X.S. Tang, J.M. Xue, S. Wang, Gene transfer using self-assembled ternary complexes of cationic magnetic nanoparticles, plasmid DNA and cell-penetrating tat peptide, Biomaterials. 31 (2010) 769-778. https://doi.org/10.1016/j.biomaterials.2009.09.085

[96] W. Huang, Z. Liu, G. Zhou, J. Ling, A. Tian, N. Sun, Silencing Bag-1 gene via magnetic gold nanoparticle-delivered siRNA plasmid for colorectal cancer therapy in vivo and in vitro, Tumor Biol. 37 (2016) 10365-10374. https://doi.org/10.1007/s13277-016-4926-0

[97] Q.A. Pankhurst, J. Connolly, S.K. Jones, J. Dobson, Applications of magnetic nanoparticles in biomedicine, J. Phys. D. Appl. Phys. 36 (2003) 167-181. https://doi.org/10.1088/0022-3727/36/13/201

[98] S.D. Xiang, C. Selomulya, J. Ho, V. Apostolopoulos, M. Plebanski, Delivery of DNA vaccines: an overview on the use of biodegradable polymeric and magnetic nanoparticles., Wiley Interdiscip. Rev. Nanomed. Nanobiotechnol. 2 (2010) 205-218. https://doi.org/10.1002/wnan.88

[99] F.M.N. Al-Deen, S.D. Xiang, C. Ma, K. Wilson, R.L. Coppel, C. Selomulya, M. Plebanski, Magnetic Nanovectors for the Development of DNA Blood-Stage Malaria

Vaccines, Nanomater. (Basel, Switzerland). 7 (2017) 30.
https://pubmed.ncbi.nlm.nih.gov/28336871. https://doi.org/10.3390/nano7020030

[100] X.F. Zhou, B. Liu, X.H. Yu, X. Zha, X.Z. Zhang, X.Y. Wang, Y.H. Jin, Y.G. Wu, C.L. Jiang, Y. Chen, Y. Chen, Y.M. Shan, J.Q. Liu, W. Kong, J.C. Shen, Using magnetic force to enhance immune response to DNA vaccine, Small. 3 (2007) 1707-1713. https://doi.org/10.1002/smll.200700151

[101] C. Plank, O. Zelphati, O.M. Mykhaylyk, Magnetically enhanced nucleic acid delivery. Ten years of magnetofectio-Progress and prospects_, Adv. Drug Deliv. Rev. 63 (2011) 1300-1331. https://doi.org/10.1016/j.addr.2011.08.002

[102] F.N. Al-Deen, J. Ho, C. Selomulya, C. Ma, R. Coppel, Superparamagnetic nanoparticles for effective delivery of malaria DNA vaccine., Langmuir. 27 (2011) 3703-3712. https://doi.org/10.1021/la104479c

[103] C. Selomulya, Y.Y. Kong, S.D. Xiang, C. Ma, R.L. Coppel, M. Plebanski, Design of magnetic polyplexes taken up efficiently by dendritic cell for enhanced DNA vaccine delivery FM Nawwab AL-Deen1, Gene Ther. 21 (2014) 212-218. https://doi.org/10.1038/gt.2013.77

[104] V.I. Mobarakeh, M.H. Modarressi, P. Rahimi, A. Bolhassani, E. Arefian, Modification of SPION nanocarriers for siRNA delivery: A therapeutic strategy against HIV infection, Vaccine Res. (2019). https://doi.org/10.29252/vacres.6.1.43

[105] S. Kamalzare, Z. Noormohammadi, P. Rahimi, F. Atyabi, S. Irani, F.S.M. Tekie, F. Mottaghitalab, Carboxymethyl dextran-trimethyl chitosan coated superparamagnetic iron oxide nanoparticles: An effective siRNA delivery system for HIV-1 Nef., J. Cell. Physiol. 234 (2019) 20554-20565. https://doi.org/10.1002/jcp.28655

[106] J.H. Erasmus, A.P. Khandhar, M.A. O'Connor, A.C. Walls, E.A. Hemann, P. Murapa, J. Archer, S. Leventhal, J.T. Fuller, T.B. Lewis, K.E. Draves, S. Randall, K.A. Guerriero, M.S. Duthie, D. Carter, S.G. Reed, D.W. Hawman, H. Feldmann, M.J. Gale, D. Veesler, P. Berglund, D.H. Fuller, An Alphavirus-derived replicon RNA vaccine induces SARS-CoV-2 neutralizing antibody and T cell responses in mice and nonhuman primates., Sci. Transl. Med. 12 (2020). https://doi.org/10.1126/scitranslmed.abc9396

[107] M. Behzadi, B. Vakili, A. Ebrahiminezhad, N. Nezafat, Iron nanoparticles as novel vaccine adjuvants, Eur. J. Pharm. Sci. 159 (2021) 105718. https://www.sciencedirect.com/science/article/pii/S0928098721000208. https://doi.org/10.1016/j.ejps.2021.105718

[108] A.J. Grippin, B. Wummer, T. Wildes, K. Dyson, V. Trivedi, C. Yang, M. Sebastian, H.R. Mendez-Gomez, S. Padala, M. Grubb, M. Fillingim, A. Monsalve, E.J. Sayour, J. Dobson, D.A. Mitchell, Dendritic Cell-Activating Magnetic Nanoparticles Enable Early Prediction of Antitumor Response with Magnetic Resonance Imaging, ACS Nano. 13 (2019) 13884-13898. https://pubmed.ncbi.nlm.nih.gov/31730332. https://doi.org/10.1021/acsnano.9b05037

Materials Research Forum LLC
https://doi.org/10.21741/9781644902332-10

[109] B. Yoo, V.C. Jordan, P. Sheedy, A.-M. Billig, A. Ross, P. Pantazopoulos, Z. Medarova, RNAi-Mediated PD-L1 Inhibition for Pancreatic Cancer Immunotherapy, Sci. Rep. 9 (2019) 4712. https://doi.org/10.1038/s41598-019-41251-9. https://doi.org/10.1038/s41598-019-41251-9

[110] Y.-S. Tang, D. Wang, C. Zhou, W. Ma, Y.-Q. Zhang, B. Liu, S. Zhang, Bacterial magnetic particles as a novel and efficient gene vaccine delivery system, Gene Ther. 19 (2012) 1187-1195. https://doi.org/10.1038/gt.2011.197. https://doi.org/10.1038/gt.2011.197

[111] H.A. Schreiber, J. Prechl, H. Jiang, A. Zozulya, Z. Fabry, F. Denes, M. Sandor, Using carbon magnetic nanoparticles to target, track, and manipulate dendritic cells, J. Immunol. Methods. 356 (2010) 47-59. https://pubmed.ncbi.nlm.nih.gov/20219468. https://doi.org/10.1016/j.jim.2010.02.009

[112] J. Ho, F.M.N. Al-Deen, A. Al-Abboodi, C. Selomulya, S.D. Xiang, M. Plebanski, G.M. Forde, N,N'-Carbonyldiimidazole-mediated functionalization of superparamagnetic nanoparticles as vaccine carrier., Colloids Surf. B. Biointerfaces. 83 (2011) 83-90. https://doi.org/10.1016/j.colsurfb.2010.11.001

[113] S. Sungsuwan, Z. Yin, X. Huang, Lipopeptide-Coated Iron Oxide Nanoparticles as Potential Glycoconjugate-Based Synthetic Anticancer Vaccines, ACS Appl. Mater. Interfaces. 7 (2015) 17535-17544. https://doi.org/10.1021/acsami.5b05497. https://doi.org/10.1021/acsami.5b05497

[114] X.-B. Liu, G.-W. Yu, X.-Y. Gao, J.-L. Huang, L.-T. Qin, H.-B. Ni, C. Lyu, Intranasal delivery of plasmids expressing bovine herpesvirus 1 gB/gC/gD proteins by polyethyleneimine magnetic beads activates long-term immune responses in mice, Virol. J. 18 (2021) 60. https://doi.org/10.1186/s12985-021-01536-w. https://doi.org/10.1186/s12985-021-01536-w

[115] C. Hüttinger, J. Hirschberger, A. Jahnke, R. Köstlin, T. Brill, C. Plank, H. Küchenhoff, S. Krieger, U. Schillinger, Neoadjuvant gene delivery of feline granulocyte-macrophage colony-stimulating factor using magnetofection for the treatment of feline fibrosarcomas: a phase I trial., J. Gene Med. 10 (2008) 655-667. https://doi.org/10.1002/jgm.1185

Keyword Index

About the Editors

Dr. Rajshree B. Jotania is a professor of Physics, Department of Physics, Electronics and Space science, at Gujarat University, Ahmedabad, India. She obtained her B.Sc., M.Sc. and Ph. D from Saurashtra University, Rajkot, India. She was Junior Research Fellow (DAE-BRNS project) during 1987 to 1989 at Physics Department, Saurashtra University, Rajkot, India. She obtained a few regional and national awards for contribution toward scientific research. She worked at National Chemical Laboratory, Pune, India for few months as a Summer Visiting Teacher Fellow in 2005 and as a Visiting Scientist fellow in 2011. She possesses 32 years of teaching experience at UG and PG level. She is a member of board of studies at few Universities of Gujarat, India and a Mentor of DST-INSPIRE (Department of Science and Technology-Innovation in Science Pursuit for Inspired Research) program. She has published more than 100 papers in various research journals and conference proceedings. She has delivered more than 25 invited talks at various DST-INSPIRE Internship science camp in India. She has edited four books entitled 'Ferrites and ceramic composites' (Vol. I & II, Trans Tech Publisher (TTP), Switzerland), Magnetic Oxides and Ceramic Composites (Vol. I & II, Materials Research Forum, LLC, USA) and is working on one more book (Green Nanomaterials for Clean and Sustainable Environment). She has visited Singapore, Malaysia, USA (New York), and North Africa (Tunisia) for research work. She has attended more than 55 international, national conferences/symposiums/seminars/academy meeting and worked as a chair person as well as delivered invited talks in a few international and national conferences. She possesses a life membership of eight professional bodies and she has guided seven Ph. D, twelve M. Phil students. At present five students are working under her guidance for Ph. D. To date she has completed five research projects of various agencies. She has worked as deputy co-coordinator, DRS (SAP-I) program. She was a secretary, Gujarat chapter RC-07, Indian Association of Physics Teacher (IAPT). Recently she has been associated with a few universities for scientific co-operation in nanomagnetic and nanotechnology field.

Prof. Rajshree B. Jotania
Department of Physics, Electronics and Space science,
University school of sciences,
Gujarat University,
Ahmedabad, Gujarat, India,
Email: rbjotania@gujaratuniversity.ac.in

Dr. Martín F. Desimone is Professor in the Department of Chemical Sciences in the Faculty of Pharmacy and Biochemistry of the University of Buenos Aires and holds a Principal Researcher position at CONICET, Argentina. He studied Pharmacy, Biochemistry and received his Ph. D. from the University of Buenos Aires, Argentina. He was a postdoctoral visiting scientist at the University of Basque Country in Spain and held a Maître de conferences position in the Collège de France working in the Laboratoire de Chimie de la Matière Condensée de Paris (UMR 7574 Sorbonne Université, CNRS, Collège de France). He leads a research team working on nanocomposites and hybrid materials combining biomaterials with nanomaterials and biological active molecules for biomedical applications such as wound healing, tissue engineering, drug delivery, 3D printing and stimuli-responsive materials. He has established fruitful collaborations with national and international teams contributing to the work of ten Ph D. and two Master students who concluded their work with his direction. He is the recipient of several awards including awards from the National Academy of Pharmacy and Biochemistry, the Award "Innovar" in the category applied research received in two consecutive years (2016 and 2017) from the Ministry of Science, Technology and Productive Innovation (MINCYT, Argentina), the distinction "Dr. José A. Balseiro" XV edition, received in The Argentine Senate, which is the upper house of the Argentine National Congress, International Association of Advanced Materials medal of the year 2018 (Sweden), and the "Academic Excellence" from the University of Buenos Aires in four consecutive years, among others. He is cofounder of the start-up enterprise Hybridon which is producing nanotechnological applications. He has coauthored more than 120 international scientific publications, 17 book chapters and 4 patents.

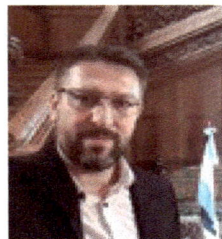

Dr Martín F. DESIMONE
Cátedra de Química Analítica Instrumental
IQUIMEFA-CONICET
Facultad de Farmacia y Bioquímica, Universidad de Buenos Aires
(1113) Junin 956 Piso 3. Buenos Aires. Argentina.
E-mail: desimone@ffyb.uba.ar / martinfdesimone@gmail.com
Tel: +54-11-52874332